The Illustrated Buyer's Guide to Used Airplanes

The Illustrated Buyer's Guide to Used Airplanes

Sixth Edition

Bill Clarke

McGraw-Hill

New York   Chicago   San Francisco   Lisbon   London   Madrid
Mexico City   Milan   New Delhi   San Juan   Seoul
Singapore   Sydney   Toronto

The McGraw·Hill Companies

Cataloging-in-Publication Data is on file with the Library of Congress

1 2 3 4 5 6 7 8 9 0   DOC/DOC   0 1 0 9 8 7 6 5

ISBN 0-07-145427-6

The sponsoring editor for this book was Stephen S. Chapman, the editing supervisor was Stephen M. Smith, and the production supervisor was Richard C. Ruzycka. It was set in ITC Slimbach by Wayne A. Palmer of McGraw-Hill Professional's Hightstown, N.J., composition unit. The art director for the cover was Anthony Landi.

Printed and bound by RR Donnelley.

McGraw-Hill books are available at special quantity discounts to use as premiums and sales promotions, or for use in corporate training programs. For more information, please write to the Director of Special Sales, McGraw-Hill Professional, Two Penn Plaza, New York, NY 10121-2298. Or contact your local bookstore.

This book is printed on recycled, acid-free paper containing a minimum of 50% recycled, de-inked fiber.

*This book is dedicated to my wife,
for keeping me on course.*

Contents

Appendixes 463

Preface

ARE YOU THINKING ABOUT AIRPLANE OWNERSHIP? At one time or another, nearly every pilot considers owning his or her own airplane. Airplane ownership is not difficult, although, like everything else, you must first know and understand certain complexities. Since most of these complexities involve money, you must thoroughly understand airplane ownership to prevent later financial grief.

In 1980, a new Cessna 172 sold for about $39,000. By 1986, the same model of airplane carried a price tag of nearly $75,000. Later that year, the 172 marched off into history as production stopped. The world's most popular four-place airplane had been priced out of existence. At the same time, top-of-the-line airplanes such as the Bonanza A36 were sporting prices of nearly $250,000. In 1996 Cessna reintroduced the 172—at a price of $124,500—and by 1998 the A36 was sticker-priced at over $425,000. By 2004, the Cessna was selling for in excess of $170,000 and the A36 for more than a half-million dollars.

Although high prices are a way of life in our society, it appears that aviation costs have soared when compared to the average cost of living. Many factors contribute to the phenomenon, including the high cost of labor, ever-increasing premiums paid for product liability insurance, and increased charges for materials. Regardless of the reasons, most pilots are financially excluded from the new airplane market. A good used airplane, however, is a viable alternative to an expensive new airplane. As a matter of record, in 2003 the Aircraft Owners and Pilots Association (AOPA) reported that 625,011 pilot certificates were held and 160,850 piston-powered general aviation airplanes were active (most over 30 years old).

The Illustrated Buyer's Guide to Used Airplanes, 6th edition, is written to help you, the prospective used airplane buyer, to successfully search for a suitable and cost-effective airplane (and not get financially burned in the process). Part 1 discusses the pros and cons of airplane ownership, and explains how to determine the size and type of aircraft most suitable for your flying needs. It then explains how to locate and evaluate a used airplane and read those cryptic used airplane advertisements.

You will learn about the government paperwork required for airplane ownership, how to select a home base, and how to save money by doing your own preventive maintenance. Care and protection of your aviation investment is explained.

Although you may have a basic idea of what a particular make and model of airplane looks like or what some of the specifications are, you must have a ready source of additional, specific, and accurate information. Part 2 is such a source, containing historical information, photographs, and specifications for most airplanes that can be found on today's used market. Alternative aircraft are also investigated, including floatplanes, personalized homebuilts, serene gliders, and the fast, powerful war birds.

Part 3 provides stories and other pieces of information from the hangar and airport that can provide insight into particular makes and models of airplanes. Then follows the airworthiness directive information, often requiring expensive and complex maintenance and repairs, that is the word of law from the government. The airworthiness directives that apply to airplanes described in this book are listed and briefly explained.

A new chapter, Light-Sport Aircraft, has been added to this edition. Here you will find information about a new category of airplanes and the pilots who can fly them.

The appendixes contain the NTSB aircraft accident chart, which rates the various makes and models of airplanes. They also list points of contact for FAA offices, state aviation agencies, manufacturers, and airplane type clubs. This includes addresses, telephone and fax numbers, and Internet information (where available).

Also contained in the appendixes is the all-important used airplane price guide. This is an up-to-date listing of typical used airplane selling prices and a means for determining real value for a specific airplane. Following the price guide is a list of the currently registered makes and models from the FAA Registry. These numbers indicate the potential numbers of a specific make and model that can be found on the used airplane market.

In summary, *The Illustrated Buyer's Guide to Used Airplanes*, 6th edition, is written to help you in economical decision making and to guide you around the many pitfalls you might encounter when buying and owning a used airplane.

Bill Clarke

The Illustrated Buyer's Guide to Used Airplanes

Part I

The Art of Buying and Owning an Airplane

1

The General Aviation Market

WHERE IS THE GENERAL AVIATION MARKET GOING? That certainly is a very good question, but unfortunately there are no solid answers. Although there is a fairly accurate history of the general aviation market, only forecasts exist for the future.

The general aviation market, as analyzed by itself, has experienced a general increase in airplane prices over the past years and a very sharp price increase in the last 10 years. The recent increases are steep enough that quality used airplanes have often been purchased, operated, and properly maintained for several years and then sold for a profit. The overall general aviation market is fueled by *supply and demand.*

What is supply and demand? In simple terms, it determines market prices for everything—including used airplanes. It consists of the immediate supply of a sought-after product and the prices that purchasers are willing to pay for that product.

No manufacturer of small airplanes produces an appreciable number of aircraft on an annual basis. Some build fewer than one plane per month and others build nothing at all. This has caused a short supply of new airplanes and a dwindling supply of used airplanes for a market that is very alive with prospective purchasers. The supply of used airplanes gets smaller each year because of attrition (permanent removal from the fleet because of age, damage, and maintenance expense).

The only major exception to low production numbers is Cessna. Since reentering the single-engine piston market, it has produced more than 1000 model 172 and 182 airplanes.

PRODUCTION LEVELS

During the mid- to late 1970s, U.S. production levels of single-engine general aviation airplanes ran from 10,000 to over 14,000 planes annually. This was the heyday of general aviation (Fig. 1-1). The production levels of the 1970s were exceeded only by the year 1946, when some small airplane factories turned out an airplane every hour and the annual production level exceeded 25,000 airplanes.

Records for the most recent two years show current total production levels of the same types of airplanes to be increasing—increasing, that is, over the "under 500 airplanes yearly (only 444 in 1994)" reported for the four-year period just prior, which was the lowest since World War II, when production was not allowed, due to the war effort. The total numbers may be somewhat inaccurate, as many of the airplanes included are not suitable for pilot ownership (Cessna Caravan, Piper Malibu, and twin engine airplanes, for example, which are meant for charter work), but do show the trend.

During 1978, Cessna manufactured 7423 piston-powered, single-engine airplanes. From 1988 through 1996, none were produced. In 1996, small, active airplane makers in the United States produced (or estimated production of) a total of 575 new airplanes of the types covered in this book. Subtract the approximate 145 Caravans and Malibus built for a total of 430 airplanes considered to be owner/pilot airplanes. The active manufacturers included American Champion, Bellanca, Cessna, Commander, Maule, Mooney, Raytheon (Beech), and Piper.

What caused this situation? Why did production nearly stop? It's hard to say, and the finger of blame doesn't point in only one direction. However, there is little doubt that product liability on the part of the manufacturers was the major contributing factor.

How bad are the effects of product liability? The NTSB (National Transportation Safety Board) found that the causes of 203 crashes of Beech airplanes between 1989 and 1992 included weather, poor maintenance, and

Fig. 1-1. *Heyday shot of a Cessna 172, when these planes were being built in large numbers.*

pilot or air-control errors. None were blamed on faulty airplane design. Yet Beech spent an average of $530,000 per incident defending itself from product liability suits. The other airplane manufacturers suffered a similar fate under product liability claims.

In the period of 15 years after the bubble burst, it is estimated that 100,000 jobs in the general aviation market were lost as a direct result of product liability. For example, Cessna stopped producing piston-engine airplanes because of the high cost of defending itself in product liability suits. Other factors affecting the market include the lack of an economical means to certify new models of airplanes, which prevents startup manufacturers from entering the market, and the ever-increasing cost of labor and materials. These factors have resulted in the continued increase in selling prices of new airplanes and, therefore, used airplanes.

Exports

A less-known factor involving the high price of used airplanes is the overseas market. Suffering from the lack of American production of small airplanes, the overseas market has added to the supply and demand problem.

According to the U.S. Department of Commerce, during the most recent years of record (1990 to 1995) an average of over 180 general aviation single and multiengine airplanes (the type of planes covered in this book) were exported from the United States annually. Although these numbers don't sound alarming, as there are over 100,000 small airplanes active in the U.S. registry, only the best used airplanes are shipped overseas. Ercoupes, old rag wings, and even many of the planes built in the 1960s and 1970s don't qualify; only the best of the best airplanes are finding homes overseas.

An additional export category is the new airplane. Nearly 25 percent of all new single-engine airplanes are exported.

Fact: The average age of all the two- and four-place small airplanes in the U.S. general aviation fleet is over 30 years.

More numbers

According to the figures published by GAMA (General Aviation Manufacturers Association) and the AOPA (Airplane Owners and Pilot's Association), there were 165,073 active single-engine airplanes in the U.S. general aviation fleet in 1990. By 1998, that number had dwindled to 157,056 airplanes. This drop in number is consistent with the low production levels of airplane manufacturers. As older airplanes become uneconomical for further use, they are not being replaced with new models; they are simply being removed from the supply of used airplanes. The 2003 total was 156,098 airplanes.

Over the past several years, approximately 25 percent of the small-airplane fleet has changed hands yearly.

GAMA forecasted that the drop in the annual production number would end after 1996 and begin to rise slightly starting in 1997. In early 1999, it appeared they were correct.

As an interesting side note, the total number of reported general aviation accidents has dropped significantly since 1980, when the production levels also dropped. The drop in accident numbers is therefore more a result of fewer airplanes and less flying than a result of improved safety. The fatal accident levels fluctuated from 1.69 to 1.87 incidents per 100,000 hours flown during the same time period, indicating no significant change.

The only increase

The single category of airplanes whose activity numbers are showing an increase are the experimental amateur-built (homebuilt) airplanes. There were 6854 active homebuilt airplanes in 1993 and 21,950 by 2003. Homebuilding is done for reasons such as personal accomplishment, learning experience, and creating a specialized airplane. Homebuilding is seldom considered for economic reasons, although it can, in some cases, make for inexpensive flying.

TURNAROUND

Public law 103-298, the GARA (General Aviation Revitalization Act of 1994), reduced the exposure of airplane manufacturers to product liability lawsuits. Rather than a manufacturer being held responsible for possible or alleged product flaws through the entire life of an airplane, the exposure period is now limited to 18 years from the date of first delivery.

This means that airplanes built prior to 1987 have been removed from product liability exposure. Over the next few years, the vast majority of the small-airplane fleet will be removed from the product liability arena due to the small number of aircraft being produced. Remember that the average small airplane was built in 1967 and is, therefore, well over 30 years old.

The immediate effect of GARA was the industry-claimed resurgence of airplane manufacturing. Cessna Aircraft again started production and between 1996 and 1999 built over 1000 small piston-powered airplanes for the general aviation fleet. Beech (Raytheon), Mooney, and New Piper continued and saw increases in production. The smaller companies of American Champion, Aviat, Lake, and Maule provided for their specialized markets. Other familiar old names planned comebacks, including Luscombe, Meyers, and Swift. A few new names entered the mix, including Cirrus, Diamond, Lancair, and Zenith. As of this writing, only the Katana is actually seen on the used market. The viability of the others remains to be seen (Fig. 1-2).

Courtesy of Cessna

Fig. 1-2. *The 1996 Cessna 172, signifying the restart of general aviation.*

1996 production numbers

In the first quarter of 1996, the heavily touted resurgence of smallairplane production amounted to only 119 airplanes. However, the trend held and about 534 airplanes were built for the general aviation market in 1996. Numerically, this didn't show any significant gain over immediately prior years.

2004 production numbers

In early 2005, the General Aviation Manufacturers Association (GAMA) released its final production figures for 2004. The following is a breakdown of two- and four-place single-engine and light twin production by manufacturer and model:

AMERICAN CHAMPION

7ECA	2
7GCAA	12
7GCBC	24
8GCBC	18
8KCAB	38

AVIAT

Pitts	9
Husky	33

CESSNA

172	236
182	329
206	89

CIRRUS

SR20	91
SR22	459

DIAMOND

DA20	58
DA40	203

MAULE

M-7	11
MT-7	1
MX-7	5
MXT-7	8

MOONEY

M20M	9
M20R	28

RAYTHEON (BEECH)

A36	62
58	31

SOCATA

TB-10	3
TB-21	2

THE NEW PIPER AIRCRAFT

PA-28-161	18
PA-28-181	19
PA-28R	12
PA-32	38
PA-32R	40
PA-34	10
PA-44	11
PA-46	15

TIGER AIRCRAFT

AG5B	19

FORECAST

Will we ever see small-airplane production numbers at past record levels? It is doubtful this will happen in the foreseeable future. Too many financial restrictions are currently imposed on the market. New airplanes are too expensive for the majority of the buying public, but used airplanes are affordable and can, under some circumstances, actually represent an investment for profit—as well as fun and purpose. For the latest twist to the general aviation market, read Chap. 16, "Light-Sport Aircraft."

Used values rising

In 1985, a Piper J3's value averaged $10,500. The same J3 today will fetch $30,000 (or more). This represents an increase of nearly 3 times. An average 1978 Cessna 182 shows $35,000 as its 1985 value and $98,000 today, for an increase of nearly 3 times. A similar 1980 Piper PA-28-236 Dakota shows $44,000 and $100,000, respectively, more than doubling in value.

Those numbers are great for the seller, when you look at the CPI (Consumer Price Index) for the same period and note its approximate 75 percent rise for the same time span. The high prices are tough for the purchaser, but that is the law of supply and demand at work.

CPI chart

The U.S. Department of Commerce obligingly provides an accurate means of comparing costs from one year to another in the form of the Consumer Price Index (CPI). Using the numbers provided in Table 1-1, any year can be compared to another.

ON THE UPBEAT

In March of 1996, Charles Suma, CEO of The New Piper Company, gave a very optimistic speech at the FAA forecast Conference:

> Many had predicted the slow, agonizing death of the piston engine powered general aviation industry by mid-1995. How wrong could they be? Very wrong! Today we see a new level of confidence in our customers and new interest in general aviation. Why? The General Aviation Revitalization Act of 1994, known as GARA, was passed on August 17, 1994. This was perceived and accepted by our customers as having immediate and long-term positive effects in our industry.
>
> Today you will hear of new technology, new aircraft, increases in production, and the rebuilding of the general aviation workforce. This morning the GA Team 2000 was announced. This program is

Table 1-1. Consumer Price Index (CPI) Conversion Factors, 1947 through 2004

Year	Conversion factor	Year	Conversion factor
1947	0.118	1976	0.301
1948	0.128	1977	0.321
1949	0.126	1978	0.345
1950	0.128	1979	0.384
1951	0.138	1980	0.436
1952	0.140	1981	0.481
1953	0.141	1982	0.511
1954	0.142	1983	0.527
1955	0.142	1984	0.550
1956	0.144	1985	0.570
1957	0.149	1986	0.580
1958	0.153	1987	0.601
1959	0.154	1988	0.626
1960	0.157	1989	0.656
1961	0.158	1990	0.692
1962	0.160	1991	0.721
1963	0.162	1992	0.743
1964	0.164	1993	0.765
1965	0.167	1994	0.785
1966	0.172	1995	0.807
1967	0.177	1996	0.831
1968	0.184	1997	0.850
1969	0.194	1998	0.863
1970	0.205	1999	0.882
1971	0.214	2000	0.912
1972	0.221	2001	0.938
1973	0.235	2002	0.952
1974	0.261	2003	0.974
1975	0.285	2004	1.000

To convert from a prior year: Divide the dollar amount by the conversion factor. Example: 1950 price of $4250 is divided by the conversion factor of 0.128 to get the 2004 dollar value of $33,203.

To convert from the present: Multiply the dollar amount by the conversion factor. Example: Present price of $19,950 is multiplied by the conversion factor of 0.177 to get the 1967 dollar value of $3531.

directed toward increasing our pilot population by attracting more new men and women into aviation. This is a very important goal. We are extremely pleased by the support from Administrator Hinson and the entire aviation community. This program will only succeed if it has broad participation and commitment from all segments of the aviation community.

There is not a single company, government agency, or individual at this conference that knows the significance of GARA more than The New Piper Aircraft, Inc. and myself. If there is a doubt in anyone's mind of the effect of this landmark legislator, we are living proof. We are The New Piper because of GARA and its limiting effect on the enormous product liability tail.

We call this the rebirth of an industry. If we believe this, then let's call the 1980s and early 1990s the embryonic stage of the birth process and understand that this process has developed a new industry. One that as a newborn is fragile, requires nurturing, protection and the wisdom of its parents to learn how to survive in, and contribute to, the world of the future. As parents it is our obligation to pass to this child the lessons we learned as an industry in the 1970s so history does not repeat itself.

I believe we all realize that product liability was a major factor in the decline of this sector of general aviation; however, do not forget there were other factors that when combined also have had a major impact on the marketplace. Today's marketplace and economy are substantially different than the 1970s, when this sector of general aviation produced in excess of 145,000 aircraft in 10 years. The majority of these are still in use today, worldwide. Business factors are significantly different, such as:

- The loss of investment tax credits.
- Elimination of the accelerated depreciation.
- Insurance costs for owners, operators, and manufacturers are substantially higher and do not track with inflation.
- Fuel costs have escalated at a higher rate than inflation.
- Decline in middle-class consumers and their ability to purchase products.
- Cessation of the G.I. Bill.
- Long life of the product.
- Risk of overproduction.

These factors have forced a decline in the consumer base with the income and desire to purchase new aircraft products. The decision to purchase an aircraft is one based on need, price, emotional appeal, performance, operating costs, and residual values. Companies such as New Piper, which did not recede from the market, have over the past four years revised their sales and marketing strategies to identify new sales opportunities by addressing these issues.

Manufacturing and sales practices of the 1970s—such as build them, tie them down and then find someone to buy them at discounted prices—will not work in today's market. Consumers are value driven and strongly dislike their investments being depreciated. This is a fragile market that will be affected strongly by sales gimmicks and overproduction of products, which result in diminished returns on investments to consumers. Such statements as "we did it before, we can do it again" show a lack of understanding of today's consumer, which falls under the philosophy of "build and they shall buy." This may work in Hollywood movies or in Disney World, but in the real world these types of sales tactics will only hurt our industry, not help revitalize it.

The majority of current sales are replacement of existing products or upgrades to higher-level products. These are not on a one-for-one basis, meaning if an FBO (fixed-base operator) currently operates three aircraft, he or she may decide to buy one or possibly two new aircraft and sell the old aircraft on the open market. This has a ripple effect in the market as the old aircraft are refurbished or sold to a consumer to use "as is," depending on the condition. These aircraft will not go away, but will become an alternative to buying new. Remember our consumers are sophisticated in their knowledge of new products available in the market and the long-term value of their investment. To succeed, you require a true understanding of the customer.

In no way are we advocating a negative or down market; however, we do advocate a growth-oriented, conservative approach to the piston engine segment of general aviation. The success of this market depends largely on the approach to the marketplace and the integrity we build with customers.

As The New Piper, we have a vision that is based on technology driving innovations in current products and providing the foundation for development of new platforms. To achieve that vision we feel it imperative to maintain a responsible, growth-oriented, but bottom-line approach to the marketplace.

The message I would like each of you to leave this luncheon with today is one of responsible, conservative growth. As an industry we have a unique opportunity to rebuild the piston engine sector of general aviation. This will require the patience of the manufacturers, consumers, and industry support groups—as it will not happen overnight. Realistic growth patterns based on true customer needs and pacing production levels to meet the demand, not exceed the demand, are goals we as an industry should strive to achieve. We must moderate the cyclical nature of our industry. It is always easier to increase production levels than reduce and face the negative consequences that result.

Let's not get caught up in the past promises, but look to the future and remember our responsibility as an industry is to our customers.

What really counts

First and foremost, purchasing a small airplane cannot be examined only in the light of production numbers and prices. Airplane ownership cannot be measured with a ruler or a chart. A personally owned airplane is a pleasure item with possible business applications.

Edward W. Stimpson, then President and now Vice Chairman of GAMA, in an essay written for the Smithsonian's Air and Space Museum, summed up general aviation by saying, "But general aviation is more than just manufacturing and employment statistics. In fact, it's not about transportation at all. It's about flying."

2

To Own or Not to Own

SO YOU WANT YOUR OWN AIRPLANE. Well, sooner or later nearly everyone who earns a pilot's license has the desire to own an airplane. The fact that you're reading this book means you at least dream of having your own plane (Figs. 2-1 and 2-2).

Although new airplane prices have gone straight through the roof, the prudent purchase of a used airplane can, in many cases, actually reduce the cost of flying. This chapter examines the financial justification. In other words, is owning your own airplane worth what it will cost?

Owning an airplane gives you pride and responsibility of ownership, the freedom to fly anytime and anywhere you choose, and the opportunity to modify and equip your airplane as you like. In return you protect, care for, cherish, and love your airplane. But most of all you pay for it.

THE EXPENSE OF OWNERSHIP

Ask yourself the following question: "Can I really afford to own an airplane?" Before you answer, consider the following:

- Do you fully understand the total costs of airplane ownership?
- Will you use the airplane often enough to justify ownership?
- Have you thoroughly considered the alternatives to ownership?

Let's examine the real costs of ownership, costs not included in the price of purchase. In the airplane ownership examples later in this chapter, assume that cash is paid for the airplanes and that loans only serve to increase the total expense of ownership.

Vocabulary

The language of airplane ownership and operation abounds with specialized words and phrases:

Courtesy of Beechcraft

Fig. 2-1. *Airplanes can be used for family and business traveling, requiring speed, equipment, and considerable expense.*

Courtesy of FletchAir, Inc./photo by G. Miller

Fig. 2-2. *Airplanes can also represent loads of economical fun.*

Fixed costs: The expenses of ownership regardless of the amount of use the airplane gets. Fixed costs are incurred even if the airplane sees no use at all. Included in fixed costs are hangar and tiedown fees, state and local property taxes on the aircraft, insurance premiums, and the cost of the annual inspection (difficult to estimate because of the mechanical variables of different aircraft).

Operating costs: Include the price of fuel and oil used per hour, an engine reserve, and a general maintenance fund.

Hourly cost: A calculated figure based on the annual total of fixed and operating costs divided by the total hours of operation. Hourly cost is the all-important number that shows whether owning your own airplane is worth what it costs.

Engine reserve: A monetary fund, built up on an hourly basis, to pay for the eventual engine overhaul or rebuild. Figure this hourly rate as the estimated cost of an engine overhaul or rebuild divided by the recommended overhaul time limit (generally called the TBO, or time between overhaul).

General maintenance fund: A cumulative savings used to pay for minor repairs and service (normally $8 to $25 per hour of operation, depending on aircraft complexity) that often need a mechanic's or technician's attention, including oil changes, periodic Airworthiness Directives (ADs) compliance, minor mechanical defects, and avionics problems.

Worksheets

A worksheet will help you calculate the cost of ownership (Fig. 2-3). The sheet asks for considerable specific information, so you may want to call your local airport to obtain some of the data. After filling in the blanks, follow the instructions for computations.

An example

Let's examine a hypothetical case of ownership. The airplane is a 1975 Cessna 150 valued at $20,000 to be flown 100 hours annually (see Table 2-1). The costs are $45 per month for outside tiedown (very modest), a 1 percent local tax applied to the value of the airplane, and a state aircraft annual registration of $10. The insurance cost is based on the value of the airplane and the pilot's experience. The annual inspection cost is based on the average of estimates made by several FBOs (assuming no serious problems are found).

The GPH is an estimate of fuel usage at a cost of $3.35 per gallon, and the engine reserve of $6 per hour should be adequate for overhaul or rebuild. The general maintenance rate of $8 per flying hour will allow a reserve to build up for the repair of routine mechanical difficulties (brakes, tires, nosewheel shimmy, flap actuator jack problems, radio failure, and so on). As a rule of thumb, the hourly cost for general maintenance (not including engine overhaul) can be estimated as one-half the hourly fuel cost.

The example shown in Table 2-1 is for an airplane flown 100 hours during the year. Notice that the total hourly cost is less than the cost of renting a similar airplane, making a good argument for ownership. Now let's try other operating times and see how usage, or the lack of it, affects the total hourly cost.

One hour of annual use

A worst-case scenario of flying for only one hour during the entire year shows that usage is necessary to make ownership practical (see Table 2-2). Notice how these fixed costs do not change to reflect the lack of usage.

OWNERSHIP WORKSHEET

Fixed Costs

1. Storage (12 months) .... _____
2. Annual taxes .... _____
3. Annual state registration .... _____
4. Insurance premium (12 months) .... _____
5. Annual inspection (estimate) .... _____
6. Total fixed costs (add lines 1–5) .... _____

Operating Costs

7. Fuel cost per gallon X GPH .... _____
8. Engine reserve per hour .... _____
9. General maintenance per hour .... _____
10. Total operating costs (add lines 7–9) .... _____

Hourly Cost

11. Total fixed cost (from line 5) .... _____
12. Total operations (line 10 X hours flown) .... _____
13. Total costs (add lines 11 and 12) .... _____
14. Total hourly cost (line 13 / hours flown) .... _____

Practicality of Ownership

15. Rental price for comparable airplane .... _____
16. Total hourly cost (from line 14) .... _____

Fig. 2-3. *Use this worksheet to figure if airplane ownership is economically practical for you.*

Table 2-1. Hypothetical Case of Ownership of 1975 Cessna 150

Fixed costs	
12 months of storage	$540.00
Annual taxes	$200.00
Annual state registration	$10.00
Insurance premium (12 months)	$950.00
Annual inspection (estimate)	$525.00
Total fixed costs	$2225.00
Operating costs	
Fuel cost per gallon × GPH	$16.75
Engine reserve per hour	$6.00
General maintenance per hour	$8.00
Total operating costs	$30.75
Hourly cost	
Total fixed costs	$2525.00
Operating costs × hours flown	$3075.00
Total costs	$5600.00
Total hourly cost	$56.00
Practicality of ownership	
Rental price for comparable airplane	$65.00

They are fixed for the year, even if the plane is never used, which certainly makes that one hour of use expensive!

50 hours

Many privately owned airplanes see only 50 hours of operation a year (see Table 2-3). Unfortunately, this is a reasonably accurate average. In this example of only 50 hours of annual usage, the practicality of ownership has disappeared against the rental cost of $65.00 per hour.

200 hours

The fortunate owner who can fly a plane 200 hours a year gets the most return of any examples cited thus far (see Table 2-4). It is very simple: the more use the airplane gets, the lower the total hourly cost.

Table 2-2. Effect of Usage on Total Hourly Cost

Fixed costs

12 months of storage	$540.00
Annual taxes	$200.00
Annual state registration	$10.00
Insurance premium (12 months)	$950.00
Annual inspection (estimate)	$825.00
Total fixed costs	$2525.00

Operating costs

Fuel cost per gallon × GPH	$16.75
Engine reserve per hour	$6.00
General maintenance per hour	$8.00
Total operating costs	$30.75

Hourly cost

Total fixed costs	$2525.00
Operating costs × hours flown	$30.75
Total costs	$2555.75
Total hourly cost	$2555.75

Practicality of ownership

Rental price for comparable airplane	$65.00

Table 2-3. Hourly Cost for Annual Usage of 50 Hours

Practicality of ownership

Total fixed costs	$2225.00
Operating costs × hours flown	$1537.50
Total costs	$4062.50
Total hourly cost	$81.25

More examples

Let's see what a four-place airplane with a 225-hp engine and retractable landing gear will cost (see Table 2-5). The example airplane is based near a large metropolitan area.

The charge for outside tiedown is $150 per month, which is relatively typical near major cities, and indoor storage could be as high as $500 per

Table 2-4. Hourly Cost for Annual Usage of 200 Hours

Total fixed costs	$2225.00
Operating costs × hours flown	$6150.00
Total costs	$8675.00
Total hourly cost	$43.38

Table 2-5. Cost of Operating a Four-Place Airplane

Fixed costs	
12 months of storage	$1,800.00
Annual taxes	$3,000.00
Annual state registration	$60.00
Insurance premium (12 months)	$2,400.00
Annual inspection (estimate)	$1,450.00
Total fixed costs	$8,710.00
Operating costs	
Fuel cost per gallon × GPH	$53.60
Engine reserve per hour	$10.00
General maintenance per hour	$25.00
Total operating costs	$88.60
Hourly cost	
Total fixed costs	$8,710.00
Operating costs × hours flown	$8,860.00
Total costs	$17,570.00
Total hourly costs	$175.70

month. A 4 percent local tax applies to the $75,000 value of the airplane. A state registration, based on the plane's weight, costs $60 (state registration costs vary from nothing to being based on either weight or value, depending on the particular state the airplane is based in or where the owner resides). The insurance cost is an estimate based on the value of the airplane and the pilot's experience (or lack of experience) in a retractable aircraft. The annual inspection reflects the increased cost of a more complex airplane over that of the original example (Cessna 150). Fuel usage is calculated at 16 gallons per hour at $3.35 per gallon. The engine reserve is $10 per hour, based on the current overhaul costs for a 225-hp engine.

The general maintenance rate of $25 per flying hour allows for a repair reserve. The example is based on 100 hours annual usage.

What now?

Compare these figures or figures based on the airplane you want to own with the straight hourly rental charged for a similar airplane at the local FBO. Keep in mind that you can sometimes purchase blocks of time at a reduced rate, often as much as 20 percent below the posted rate. None of these figures takes into consideration the purchase price of an airplane or the costs of financing in cases where a loan exists for purchase funds.

AGAINST OWNERSHIP

As the owner of the airplane, all maintenance is your responsibility. You won't have a squawk book to write pilot/renter complaints into and expect the FBO to address them before the next flight. You will have to either fix your problems yourself or pay out of your own pocket to have them repaired.

Do you enjoy the clean airplane rented from the FBO? If you own an airplane, you will have to keep it clean yourself. Consider the size of an airplane. The top surface of the wings can exceed 200 square feet, which is a lot of area when you are the one doing the washing and polishing. And that doesn't even include the windshields, upholstery, carpets, and oil-stained belly!

But don't get glum. Many families enjoy flying as a group activity. Maintenance and cleaning the airplane can be just part of the fun. Often, small airports operate in a country-club fashion with cookouts and other get-togethers.

It is important to point out, however, that an airplane must be flown on a regular basis. As seen in the ownership examples, too few hours of annual use results in high total hourly costs—eventually negating any financial reasons for ownership.

A PRIMER ON AIRCRAFT FINANCING

Financing any big-ticket item requires quite a bit of money. The exact cost is usually well hidden within the fine print of a loan contract, and most purchasers only concern themselves with the monthly payment (a fact well known by profit-making bankers). Monthly payments amortize (pay back) a loan and include a portion of the principal and a portion of the interest of the entire loan. How much of each is determined by the type of loan.

The *principal* is the amount borrowed and the *interest* is the price paid for use of another person's money. This applies whether you borrow from a bank, loan company, or individual. Generally, two forms of interest are available: fixed rate and variable rate. A *fixed interest rate* is a single interest rate set by the loan contract that remains unchanged for the life of that contract. *Variable interest rates* can go up or down during the life of the loan contract, normally tracking the prime lending rate.

When purchasing an airplane or any other expensive item, always plan for the worst if the loan is based on a variable interest rate. A variable-rate loan could produce higher monthly payments, because of a rise in the prime interest rate, than was originally anticipated (or budgeted for). A fixed-rate loan is best because you know from month to month what the payment is going to be.

A mixture of fixed and variable interest rates is a *variable-term loan*, which sets a never-changing monthly payment for the life of the contract. The length of the contract, however, can vary depending on the prime interest rate. If the interest rate rises, the total of your loan will increase (principal + interest). You will owe more money, hence your loan payment plan will be extended at the set monthly payment for the number of months necessary to pay all the additional accrued interest.

Some loan contracts have a balloon payment tacked onto the end (last payment) to keep monthly payments artificially low. Take, for example, a $20,000 loan for which, during the life of the contract, the sum total of the payment, less the interest, amortizes only $5000 of the principal (because of the small size of the monthly payment). The remaining $15,000 of the principal becomes the last payment, thus the term *balloon payment*.

A rather nasty surprise often found in loan contracts is the *prepayment penalty clause*, which means you have to pay an additional premium (penalty) to complete the loan contract at a date earlier than agreed. Prepayment clauses often demand such a stiff penalty for early payment that there will be no savings for early payoff. A prepayment penalty clause can be very expensive if you must satisfy a loan contract before completing the payment schedule, as would be the case when selling or trading an airplane.

A fact often overlooked when considering the cost of airplane ownership is lost income. In this case, lost income refers to interest not earned on money used to purchase an airplane. For example, a certificate of deposit or similar investment can safely return five or more percent on deposited funds (savings). On a $50,000 cash airplane purchase, this could mean lost income of about $2500 per year.

On the bright side is the fact that most airplanes appreciate in value each year. Of course the amount of appreciation depends on the make and model and the continued appearance and mechanical condition of the airplane.

Splitting the costs

It is possible, under some circumstances, to split the costs of airplane ownership among more than just the owner's pocket. Two methods, partnerships and leasebacks, have been successfully used by many. Both have positives and both have drawbacks—and neither is for everyone.

LEASEBACKS

Most privately owned airplanes fly fewer than 100 hours a year. As was seen when the costs of ownership were discussed earlier, it is sometimes difficult to justify the cost of owning an airplane for such low usage.

Some pilots want to own an airplane, knowing full well that they will not really use it enough to financially justify ownership. Others need financial help to make the initial purchase and are looking for the generation of income from the airplane as an offset to the expense of individual ownership.

It is not just individuals that can have a need for financial assistance making airplane purchases. Most financial institutions want FBOs to have a 15 to 20 percent cash investment in any airplanes they finance. Down payments can be a difficult problem for many small FBOs. So, some FBOs seek individuals to buy an airplane and lease it to back to them. Enter the leaseback.

A leaseback can sound like a good deal, because both parties get what they want. However, problems with leasebacks can start in how the contract of lease is written, who controls use of the airplane, and who pays the storage and maintenance bills.

Remember, a leaseback airplane is a rental airplane and rental airplanes live hard lives. They often are not well cared for—certainly not with TLC (tender loving care)—and suffer for it. The airplane's suffering converts to maintenance expense for the owner.

The most common leaseback arrangement calls for the FBO to rent the airplane to customers at a set hourly rate. The FBO keeps part of the rental and gives the remainder to the owner. The latter is responsible for all the operational and fixed costs including: fuel, maintenance, insurance, storage, engine overhaul, paint, and the plane's interior.

The actual owner will not, in most cases, be able to use the airplane at any time desired. The owner will need to schedule for it, just like a renter. If it is already booked, you can't use it. A leaseback is primarily for the benefit of the FBO, not the airplane owner.

Be very wary of any leaseback contract you are asked to sign. Have an experienced aviation attorney review the contract and be sure you understand it completely. Maintenance can be a real problem and must be spelled out very carefully, or you will be very much taken advantage of.

Leasebacks are not for everyone and must be completely explored for pitfalls before making any legal or financial commitment. Something else to think about is: What happens if the FBO you lease your airplane to goes out of business?

Before you become involved with a leaseback, consider that you might be better off continuing to rent while saving money for the eventual purchase of an airplane.

PARTNERSHIPS

A partnership can be *the* solution to aircraft ownership for some—for others, partnerships are nothing but trouble from the very first day. In general, poor planning involving money and scheduling the use of the airplane cause most partnerships to self-destruct.

Partnerships are not limited to just two individuals, but often have four, five, or even more partners. Although the expenses get spread over more pockets as the number increases, the problems seem to increase.

Often, partnerships run into difficulties when one of the partners uses the airplane more than the other(s). This is particularly true if the costs are split evenly among the partners and not through some form of prorating based on hourly usage. A limited solution to this inequity might mean that the fixed and maintenance costs are divided equally among the partners, but each is responsible for his or her own direct operating costs (fuel).

A more workable, yet also a simple, means of equitably spreading the costs among the partners is to set up an hourly use plan. This plan would charge each partner for the airplane's hourly operating costs, including maintenance reserves, and only the fixed costs would be spread equally among the partners. Some successful partnerships actually split all the expenses based upon hours of use; however, this may mean an adjustment at some point each year, as maintenance expenses are incurred.

PROBLEM AREAS

Insurance coverage can be a very real problem to partnerships, particularly due to differing pilot experience levels. For example, several high-time pilots take in a new low-time pilot as a partner. The insurance costs will probably increase. The high-time partners might feel this additional expense should be borne exclusively by the new low-time partner.

Scheduling conflicts are often the most contentious problem in airplane partnerships. The best arrangement is the two partner system, with one flying mostly during the week and the other mostly on weekends. When all partners want to fly on the weekend, you have problems.

Formal scheduling might help, but does not take into consideration weather, family, maintenance, etc. Flexibility must be built into the partnership to insure that each partner gets a fair amount of airplane use.

CHOOSING PARTNERS

The best partners are flying buddies with common interests, not just several individuals with a common interest of reducing the cost of flying.

All agreements must be in writing. This is not the day and age for verbal partnerships, even among friends. If push comes to shove and the entire agreement is in writing, everything is clear to all partners. Repeat, the agreement should be as clear and straightforward as possible and leave no ambiguity about who is responsible for what. It is recommended that a lawyer write the agreement.

Plans should be made ahead for dissolution of the partnership. When one or more partners desire to sell, some predetermined means should apply. Either the remaining partners can do a buyout or the plane can be sold and the proceeds split among the partners. Of course, as an alternative, a new partner could be found.

QUESTIONS AND ANSWERS

Before purchasing an airplane, the prospective airplane owner needs to give serious thought to how the plane will be used. You must be completely objective and honest in your thinking. Remember that you plan to have this airplane for a long time and you want to be very satisfied with it.

In order to determine what airplane is best suited to your particular needs, you need to answer some questions about your flying. A good way to answer these questions is to talk to other pilots or owners of planes you are considering. You may also want to contact the type of club supporting a specific make and model you are considering (see Appendix D), or you could purchase the airplane's flight, owner, service, parts, and engine operation manuals. Type clubs are warehouses of information from owners, and the manuals can provide factual information right at your fingertips. Some manuals are available directly from manufacturers and sometimes from an FBO, or you might be able to borrow them from another pilot. Copies of airplane manuals are available for purchase from:

Essco, Inc.
378 S. Van Buren Avenue
Barberton, OH 44203
(230) 644-7724
www.esscoaircraft.com

or

The Airport Shoppe
2635 Cunningham Avenue
San Jose, CA 95148
(800)634-4744
www.airportshoppe.com

Planned use

Why do you fly? Is your flying for sport and relaxation, for business or family transportation, or for hauling cargo?

Sport pilots fly for recreational purposes, generally during the evenings and on weekends. Sometimes forays are made to other airports, to fly-ins or other social gatherings, or just over the local area.

Pilots who use their airplanes for serious business or family transportation want reliable, speedy transportation with good load-carrying characteristics. Their reason for flying is to get from point A to point B.

Small-cargo pilots, usually associated with bush flying in Alaska, in Canada, or on large ranches, will generally be interested in airplanes capable of operation from unimproved areas and having large carrying abilities.

Flying skills

What are your real flying skills? Are you most comfortable with a low-performance tricycle-gear airplane or are you well qualified in a high-performance retractable or small twin?

Review your current piloting skills and the types of airplanes you most often fly. Your proficiency and flying comfort level (where you are most secure) are good indicators for selecting an airplane. Also consider skills you plan to acquire.

Passengers

How many people fly with you? Is most of your flying solo or perhaps with your spouse or a flying buddy? Do you only occasionally take the entire family for a day of flying adventure?

You buy an airplane to do a job. There is no real reason to select an airplane with capabilities far exceeding the requirements of that job. If you usually rent a two-place plane and it suits your needs, then stay in the two-place category. There is no financial justification for owning something larger. When the infrequent need for a larger airplane does arise, you can always rent one.

Distance traveled and speed

How far do you generally fly? Are most of your flights fewer than 100 miles or are they frequently hundreds of miles? Closely related to the issue of distance traveled is the question of speed. How quickly do you really need to get there? Be tough with the answers and remember to be objective.

Where you are planning flights to and how quickly you must get there are valid points for consideration. Pilots flying from point A to point B for a business meeting are interested in speed. A 300- to 400-mile morning trip could take as little as two hours or as long as four hours. A "hurry up, I want to get there and get it done" attitude requires a fast airplane.

For the family on vacation, seeing the countryside is important and speed will not be as much of a consideration. High speed could even be a detractor. Slow down and enjoy the ride. An important point to remember is that speed equates to larger engines, more complex airplanes, and the resulting higher costs of ownership. Speed is expensive.

VFR/IFR

Related to speed and distance are the problems associated with all-weather flying. The airplane used for business transportation from point A to point B should most probably be equipped for IFR. This allows for travel during poor weather conditions and eliminates most weather-caused delays. Do you need to be able to leave immediately, or can you afford to wait for good weather?

Avionics add to the value of a used airplane, and they also add to maintenance expenses. Radio navigation equipment is expensive to purchase, install, and maintain. If you need IFR, then be equipped for it; if you are not instrument rated or don't fly in bad weather, however, save your money. Buy only what you need.

Airports

The airports you regularly fly in and out of will influence your choice of airplanes. If all your flying is from paved runways, the choice is wide open. However, if you find yourself flying from rough grass strips or unimproved areas, you must select an airplane that will stand up to the use and abuse it will receive.

Put it together

Studying and answering these questions, objectively of course, will allow you to determine your level of piloting skill and identify your hardware requirements. You are well on your way to selecting an airplane appropriate for you. Merely having a basic airplane in mind, however, will not do for very long. You must consider many more factors before making a final choice.

MAINTENANCE

Maintenance is one of aviation's biggest expenses. All airplanes require maintenance, but some require more maintenance than others. And the more maintenance required, the more expenses incurred.

Basic automobile maintenance usually means filling it with gas, checking the oil and water, and driving away. Airplanes are quite different; annual inspections (required by regulation), minor repairs, major repairs, and equipment installations are all part of the airplane maintenance picture. When selecting a used airplane, maintenance should weigh heavily in the decision-making process. The more complex an airplane, the more maintenance dollars you will spend. Consider the following maintenance requirements:

- Airframe coverings can be metal or fabric. Fabric needs periodic replacement and metal is for life.

- Landing gear can be fixed or retractable. Retractable gear has many moving parts. These parts wear and will eventually need repair or replacement.

- Propellers are either fixed-pitch or constant-speed (variable-pitch). The latter may have as many as 200 moving parts. Fixed-pitch propellers have no moving parts.

- Engines are often modified for performance reasons, for example, adding a turbocharger. Such modifications are expensive in themselves and often lead to further maintenance expenses and, in some instances, shorter engine life.

- Aircraft age is a prime aspect of maintenance. Older airplanes require far more maintenance than newer planes. This often makes purchasing an older, inexpensive airplane less attractive than buying a newer, higher-priced plane.

These five different maintenance requirements illustrate that simple generally means less expensive. I won't use the word *cheap*, as there is nothing cheap about airplane maintenance. All airplane maintenance is, in varying degrees, expensive. When it comes to airplane equipment, keep the following in mind: "If you don't have it, it can't break, and therefore it will cost nothing to maintain."

Along with routine maintenance, you will find Airworthiness Directive (ADs) and General Aviation Airworthiness Alerts of prime concern, as they can mandate immediate maintenance expense.

Airworthiness directives

Unfortunately, airplanes are not perfect in design or manufacture. They will, from time to time, require inspection, repairs, or service as a result of unforeseen manufacturing problems or defects. These faults may not be identified until years after the actual date of manufacture, but the faults often affect a large number of a particular make and model of airplane. The required inspection or maintenance procedures are set forth in the ADs, as described in Federal Aviation Regulations Part 39, and must be complied with.

An Airworthiness Directive may require only a simple one-time inspection to assure that the defect does not exist, a periodic inspection (every 50 hours of operation, for example) to watch for an impending problem, or a major airframe or engine modification to remedy a current difficulty. Compliance with some ADs can be relatively inexpensive, particularly those involving only minor inspections. However, complying with an AD requiring extensive engine or airframe modifications and repairs can devastate a bank account.

Although ADs correct deficient design or poor quality control of parts and workmanship, manufacturers are not normally considered responsible for the costs incurred in AD compliance. Unlike automobile recalls, the financial burden of meeting AD requirements is routinely met by the airplane owner. Unfortunately, the manufacturer can even profit from selling the parts necessary to comply with the AD. There is no large consumer voice involving aircraft manufacturer responsibility.

Mechanics, inspectors, and the FAA maintain Airworthiness Directive files. An Airworthiness Directive compliance check is part of the annual inspection, thereby assuring continued compliance. Records of AD compliance become a part of the aircraft logbooks. When looking to purchase an airplane, always check for AD compliance.

You can find a listing of Airworthiness Directives applicable to airplanes covered in this book in Chap. 14.

General aviation airworthiness alerts

On a monthly basis, the FAA compiles the Malfunction or Defect Reports, or MDRs (Fig. 2-4), submitted by owners, pilots, and mechanics from all over the nation into a single listing. Called General Aviation Airworthiness Alerts, the list is published and sent to interested parties, including manufacturers, mechanics, and inspectors. Alerts are not the law, as ADs are, but they can indicate areas of mechanical concern and are often the basis for later ADs.

Some aircraft are so laden with ADs and reported mechanical difficulties that they scream out, "Spend money on me!" Check your proposed airplane selection carefully against AD and alert lists. It is better to be discouraged now, at decision-making time, than later when it's time to spend your hard-earned dollars for all the required service work.

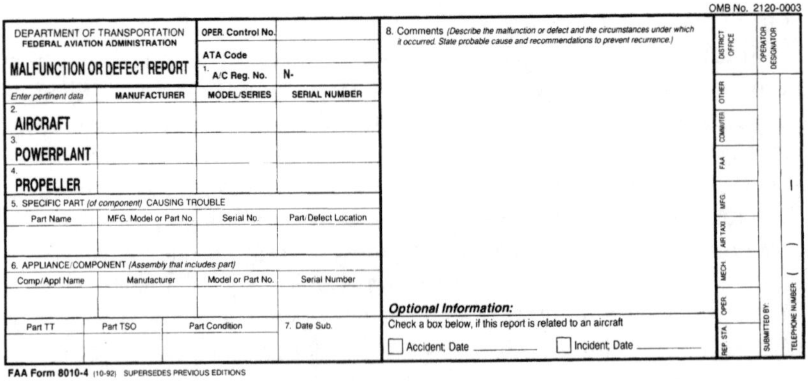
OMB No. 2120-0003

DEPARTMENT OF TRANSPORTATION
FEDERAL AVIATION ADMINISTRATION
MALFUNCTION OR DEFECT REPORT

OPER. Control No.
ATA Code
1. A/C Reg. No. N-

Enter pertinent data	MANUFACTURER	MODEL/SERIES	SERIAL NUMBER
2. AIRCRAFT			
3. POWERPLANT			
4. PROPELLER			

5. SPECIFIC PART (of component) CAUSING TROUBLE

Part Name	MFG. Model or Part No.	Serial No.	Part/Defect Location

6. APPLIANCE/COMPONENT (Assembly that includes part)

Comp/Appl Name	Manufacturer	Model or Part No.	Serial Number

Part TT	Part TSO	Part Condition	7. Date Sub.

8. Comments (Describe the malfunction or defect and the circumstances under which it occurred. State probable cause and recommendations to prevent recurrence.)

Optional Information:
Check a box below, if this report is related to an aircraft
☐ Accident; Date ______ ☐ Incident; Date ______

DISTRICT OFFICE
OPERATOR DESIGNATOR
OTHER
COMMUTER
FAA
MFG.
AIR TAXI
MECH.
OPER.
REP. STA.
SUBMITTED BY:
TELEPHONE NUMBER ( ) -

FAA Form 8010-4 (10-92) SUPERSEDES PREVIOUS EDITIONS

Fig. 2-4. *A Malfunction or Defect Report.*

Information sources

AD and General Aviation Airworthiness Alert information specific to a particular make, model, and serial number of aircraft is available through sources such as the Aircraft Owners and Pilots Association, various type clubs, AD search services advertising in aviation publications, and on CD-ROM from specialized vendors. Warning: Don't buy more airplane than you can afford to properly maintain. Good maintenance not only promotes flying safety, it protects your investment.

INSURANCE

To protect an airplane investment, you must insure it against loss. Loss means costs of repair or replacement in the event of damage. If you are considering buying a specific airplane, make sure to examine the insurability of that airplane.

Insurance companies are in business to make money by providing a service called *coverage*. The *insurance premium*, the amount charged for this coverage, is based on two main factors (sometimes called risk factors): the pilot and the airplane.

The pilot risk factor

The pilot risk factor is determined by the total number of hours flown, types of airplanes flown, ratings, and violation/accident history. Simply stated, it is the competency level of the pilot. To the insurance company, it is the potential for the pilot to cause a financial loss to the insurance company.

The airplane risk factor

Examining the loss ratio history for a particular make and model of airplane and the current availability of replacement parts, in the event of a loss, determines the airplane risk factor.

Parts availability is directly related to the total number of the make and model of airplane manufactured. Orphan airplanes, those long out of production with no parts supply available, are expensive to insure and maintain because of the parts problems.

RESALE

What is the possibility of quickly reselling the airplane at a sum near the purchase price? If you are concerned about resale, it is very important to consider current values and, in the event you need money in a hurry, how quickly the airplane will sell.

An orphan airplane can be a very inexpensive purchase, but it is usually not a hot seller. As already described, most orphans were built in small numbers and are no longer produced or supported by any manufacturer or parts supplier. If you purchase an orphan, understand that when the time comes to sell it, you may have to either sell cheap or wait a long time to find a buyer. This is not to condemn orphans, for they can represent some very affordable flying. Rather, you have to realize what you are getting into.

If being able to resell an airplane is a prime concern, then buy a Cessna or Piper basic four-place fixed-gear airplane. There is always a market for them.

WATCH OUT FOR BARGAINS

There are few bargains in the world, and the field of used airplanes is no exception to the rule. Any plane selling for a price that seems much lower than it should be probably has a serious flaw. Even if you know of the flaw and feel it is not as bad as it sounds, get some professional advice before buying a bargain-priced airplane. It may very well be no bargain at all!

Repairing even simple things on an airplane generally requires a lot of money. For example, replacing a typical four-place airplane's interior can cost more than $3500. Of course, you could get the materials and do the work yourself for about $1000. This might appear to be a good way to save money, until you find that it takes the better part of the summer to get the job done—the part of summer during which you expected to take a flying vacation.

Don't even consider one of those back-of-the-hangar airplanes, the one that doesn't have an engine in it or has an inch of dirt and dust on it. Mechanics don't even want planes long in disuse or in pieces, except for parts. They don't represent an economical entry into airplane ownership for the average person.

When it comes to spending money on an airplane, consider actually purchasing the airplane as the tip of the iceberg. You can either pay a higher price for a plane in good shape, or pay less for a plane requiring work and additional money for maintenance. In the end, the total spent will be nearly equal.

MAKE SURE TO HAVE FUN

If you are still being objective and honest with yourself, you are now ready to look at the specific makes and models of airplanes currently available on the used market and evaluate them with respect to your individual needs. Remember that the final objective of purchasing a used airplane is to find an airplane that is affordable to own and operate and that does not overtax your flying skills.

Some years back a friend of mine earned his private pilot's license, then bought an airplane. He learned to fly in a Cessna 150 but he bought a Cessna 180, a taildragger with a big engine that can be a fire-breathing dragon in inexperienced hands. After the airplane dealer checked him out in the 180 and after he paid high insurance rates based on his lack of proficiency in that type of airplane, he went out to play bush-pilot.

It took only a couple of flights for his inexperience to catch up with him, and the plane took him for an unforgettable ride across the infield during a landing. The plane was undamaged, but the experience scared him so badly that the airplane sat unused for months, costing tiedown fees, insurance premiums, and all the other things known as *fixed costs*.

Finally, I flew it a few times, then demonstrated and sold it for him. He later bought a Cessna 172 and I guess lived happily ever after.

The point is: Don't buy more airplane than you can handle, because if you can't handle it, you won't enjoy it. And if you don't enjoy it, you won't fly it.

3

Determining Airplane Values

PLACING A CASH VALUE ON A USED AIRPLANE is not as simple as valuing a used automobile. Airplane valuation is not a simple matter of make, model, and year. There are four areas of high-dollar items that must be judged very carefully to place an accurate value on an airplane:

- Airframe, which is subject to fatigue, damage, and corrosion
- Engine (the engine's life expectancy is limited)
- Modifications for better performance or particular use
- Avionics, which become outdated as electronic technology improves

The sum of the values of these areas determines an airplane's overall value.

AIRFRAMES

For most purposes, airplanes have one of two types of airframes: all-metal or tube-and-fabric. Most modern airframes have an all-metal construction. Two notable exceptions to this rule that are currently being manufactured are Aviat and Maule.

The advantages of all-metal construction are easier outdoor storage and a much longer life. The disadvantages are expensive repairs and outrageous painting costs. Painting a typical four-place airplane will cost between $3000 and $6000, with some high-dollar finishings priced above $8000. Repainting airplanes is a labor-intensive job that requires knowledge and experience, hence the high prices (Fig. 3-1).

Fig. 3-1. *A typical all-metal airplane.*

Tube-and-fabric airplanes are very expensive to re-cover and do not weather well outside. Exceptions are airplanes covered with synthetic coverings, which can last for many years, and those covered with fiberglass, which is considered permanent. Regardless of the material, recovering is an expensive art and every tube-and-fabric airplane will at some point require re-covering, even those considered permanent. Currently, a re-covering job costs from $8000 to $12,000 (Fig. 3-2).

The total hours of operation affect the value of all airframes. It stands to reason that anything exposed to the stress and strain that airframes experience during operation will fail sooner or later. Additionally, corrosion of the structure may become a problem, and is usually very expensive to repair.

If an airframe has a history of severe damage, the airplane's value will be reduced. The method used to repair damage is the determining factor of the amount of value reduction. Was the airplane repaired to be like new or was it just a "make-do" job? Proper reconstruction of a severely damaged airplane requires considerable time, skill, and expense. The work must be done in a properly equipped shop using jigs to ensure the alignment of all parts.

Along the lines of damage to the structure comes weather-related damage such as from a flood. Flooding not only soaks the airplane with water, it leaves behind sediment, mud, and other debris that can later cause trouble. This means that an airplane submerged in a flood most likely have not only a damaged engine and engine accessories, but also rust and corrosion of metal parts in the airframe, destruction of wooden airframe parts, and destruction of instruments and avionics. Although

Fig. 3-2. *A tube-and-fabric airplane.*

technically repairable, a flood-ravaged airplane should be at the very bottom of most purchasers' lists.

Consider the airplane's past usage. Was it a trainer, crop duster, patrol, or rental airplane? If so, the hours are likely high and the usage rough. Yet commercial or rental use can also denote very good maintenance. After all, maintaining what you have is cheaper than running it into the ground and replacing it.

Although often overlooked, ascertain if the fuel tanks have recently been repaired or restored. Many older airplanes have problems with fuel tanks or bladders, ranging from leakage to contamination. They can be repaired by several methods or replaced. The cost of fuel tank or bladder work ranges from several hundred dollars to several thousand dollars, depending on the specific airplane.

Recognize that there is no such thing in aviation as "a little fixing up" to remedy small problems. Upgrading a complete interior, repainting the plane's outside, replacing windows, and the like is very expensive. Therefore, a plane that needs "a little fixing up" is worth quite a bit less money than a plane needing no fixing up.

ENGINES

Engines are the most costly single item, maintenance-wise, attached to an airplane. An engine failure involving internal breakage can cost thousands

Fig. 3-3. *The very expensive heart of a power airplane.*

of dollars to repair. Therefore, the condition and history of the engine is a very important part of the overall airplane's value (Fig. 3-3).

Rebuilds, overhauls, and confusion

Advisory Circular AC-43-11 explains the differences between engine rebuilding and engine overhauling. The circular is followed by some popular phrases and words that are part of the airplane vocabulary you must learn when searching for, owning, and selling an airplane:

AC-43-11

SUBJECT: Reciprocating Engine Overhaul Terminology and Standards

Date: 4/7/76

Initiated by: AFS-830

1. PURPOSE.
This advisory circular discusses engine overhaul terminology and standards that are being used in the aviation industry:

a. To inform the owner or operator of the variety of terms used to describe types of reciprocating engine overhaul.

b. To clarify the standards used by the industry during reciprocating engine overhaul.

c. To review the Federal Aviation Regulations (FAR) regarding engine records and standards.

2. REFERENCES.

FAR 43, Sections 43.9, 43.13(a), and 43.13(b); FAR 91, Sections 91.173 and 91.175.

3. BACKGROUND.

In the maintenance of aircraft engines, terms such as top overhaul, major overhaul, etc., are used throughout the aviation industry. The standard to which an engine is overhauled usually depends on the terms used by the person who is performing the engine overhaul. These terms are familiar to the aviation community, but their specific meanings are not fully understood. This could result in similar engines being overhauled to different tolerances. We believe that through the discussion that follows, owners or operators and engine overhaul facilities will have a better understanding of the terms and standards relating to those terms.

4. DISCUSSION.

a. The selection of an overhaul facility by the average aircraft owner is usually determined by the cost quoted by the engine overhauler. Engine overhauls can be accomplished to a variety of standards. They can also be accomplished by many different facilities, ranging from engine manufacturers, large repair stations, or individual powerplant mechanics. The selection of an overhaul facility can and does, in most cases, determine the standards that are used during overhaul. The FAR requirement in Section 43.13(a) is that the person performing the overhaul shall use methods, techniques, and practices that are acceptable to the Administrator. In most cases, the standards that are outlined in the Engine Manufacturer Overhaul Manuals are standards acceptable to the Administrator.

b. These manuals clearly stipulate the work that must be accomplished during engine overhauls and outline limits and tolerances used during the inspections. There is no dictionary that provides a commonly accepted standard definition of all the terms used in the aviation industry. The terms discussed in this advisory circular are offered for information purposes only and are not to be considered as definitions set forth in the Federal Aviation Regulations.

c. The only definition regarding engine overhaul is the word "rebuilt." This is defined in FAR 91.175 and refers to rebuilt engine maintenance records.

5. ENGINE OVERHAUL TERMINOLOGY.

a. "Rebuilt."

(1) The term "rebuilt" is defined in FAR 91.175. The definition allows an owner or operator to use a new maintenance record without previous operating history for an aircraft engine rebuilt by the manufacturer or an agency approved by the manufacturer.

(2) A rebuilt engine as defined in FAR 91.175 "is a used engine that has been completely disassembled, inspected, repaired as necessary, reassembled, tested, and approved in the same manner and to the same tolerances and limits as a new engine with either new or used parts." All parts used must conform to the production drawing tolerances and limits for new parts or be of approved oversized dimensions for a new engine.

b. Overhaul. In the general aviation industry, the term engine overhaul has two identifications that make a distinction between the degrees of work done on an engine:

(1) A major overhaul consists of the complete disassembly of an engine, inspected, repaired as necessary, reassembled, tested, and approved for return to service within the fits and limits specified by the manufacturer's overhaul data. This could be to new fits or limits or serviceable limits. The determination as to what fits and limits are used during an engine overhaul should be clearly understood by the engine owner at the time the engine is presented by overhaul. The owner should also be aware of any parts that are replaced, regardless of condition, as a result of a manufacturer's overhaul data, service bulletin, or an airworthiness directive.

(2) A top overhaul consists of the repair of parts outside of the crankcase and can be accomplished without completely disassembling the entire engine. It can include the removal of cylinders, inspection and repair to cylinders, inspection and repair to cylinder walls, pistons, valve-operating mechanisms, valve guides, valve seats, and the replacement of pistons and piston rings. A top overhaul is not recommended by all manufacturers. Some manufacturers indicate that if a powerplant requires work to this extent, it should be given a complete overhaul.

6. FITS AND LIMITS.

As discussed above, two kinds of dimensional limits are observed during engine overhaul. These limits are outlined in the engine overhaul manual as a "Table of Limits" or a "Table of Dimensional Limits." These tables, listing the parts of the engine that are subject to wear, contain minimum and maximum figures for the dimensions of those parts and the clearances between mating surfaces. The lists specify two limits as follows:

a. Manufacturer's Minimum and Maximum. These are also referred to by some manufacturers as new parts or new dimensions. These are the dimensions that all new parts meet during manufacture and are held to specific quality control standards as required by the FAR in the issuance of an Engine Type Certificate to a manufacturer. It is important to note that new dimensions do not mean new parts are installed in an engine when a manufacturer or his authorized representative presents zero time records in accordance with FAR Section 91.175. It does mean that used parts in the engine have been inspected and found to meet the manufacturer's new specifications.

b. Service Limits. These are the dimensions that represent limits that must not be exceeded and are dimension limits for permissible wear.

(1) The comparative measurements of parts will determine their serviceability; however. it is not always easy to determine which part has the most wear. The manufacturer's new dimensions or limits are used as a guide for determining the amount of wear that has occurred during service. In an engine overhaul certain parts must be replaced regardless of condition. If an engine is overhauled to "serviceable" limits, the parts must conform to the fits and limits specifications as listed in the manufacturer's overhaul manuals and service bulletins.

(2) If a major overhaul is performed to serviceable limits or an engine is top overhauled, the total time on the engine continues in the engine records.

7. REMANUFACTURE.

a. The general term remanufacture has no specific meaning in the FARs. A new engine is a product that is manufactured from raw materials. These raw materials are made into parts and accessories that conform to specifications for the issuance of an engine type certificate. The term "remanufactured" infers that it would be necessary to return the part to its basic raw material and manufacture it again. "Remanufactured" as used by most engine manufacturers and overhaul facilities means that an engine has been overhauled to the standards required to zero time it in accordance with FAR 91.175.

b. However, not all engine overhaul facilities which advertise "Remanufactured Engines" overhaul engines to new dimensions. Some of these facilities do overhaul to new dimensions, but may not be authorized to zero time the engine records. As outlined in FAR 91.175, only the manufacturer or an agency approved by the manufacturer can grant zero time to an engine.

8. ENGINE OVERHAUL FACILITIES.

a. Engine overhaul facilities can include the manufacturer, or a manufacturer's approved agency, large and small FAA certificated repair stations, and engine shops that perform custom overhauls or individual certificated powerplant mechanics. The services offered by these facilities vary. However, regardless of the type or size of the facility, all are required to comply with FAR 43.13(a) and 43.13(b). In this regard, it is the responsibility of the owner to assure that proper entries are made in the engine records (refer to FAR 91.165 and 91.173).

b. Engine overhaul facilities are required by FAR 43.9 to make appropriate entries in the engine records of maintenance that was performed on the engine. The owner should ensure that the engine overhaul facility references the tolerances used (new or serviceable) to accomplish the engine overhaul.

Buzzwords you should know

Airplane engine talk is rife with acronyms, abbreviations, and specialized jargon. The following will assist you in understanding engine talk:

TBO (time between overhaul): The engine manufacturer's recommended maximum engine life. It has no legal bearing for airplanes not used in commercial service, but it is certainly an indicator of engine life expectancy. Many well-cared-for engines last hundreds of hours beyond TBO, but not all. Unfortunately, many never reach TBO before requiring overhaul.

Nitriding: A method of hardening cylinder barrels and crankshafts to reduce wear, thereby extending the useful life of the part.

Chrome plating: A process that brings the internal dimensions of a cylinder back to specifications by producing a hard, machinable, long-lasting surface. Because of the hardness of chrome, break-in time for chromed cylinders is longer than for normal cylinders. An advantage of chrome plating is resistance to destructive oxidation (rust) within combustion chambers.

Magnaflux/Magnaglow: Used for examinations to detect invisible defects in ferrous metals (cracks). Engine parts normally examined by these means are crankshafts, camshafts, piston pins, rocker arms, and cases.

Cylinder codes

Modified cylinders are often installed during maintenance or repairs to an engine. These cylinders are color-coded by paint or colored metal banding. The color of the paint or band indicates the cylinder's physical properties:

Orange: Chrome-plated cylinder barrel
Blue: Nitrided cylinder barrel
Green: Cylinder barrel 0.010 inch oversize
Yellow: Cylinder barrel 0.020 inch oversize

Used engines

Many used airplane advertisements proudly state the hours on the engine, for instance 745 SMOH. The acronym SMOH means *since major overhaul*. The number should reflect accurate usage time, but is not an indicator of how the engine was used. The SMOH can be very misleading, as nothing about the overhaul is explained. The airplane's history must be examined to determine how the overhaul was done and to what standards. Few real standards exist, as seen in AC-43-11.

In some instances, engine repair facilities refer to engines overhauled to new limits (dimensions and specifications used when constructing a new engine) as *remanufactured*. Remember that this term has no official validity, hence is meaningless. Further, some engine overhaul facilities advertise in a confusing manner and can be very misleading.

If the engine was rebuilt and has a new logbook showing zero time, then the work was done by the engine's manufacturer. All other overhauls are subject to wide and varied limits and specifications.

None of this information should be construed to mean that only the manufacturer can properly overhaul an engine. Far from it! There are many very good engine overhaul shops producing high-quality work.

Time and value

The engine time (hours) since new or an overhaul/rebuild is an important factor in placing overall value on an airplane. The recommended TBO, less the hours now on the engine, is the remaining time. There are three basic terms generally used for referring to time on an airplane engine (Fig. 3-4):

Low time: First third of TBO

Mid time: Second third of TBO

High time: Last third of TBO

Each is an indication of the potential value (life remaining) of the engine. Other variables also come into play when referring to TBO (see also Table 3-1):

- Are the hours on the engine since it was new, rebuilt (by the factory), or overhauled?
- At what date was the engine new, rebuilt, or overhauled?
- What type of flying was the engine used for?

Fig. 3-4. *Modified wing tips can improve slow flight, landing, and takeoff characteristics.*

Table 3-1. Value of the Engine

Value	Time	Status	Flying type	TBO
Poor	1800	New	Training	2000
Good	1000	New	X-country	1800
Fair	800	New	Training	2000
Excellent	500	New	X-country	2000
Very poor	1800	SMOH	Any type	2000
Fair	1000	SMOH	X-country	1800
Poor	1000	SMOH	Training	2000
Good	500	SMOH	X-country	2000

- Was it used on a regular basis?
- What kind of maintenance did the engine get?

Airplanes that have not been flown on a regular basis nor maintained in a like fashion have engines that will never reach full TBO. Manufacturers refer to regular usage as 20 to 40 hours of use monthly. That equates to 240 to 480 hours yearly and means a lot of flying. Few privately owned airplanes meet the upper limits of this requirement. Let's face it; most of us don't have the time or money required for such constant use. The average flying time annually accumulated in the American general aviation fleet for a single engine is about 130 hours. Some planes in commercial service run an engine to TBO in one year, while some privately owned planes simply bake in the sun and are never flown.

When an engine isn't run, acids and moisture in the oil oxidize (rust) engine components. In addition, the lack of lubricant movement causes the seals to dry out. Left long enough, the engine will seize and no longer be operable.

Abuse is just as hard on engines as no use. Hard climbs and fast descents cause abnormal heating and cooling conditions and are extremely destructive to air-cooled engines. Training aircraft often exhibit this trait because of their intensive takeoff and landing practice.

It goes without saying that preventive maintenance, such as changing spark plugs and oil, should have been done and logged throughout the engine's life. Tracking preventive maintenance should be easy, since the FARs (Federal Aviation Regulations) require that all maintenance be logged.

Beware the engine that has only a few hours of use since overhaul! Something may be not right with the overhaul, or maybe it was a very cheap job just to make the plane more salable.

When it comes to overhauls, seek out the large shops that specialize in aircraft engine rebuilding. This is not to say that the local FBO can't do a good job; it's just that the large shops specializing in this engine work generally have more experience and equipment to work with. In addition, they have reputations to maintain and will usually back you in the event of difficulties.

Typical costs for a complete engine overhaul to new limits and installed, based on current average pricing, are shown in Table 3-2 (a more complete list can be found in App. F).

Table 3-2. Typical Costs of an Engine Overhaul

Engine	Average overhaul cost (new limits)
Teledyne-Continental	
A/C 65 to 90	$10,700
C125	$12,400
C145	$12,400
E185	$18,600
O-200	$11,900
E225	$18,600
O-300	$15,900
GO-300	$17,000
IO-346	$14,700
IO-360	$18,600
TSIO-360	$20,900
O-470	$17,000
IO-470	$18,000
TSIO-470	$19,200
IO-520	$18,600
TSIO-520	$22,000
Lycoming/Textron-Lycoming	
O-235	$13,000
O-290	$10,700
O-320	$13,000
O-360	$14,100
IO-360	$17,500
O-540	$18,000
IO-540	$22,600
TIO-540	$29,400

It is possible to spend less when overhauling an engine if some parts of the engine are still serviceable. It is also possible to spend about 30 percent more and purchase a factory-rebuilt engine.

MODIFICATIONS

Many airplanes found on today's used market have been modified to update or otherwise improve them. Modifications can take many forms, but are normally to improve performance, attain greater economy of operation, or modernize the appearance of the airplane. Most modifications require an STC (Supplemental Type Certificate) and cost hundreds or thousands of dollars. These modifications do not, however, necessarily add a similar amount to the airplane's value.

Short takeoff and landing

Short takeoff and landing (STOL) conversions are perhaps king of all the modifications available to the airplane owner. The typical STOL modification changes the wing's shape (usually by the addition of a leading edge cuff) and adds stall fences (to prevent airflow disruption from proceeding along the length of the wing), gap seals, and wing tips. A larger engine is sometimes installed as part of the modification. The results of a STOL modification can be spectacular, but they are also quite expensive.

Power

Power-increase modifications are the second most popular improvement made to airplanes. They are often done, as previously mentioned, in conjunction with STOL modifications. Engine replacement increases the useful load and flight performance figures of the aircraft. A power modification appears very costly, yet is often no more expensive than a quality engine overhaul.

Wing tips and vortice generators

Wing tips can be changed to increase flight performance. Dr. Sighard Hoerner designed a high-performance wing tip for the U.S. Navy that led to the development of improved general aviation wing tips. A properly designed wing tip can provide an increase of 3 to 5 mph in cruise speed and a small increase in climb performance, but most important are the improved low-speed handling characteristics:

- 10 to 20 percent reduction in takeoff roll
- 3- to 5-mph lower stall speed
- Improved slow-flight handling

Wing tip installation time can be as low as two to three hours, making this an unusually inexpensive modification (Fig. 3-4).

Another inexpensive approach to better slow-speed handling is installing wing vortice generators. They are small bent metal/plastic strips installed along the top of the wings.

Landing gear

Taildragger conversions have become popular among owners of Cessna 150, 152, and 172 aircraft. The conversion requires the permanent removal of the nose wheel, moving the main gear forward, and adding a tailwheel. Performance benefits are an 8- to 10-mph increase in cruise, shorter takeoff distances, and better rough-field handling.

Gap seals

Gap seals are extensions of the lower wing surface from the rear spar to the leading edge of the flaps and ailerons. They cover several square feet of open space, causing a smoother flow of air around the wing. This smoother flow, or reduction of drag, causes the aircraft to cruise from 1 to 3 mph faster and stall from 5 to 8 mph slower. Gap seals are often part of a STOL installation.

Fuel tanks

Larger or auxiliary fuel tanks are sometimes installed to increase operational range. A drawback of carrying more fuel is that you will carry less load (passengers).

Auto fuels

Considerable controversy surrounds the use of auto fuels (sometimes called *mogas*) in certified aircraft engines. There are pros and cons, but it is up to the individual aircraft owner to decide whether or not to use nonaviation fuels. Note, however, that many FBOs (fixed-base operators) are reluctant to make auto fuel available for reasons of product liability and less profit. This is slowly changing, however, as mogas become available at more airports (Fig. 3-5).

Economy is the center of mogas usage. Unleaded auto fuel is certainly less expensive than 100LL (LL for low lead) by about a dollar per gallon, and it does appear to operate well in older engines that require 80 octane fuel. This gives twofold savings: once at the pump and once again for reduced engine maintenance expenses. Unfortunately, the airplane engine manufacturers claim that the use of auto fuel will void warranty service. Realistically, this threat is limited in scope, as few airplanes using mogas have new engines that would be eligible for warranty coverage anyway.

For convenience, if you have a storage tank and pump, it may be advantageous to use auto fuel. It will be easier to locate a fuel supplier willing to

Fig. 3-5. *Pumps for both av gas and mogas are becoming more common.*

keep an auto fuel storage tank filled than it will be to find an avgas (aviation gasoline) supplier willing to make small deliveries. This is particularly true for small and/or private airstrips.

Before purchasing an airplane with an auto fuel STC (supplementary-type certificate) or acquiring an auto fuel STC for your airplane, check with your insurance carrier and get their approval in writing. For further information about the legal use of auto fuels in an airplane, contact:

Petersen Aviation, Inc.
984 K Road
Minden, NE 68959
(308) 832-2050, fax (308) 832-2311
www.autofuelstc.com

Experimental Aircraft Association
P.O. Box 3086
Oshkosh, WI 54903
(920) 426-4800
www.eaa.org

Value of modifications

Expensive modifications do not always increase a plane's value in proportion to the cost of the modification. Many modifications are of value only in the eye of the current owner. Before extensively modifying an airplane to do some new form of service for which it was not originally designed, consider changing airplanes. For example, if you need to perform rough-field operations, it may be more cost-effective to purchase a Cessna 180 or Maule rather than modifying your Cessna 172. Without doubt, the 180 or Maule would be easier to sell than a highly modified 172.

AVIONICS

New airplanes have instrument panels that reflect our move into the space age, often displaying more than necessary for simple flying. However, what appears complex is usually straightforward in operation and is designed to make flying and navigation easier and safer.

Avionics today are filled with small digital displays and computerized functions. They are about as similar to past equipment as a portable computer is to a pad and pencil. As far as prices go, the new equipment represents bargains never before seen.

A good NAV/COM cost roughly $1800 20 years ago, and provided 200 navigation channels and 360 communication channels. It was panel-mounted and the VOR display (CDI) was mounted separately. Electronics have changed in the past few years, making avionics equipment more capable and loaded with features. Today, the radio in that price class (sometimes a good deal less) is a NAV/COM with the same 200 navigation channels, plus the necessary increase to 760 communication channels, with a digital display, user-programmable memory channels, and a built-in CDI.

It seems that everything about aviation is identified by abbreviations or buzzwords. Avionics are no different, with the following abbreviations commonly used:

A-Panel	Audio panel
ADF	Automatic direction finder
CDI	Course deviation indicator
COM	A VHF transceiver for voice radio communications
DME	Distance-measuring equipment
ELT	Emergency locator transmitter
GPS	Global Positioning System receiver
GPS/COM	A combination of a GPS and a COM in one unit
HT	Hand-held transceiver
LOC/GS	Localizer/glide slope
LORAN-C	A computer/receiver navigation system
MBR	Marker beacon receiver
NAV	A VHF navigation receiver

NAV/COM	A combination of a NAV and COM in one unit
RNAV	Random-area navigation
XPNDR	Transponder

VFR flying

Equipping a plane for VFR (visual flight rules) flying is based on where you intend to fly. Are you going to large airports or small uncontrolled fields? Your equipment could limit you, particularly with today's alphabet-soup airspace requirements. At the barest minimum, VFR operation requires a NAV/COM, transponder, and ELT.

Not too many years ago, most cross-country flying involved using charts and looking out the windows for checkpoints. Today's aviator has become accustomed to the advantages of modern navigation and communication systems. Take advantage of the modern systems and the safety they can provide with a VFR installation that includes a 760-channel NAV/COM, an altitude reporting transponder, ELT, and GPS. With this installation, you can be comfortable and go pretty much wherever you want.

IFR flying

IFR (instrument flight rules) flying requires considerably more equipment than for VFR, representing a much higher cash investment. Efficient IFR operation requires the following minimum equipment installed and approved for instrument flight:

- Dual NAV/COM (760 channel)
- Clock
- MBR
- ADF
- LOC/GS
- Transponder
- Audio panel
- ELT
- DME

No value

Some NAV/COM and COM radios have no value at all. In fact, they cannot be used legally due to Federal Communications Commission (FCC) rules. They may NOT be considered "backup" radios, except for their NAV use. In general, the unaccepted radios referred to in the following FCC notice can be considered detractors of an airplane's value:

This notice responds to recent inquiries from the general aviation community concerning the frequency tolerance and channel spacing requirements for aircraft radios. As of January 1, 1997, each VHF aircraft radio used on board a U.S. aircraft must be type accepted by the FCC as meeting a 30 parts-per-million (ppm) frequency tolerance (47 C.F.R. §§ 87.133). The vast majority of aircraft radios that have been type accepted under the 30 ppm frequency tolerance utilize 25 kHz spacing and have 720 or 760 channels. Each aircraft radio has a label with an FCC ID number on the unit. The FCC ID number may be checked against the "FCC Aircraft Radio List" in order to determine whether the unit has been type accepted as meeting the 30 ppm frequency tolerance. The FCC Aircraft Radio List is available through the Commission's fax-on-demand service at (202) 418-0177 (call from the handset of your fax machine, follow the recorded instructions, and select document retrieval number 000013), via the Internet World Wide Web at http://www.fcc.gov/wtb/avmarsrv.html, or via e-mail request to mayday@fcc.gov.

This rule applies to all U.S. aircraft radio stations, including those no longer required to be licensed individually. The effect of this rule is to require a 30 ppm type accepted radio to be placed on board if the pilot intends to use a VHF aircraft radio for communications. There is no requirement, however, for an older radio to be removed from an aircraft in cases where the pilot does not intend to use it to transmit radio signals (e.g., receive-only operation, an integral part of a navigation/communications unit, or decoration in a vintage aircraft).

A radio which has not been type accepted as 30 ppm may not be returned to service by simply changing the crystals, or adjusting the unit to meet the 30 ppm frequency tolerance. The only way to bring a unit into compliance is through the installation of an FCC type accepted "upgrade kit," which may be available from the unit's manufacturer. Like the radio itself the upgrade kit will have an FCC ID number that may be verified against the FCC Aircraft Radio List. Presently, however, few manufacturers offer FCC type accepted upgrade kits. If a kit is not available for a particular model of radio, the radio may not be adjusted and used for communications purposes on board an aircraft on or after January 1, 1997. If no kit is available, the radio may be reinstalled in the aircraft so long it is not intended to be used to transmit radio signals.

The Commission adopted the 30 ppm frequency tolerance in 1984 in order to conform its rules with those adopted internationally in the Final Acts of the World Administrative Radio Conference, Geneva, 1979. At that time, this action was endorsed by the Federal Aviation Administration (FAA) and was strongly supported by Aeronautical Radio, Inc., the Air Line Pilots Association, the Air Transport Association, and the National Business Aircraft Association, Inc. This action was found to be consistent with the FAA's three-phase plan to implement 25 kHz channel spacing in the 118–137 MHz band, which creates more radio channels for use by pilots. These organizations also noted that users of older radios would have limited access to FAA air traffic control channels, would experience flight delays in FAA controlled air space, and would be unable to utilize newly available aviation frequencies in the 136–137 MHz band. Based on comments by the FAA and the other groups listed above, the Commission determined that permitting the continued operation of older radios type accepted prior to 1974 would pose a threat to safety in air navigation. (FCC Public Notice released November 13, 1996.)

Note: In general, if your aircraft COM radio has more than 700 channels, it is acceptable for use after 1/1/97.

Purchasing avionics

There are several options to examine when adding or upgrading avionics. Each offers a distinct advantage. The object of careful selection is to save money and assure long-term service.

NEW AVIONICS

New equipment is state of the art, offering the newest innovations, best reliability, and a warranty. An additional benefit of new equipment is that solid-state electronics units are physically smaller and allow a fuller panel in a small plane, draw considerably less electric power than their tube-type predecessors, and are much more reliable in operation.

USED AVIONICS

Used avionics can be purchased from dealers or individuals. Trade-A-Plane is a good source of used equipment listing advertisements from avionics dealers and individuals. However, regardless of where you buy used avionics, don't purchase anything with tubes in it, more than six years of age, or built by a manufacturer now out of business. In any of these cases, parts and service could be either really expensive or nonexistent.

Additionally, don't purchase anything that is technically out of date, for example a COM with less than 760 channels. Such equipment presents a lack of capability and can be a problem at a later date.

RECONDITIONED AVIONICS

Several companies advertise reconditioned avionics at a bargain. The equipment was removed from service and completely checked out by an avionics shop. Parts that have failed, are near failure, or are likely to fail have been replaced. Albeit reconditioned, it is not new. Everything in the unit has been used, but not everything is replaced during reconditioning. You will have some new parts and some old parts.

Reconditioned equipment can make sense for the budget-minded owner because it is often a fair-priced buy that usually comes with a limited warranty. Few pieces of reconditioned equipment exceed six or seven years in age.

Instrument panels

Some airplane owners view the instrument panel as a functional device, while others see it as a statement to be made. In either case, take care when filling up the panel. Don't install instrumentation merely for the sake of filling holes. Plan it well and make it functional and easy to use. Above all else, do it economically.

FINAL VALUATION

After you have looked at the engine, airframe, any modifications, and the avionics of an airplane, you should have a good feel for the overall worth of the airplane. Compare the airplane with other similar available airplanes, realizing that small items such as new brakes, intercoms, tires, plastic or vinyl parts, and most modifications add little to the total dollar value of the airplane. A complete guide to pricing is given in Appendix F.

The Internet way

There are Internet web sites that can be used to assist in placing a value on a particular airplane. These systems are interactive, operating from a set of formulas and variables that require the user to input certain specifics about a particular airplane. A dollar value figure will result from the calculations.

These systems are reasonably simple to use and give reasonable ballpark values. The more specific information asked about a particular airplane, the more accurate estimate of a particular airplane's value will result. The input process can take several minutes and can require a complete list of all equipment, modifications, and airplane particulars.

In the end, there is a rule of thumb about the selling price of used airplanes which says: The selling price of an airplane is set when the dollar value the seller places upon his or her valued possession is equaled by the top amount offered by the prospective purchaser for the same box of rocks.

4

Before You Purchase

SEARCHING FOR A GOOD USED AIRPLANE that suits your needs can be a frustrating experience. However, frustrating or not, it is a large part of airplane ownership and a learning experience well worth doing. Most pilots enjoy looking at airplanes.

THE SEARCH

To simplify the search, it is important to begin by having an idea of what airplane you want to purchase and setting a range of expectations for that airplane, including the acceptable overall condition, extra equipment desired, and the spending dollar limit you are comfortable with.

Usually, the search for a used airplane starts at your home airport. Discuss your search with your local FBO (fixed-base operator). FBOs will normally know of any advertised and unadvertised airplanes for sale at the airport and, being an insider to the business, will probably know about the condition and history of those planes.

If there is nothing acceptable at your airport, then broaden the search. Visit other nearby airports, check the bulletin boards, and ask around. While you're there, walk around and look for airplanes with "For Sale" signs in the windows (Figs. 4-1 and 4-2). If you find nothing, you may want to put an "Airplane Wanted" ad on the bulletin board.

If you are interested in an airplane and there is no indication of the owner, write the N-number down and check the ownership, through the FAA, on the Internet or World Wide Web, or via CD-ROM. The latter will give you complete airplane registry information from serial number and exact model number to the current owner's name and address. A CD-ROM for this purpose is the Avantext Aviation Data CD, available from:

Fig. 4-1. *Perhaps you'll see something inviting, like this clean-looking Cessna, while walking around a small airport.*

Avantext, Inc.
Green Hills Corporate Center
340 Morgantown Road, Suite 3300
Reading, PA 19611
(800) 998-8857
(610) 796-2383
www.avantext.com

The FAA maintains an accurate Aircraft Registry query web site that is free for all to use. It allows queries based on N-numbers (for those airplanes you see and are curious about), serial numbers (found on an identification plate mounted on the exterior of an airplane), or even by make and model (which can aid you in locating desired airplanes in a specific geographical area. The Aircraft Registry can be found at 162.58.35.241/acdatabase/acmain.htm.

Printed press

Sometimes you can find used airplane ads in local newspapers, and they are always in the various flight-oriented magazines (*AOPA Pilot, Flying, Private Pilot*, and so on). Unfortunately, the magazine ads are usually stale because of the 60 to 90 days of lag time between ad placement and printing. There are, however, sources of timely advertising.

Fig. 4-2. *Even though it's for sale, this is not what you are looking for! This airplane needs a complete rebuild (loads of money and time). Remember that there is no such thing in aviation as a little fixing for a few dollars.*

TRADE-A-PLANE

Trade-A-Plane has been in business for more than 50 years and publishes its yellow-colored paper three times monthly. Within those pages you will find ads for nearly everything you would ever need in the airplane business including airplanes, parts, service, insurance, avionics, and much more. If you don't see it advertised in *Trade-A-Plane*, you may never see it advertised anywhere else (Fig. 4-3). Contact:

Trade-A-Plane
P.O. Box 509
Crossville, TN 38557
U.S. and Canada: (800) 337-5263, fax (800) 423-9030
International: (931) 484-5137, fax (931) 484-2532
info@trade-a-plane.com
www.trade-a-plane.com

For those really in a hurry to see *Trade-A-Plane* before anyone else, purchase a First Class U.S. Mail subscription or, better yet, a Federal Express 2nd Day Air Priority subscription. Expensive? Yes. But, if you are looking for bargains, you must be at the head of the line.

The newest innovation from *Trade-A-Plane* places it on the Internet. In a superb web site, online viewing of advertisements, sorting, preselection,

Fig. 4-3.
Symbol of Trade-A-Plane, *those yellow sheets with all the airplanes listed for sale.* Courtesy of Trade-A-Plane

and more are available. Specialized options are available for paid subscribers. A subscription to *Trade-A-Plane*, either mail or online, is a must for anyone seeking to buy, sell, or own an airplane.

GA News

General Aviation News is a twice-monthly newspaper carrying loads of up-to-the-minute news concerning general aviation. "You read it here months before the magazines print it" is a phrase used in its advertising. Airplane classifieds are printed on pink paper and carried as an insert to the newspaper. Typically, the pink page count exceeds 20 per issue. Contact:

GA News
P.O. Box 39099
Lakewood, WA 98439-0099
(800) 426-8538
www.generalaviationnews.com

A subscription to *GA News* will keep you well informed of airplane prices as well as current flying topics and news.

Online press

The online press is a new field in the advertising of used airplanes. On the Internet you will find many sources of used airplanes. Already mentioned above is the *Trade-A-Plane* web site; other web sites of interest to the airplane purchaser follow.

www.aircraftbuyer.com: Comprehensive listings of airplanes for sale, with all specifications, easily searchable. Extensive information resources from pilot reports to performance charts to insurance information. The *Aircraft Buyer* web site is superb and highly recommended.

www.aso.com: Listings by category, make, model, year.

www.wingsonline.com: Listings are searchable by type, price, and location.

www.aopa.org: A full-featured site with classified listings and Vref™ (an airplane valuation system). The AOPA web site is very good; however, only AOPA members are able to use all of its features.

Note: This list is not all-inclusive. Some web sites are better than others, more are being added, and many will quickly come and go. Only the best sites will survive.

Where the airplanes are

The following states have the largest population of general aviation aircraft based in them (in descending order):

- California
- Texas
- Florida
- Ohio
- Washington
- Arizona
- Michigan
- Illinois
- Minnesota
- Alaska
- New York

Of course, there are small airplanes in every state.

Alternative sources

Three additional sources of used airplanes are not usually considered because they often have only airplanes that are not desirable for the individual purchaser to consider for ownership. These sources are repossessions, law enforcement seizure auctions, and government surplus sales. All three have a common thread; you can normally make only a quick cursory inspection (and sometimes not even that), and a test flight of the airplane is totally out of the question.

Of the three, repossession is the safest bet. Repossessed airplanes were legally flying in the civilian fleet and should represent no difficulty in title or registration. When considering the purchase of a repossessed airplane, remember the following rule of thumb: If the owner couldn't make the payments, then most likely there was little money spent on maintenance. No matter how scanty the maintenance was, however, it should be recorded in the logbooks. Banks are often a good starting point for searching out repossessed airplanes.

The remaining two sources are more complicated to use and the aircraft involved often have serious mechanical defects that can preclude legal flight.

Aircraft sold at a government auction may have very limited records (logbooks), placing airworthiness of the entire aircraft in question. This is particularly true with surplus military airplanes. The military often uses replacement parts that are not certified. This is not to say the parts are defective, only that they are not certified for use. The lack of parts certification can cause nightmares if you attempt to register the airplane in the civil fleet. In all cases it is the responsibility of the owner (you) to prove airworthiness.

Obtaining a clear title is not a problem because the selling agency will provide it. Cash payment is expected at auctions (unless advanced arrangements are made), and the purchaser is required to remove the airplane immediately or within a few days.

For more information, visit www.firstgov.gov.

The U.S. Marshals Service sells seized aircraft through:

Aero Mod Service, Inc.
Midland International Airport
P.O. Box 60314
Midland, TX 79711
(432) 563-1666
www.aeromodservice.com

Information about Department of the Treasury sales can be obtained from:

EG&G Services
3702 Pender Drive, Suite 400
Fairfax, VA 22030
(703) 273-7373, fax (703) 934-4099
www.treas.gov/auctions/customs

To request a Defense Reutilization and Marketing Service (DRMS) sales information kit (which lists the DRMS offices worldwide and categories of Federal Supply Groups offered by DRMS) contact:

Defense Reutilization and Marketing Service
Hart Dole Inouye Federal Center
74 Washington Avenue N
Battle Creek, MI 49017-3092
(888) 352-9333
custservice@drms.dla.mil

Purchasing an airplane at auction is not for everyone. However, if you are working with a good mechanic and have the extra finances available to correct all mechanical deficiencies, it could save you money in the long run.

READING THE ADVERTISEMENTS

Most airplane advertisements use various cryptic abbreviations to describe the airplane, indicate its condition, suggest the use it has had, and tell how it is equipped. Ads normally include a telephone number and a price. A sample follows:

68 Cessna 182, 2243TT, 763 SMOH, Oct ANN, FGP, Dual NAV/COM, GS, MB, ELT, Mode C XPNDR, NDH. $60,900 firm. (607) 555-1234.

Translation: For sale, a 1968 Cessna 182 airplane with 2243 total hours on the airframe and an engine with 763 hours since a major overhaul. The next annual inspection is due in October. It is equipped with a full gyro instrument panel, has two navigation and communication radios, a glideslope receiver and indicator, a marker beacon receiver, an emergency locator transmitter, and Mode-C transponder. The airplane has no damage history. The price is $60,900 and the seller claims there will be no haggling over the price (most do, however). A telephone number is given for contacting the seller.

Abbreviations

A lot of information is packed inside those lines by way of abbreviations. The following is a list of the more commonly used advertising abbreviations:

AD	airworthiness directive
ADF	automatic direction finder
AF	airframe
AF&E	airframe and engine
ALT	altimeter
ANN	annual inspection
ANNUAL	annual inspection
AP	autopilot
ASI	airspeed indicator
A&E	airframe and engine
A/P	autopilot
BAT	battery
B&W	black and white
CAT	carburetor air temperature
CHT	cylinder-head temperature
COM	communications radio
CS	constant-speed (propeller)

C/S	constant-speed (propeller)
C/W	complied with
DBL	double
DG	directional gyro
DME	distance measuring equipment
FAC	factory
FBO	fixed-base operator
FGP	full gyro panel
FWF	fire wall forward
GAL	gallons
GPH	gallons per hour
GPS	Global Positioning System (receiver)
GS	glideslope (receiver)
HD	heavy-duty
HP	horsepower
HSI	horizontal situation indicator
HVY	heavy
IFR	instrument flight rules
ILS	instrument landing system
INSP	inspection
INST	instrument
KTS	knots
L	left
LDG	landing
LE	left engine
LED	light-emitting diode
LH	left-hand
LIC	license
LOC	localizer
LTS	lights
L&R	left and right
MB	marker beacon (receiver)
MBR	marker beacon receiver
MP	manifold pressure
MPH	miles per hour
MOD	modification

NAV	navigation
NAVCOM	navigation/communication radio
NDH	no damage history
OAT	outside air temperature
OX	oxygen
O2	oxygen
PMA	parts manufacture approval
PROP	propeller
PSI	pounds per square inch
R	right
RC	rate of climb
RE	right engine
REMAN	remanufactured
REPALT	reporting altimeter
RH	right-hand
RMFD	remanufactured
RMFG	remanufactured
RNAV	area navigation
ROC	rate of climb
RX	receiver
SAFOH	since airframe overhaul
SCMOH	since (chrome/complete) major overhaul
SEL	single-engine land
SFACNEW	since factory new
SFN	since factory new
SFNE	since factory new engine
SFREM	since factory remanufacture
SFREMAN	since factory remanufacture
SFRMFG	since factory remanufacture
SMOH	since major overhaul
SNEW	since new
SPOH	since propeller overhaul
STC	supplemental type certificate
STOH	since top overhaul
STOL	short takeoff and landing
TAS	true airspeed

TBO	time between overhaul
TC	turbocharged
TLX	telex
TNSP	transponder
TNSPNDR	transponder
TSN	time since new
TSO	technical service order
TT	total time
TTAF	total time airframe
TTA&E	total time airframe and engine
TTE	total time engine
TTSN	total time since new
TX	transmitter
TXP	transponder
T&B	turn and bank
VAC	vacuum
VFR	visual flight rules
VHF	very high frequency
VOR	visual omni range
XC	cross-country
XLNT	excellent
XMTR	transmitter
XPDR	transponder
XPNDR	transponder
3LMB	three-light marker beacon

Location by phone number

Don't forget the value of the area code in the telephone number; it usually tells the location of the airplane, which is very important for two reasons. Airplane location can dictate whether you are interested in seeing it, as you may not want to travel a couple of thousand miles to look at one airplane. The other aspect of location is the effect that climatic conditions have on the airplane:

- Salt-laden air along a sea coast can cause serious corrosion problems.
- The abrasive sand of the southwest United States takes its toll on the gyros, moving parts, and paint.
- Cold winters in the northern United States can cause excessive engine wear during engine starts and cracking of plastic materials.

- Acid rain found around manufacturing centers can increase corrosion and paint failure.

Generally, these problems can be avoided if the airplane is well cared for and treated properly. Further, the older an airplane, the less likely it has spent its entire life in one geographical area. The following area code listings will aid in searching.

Note: Be careful when calling or returning telephone calls to off-shore area codes. Some are bases of expensive scams that are designed only to take money from you by excessive telephone charges.

AREA CODE LIST BY NUMBER

201	NJ	Hackensack, Jersey City
202	DC	Washington, DC
203	CT	New Haven, Stamford, southwestern
204	MB	Manitoba; entire province
205	AL	Birmingham, Tuscaloosa
206	WA	Seattle
207	ME	entire state
208	ID	entire state
209	CA	Modesto, Stockton
210	TX	San Antonio
212	NY	New York City/Manhattan
213	CA	Los Angeles/inner downtown
214	TX	Dallas and suburbs
215	PA	Philadelphia
216	OH	Cleveland
217	IL	Champaign, Springfield
218	MN	Duluth, northern
219	IN	Gary, northwestern
224	IL	Chicago/northern suburbs
225	LA	Baton Rouge
228	MS	Biloxi, Gulfport, extreme southern tip
229	GA	Albany, Valdosta, southwestern
231	MI	northwest Lower Peninsula, Muskegon, Traverse City
234	OH	Akron, Canton, Youngstown
239	FL	Ft. Myers, Naples

240	MD	Silver Spring; Washington, DC, suburbs; Frederick
242	—	Bahamas
246	—	Barbados
248	MI	Pontiac, Troy, northwestern Detroit
250	BC	British Columbia; Victoria, Kamloops
251	AL	Mobile, southwestern
252	NC	Greenville, Rocky Mount, Cape Hatteras
253	WA	Tacoma, southern Seattle area
254	TX	Waco
256	AL	Huntsville, northern and east central
260	IN	Ft. Wayne, northeastern
262	WI	Kenosha, Racine, Waukesha, West Bend
264	—	Anguilla
267	PA	Philadelphia
268	—	Antigua, Barbuda
269	MI	Kalamazoo, southwestern
270	KY	Bowling Green, western
276	VA	Bristol, southwestern
281	TX	Houston
284	—	British Virgin Islands
289	ON	Ontario; Hamilton, Mississauga, Toronto suburbs
301	MD	Washington, DC, suburbs
302	DE	entire state
303	CO	Denver
304	WV	entire state
305	FL	Miami, Key West
306	SK	Saskatchewan entire province
307	WY	entire state
308	NE	North Platte, Grand Island, western
309	IL	Bloomington, Peoria, Moline/Quad Cities
310	CA	Santa Monica, part of Los Angeles County
312	IL	Chicago/downtown
313	MI	Detroit
314	MO	St. Louis
315	NY	Syracuse, Utica, north central
316	KS	Wichita

317	IN	Indianapolis
318	LA	Shreveport, Alexandria, northern
319	IA	Cedar Rapids, Waterloo, Burlington
320	MN	St. Cloud, central and west central
321	FL	Orlando, Melbourne
323	CA	Los Angeles/outer downtown
325	TX	Abilene, San Angelo
330	OH	Akron, Canton, Youngstown
331	IL	Hinsdale, some western suburbs of Chicago
334	AL	Montgomery, southeastern
336	NC	Greensboro, High Point, Winston-Salem
337	LA	Lafayette, Lake Charles, southwestern
339	MA	Lexington, Boston suburbs
340	VI	U.S. Virgin Islands
341	CA	Oakland
345	—	Cayman Islands
347	NY	New York City/Bronx, Brooklyn, Queens, Staten Island
351	MA	Lowell, Leominster, northeastern
352	FL	Gainesville, Ocala
360	WA	Bellingham, Olympia, Vancouver
361	TX	Corpus Christi
386	FL	northeast
401	RI	entire state
402	NE	eastern
403	AB	Alberta; southern
404	GA	Atlanta
405	OK	Oklahoma City, central
406	MT	entire state
407	FL	Orlando
408	CA	San Jose
409	TX	Galveston, Beaumont
410	MD	Baltimore, northern and eastern
412	PA	Pittsburgh
413	MA	Springfield, western
414	WI	Milwaukee

415	CA	San Francisco, Marin County
416	ON	Ontario; Toronto
417	MO	Springfield, Joplin, southwestern
418	QC	Quebec; Quebec City, Gaspé, southeastern
419	OH	Toledo, northwestern
423	TN	Chattanooga, Johnson City, southeastern, northeastern
425	WA	Everett, Redmond, northern/eastern Seattle area
430	TX	Tyler, Texarkana, northeastern
432	TX	Midland, Odessa
434	VA	Charlottesville, Lynchburg, Emporia
435	UT	entire state except urban area
440	OH	Lorain, Ashtabula, surrounds Cleveland on three sides
441	—	Bermuda
443	MD	Baltimore, northern and eastern
450	QC	Quebec; Laval, Longueuil, suburbs of Montréal
469	TX	Dallas and suburbs
470	GA	Atlanta
473	—	Grenada
478	GA	Macon, central
479	AR	Ft. Smith, Fayetteville, northwestern
480	AZ	Scottsdale, Mesa, Tempe, east Phoenix area
484	PA	Allentown, King of Prussia, outer Philadelphia suburbs
501	AR	Little Rock, central
502	KY	Louisville, Frankfort
503	OR	Portland, Salem, Astoria, northwestern
504	LA	New Orleans
505	NM	entire state
506	NB	New Brunswick; entire province
507	MN	Rochester, southern
508	MA	Worcester, Cape Cod, southeastern
509	WA	Spokane, eastern
510	CA	Oakland
512	TX	Austin

513	OH	Cincinnati
514	QC	Quebec; Montréal, Île-Perrot
515	IA	Des Moines, Ames, Ft. Dodge
516	NY	Nassau County (western Long Island)
517	MI	Lansing
518	NY	Albany, northeastern
519	ON	Ontario; London, Windsor
520	AZ	Tucson, Nogales, southeastern
530	CA	Redding, Lake Tahoe, northeastern
540	VA	Harrisonburg, Roanoke, northwestern, central western
541	OR	Eugene, all but northwestern corner (Portland, Salem)
551	NJ	Hackensack, Jersey City
559	CA	Fresno
561	FL	Palm Beach County
562	CA	Long Beach (part of Los Angeles County)
563	IA	Davenport/Quad Cities, Dubuque, Decorah
567	OH	Toledo, northwest
570	PA	Scranton, Wilkes-Barre
571	VA	Arlington; suburbs of Washington, DC
573	MO	central and western except St. Louis
574	IN	South Bend, north central
580	OK	Lawton, panhandle, southern, western, north central
585	NY	Rochester
586	MI	Warren, Macomb County
601	MS	Jackson, Hattiesburg
602	AZ	Phoenix
603	NH	entire state
604	BC	British Columbia; Vancouver
605	SD	entire state
606	KY	Ashland, Pikeville, Somerset, eastern
607	NY	Binghamton, south central
608	WI	Madison, southwestern
609	NJ	Atlantic City, Trenton, southeast, central west

610	PA	Allentown, King of Prussia, outer Philadelphia suburbs
612	MN	Minneapolis
613	ON	Ontario; Ottawa
614	OH	Columbus
615	TN	Nashville
616	MI	Grand Rapids
617	MA	Boston
618	IL	East St. Louis, southern
619	CA	San Diego, Coronado
620	KS	southern except metro Wichita
623	AZ	Glendale, Sun City, west Phoenix area
626	CA	Pasadena, San Gabriel Valley
630	IL	Hinsdale, some western suburbs of Chicago
631	NY	Suffolk County (eastern Long Island)
636	MO	St. Charles
641	IA	central
646	NY	New York City/Manhattan
647	ON	Ontario; Toronto
649	—	Turks & Caicos Islands
650	CA	San Mateo, San Francisco Airport
651	MN	St. Paul area
660	MO	Sedalia, north central
661	CA	Bakersfield, northern Los Angeles County
662	MS	Greenville, Tupelo, northern half
664	—	Montserrat
671	GU	Guam/U.S. Pacific
682	TX	Fort Worth
701	ND	entire state
702	NV	Las Vegas
703	VA	Arlington; suburbs of Washington, DC
704	NC	Charlotte
705	ON	Ontario; Sudbury, north central
706	GA	Athens, Columbus, Augusta, Rome
707	CA	north coast
708	IL	La Grange, southwestern suburbs of Chicago

709	NL	Newfoundland & Labrador; entire province
712	IA	Sioux City, Council Bluffs, western
713	TX	Houston
714	CA	Anaheim
715	WI	Eau Claire
716	NY	Buffalo
717	PA	Harrisburg
718	NY	New York City/Bronx, Brooklyn, Queens, Staten Island
719	CO	southeastern
720	CO	Denver
724	PA	western
727	FL	St. Petersburg
731	TN	western except Memphis
732	NJ	northern Jersey Shore region
734	MI	Ann Arbor, Monroe
740	OH	southeastern
754	FL	Ft. Lauderdale, Broward County
757	VA	Norfolk
758	—	St. Lucia
760	CA	southeastern
763	MN	Brooklyn Park
765	IN	Lafayette, Muncie
767	—	Dominica
770	GA	Marietta, Atlanta suburbs
772	FL	central east coast
773	IL	Chicago/outside downtown
774	MA	southeastern
775	NV	all except Las Vegas
778	BC	British Columbia; Vancouver
780	AB	Alberta; Edmonton, northern
781	MA	Lexington, Boston suburbs
784	—	St. Vincent & the Grenadines
785	KS	north central, northwestern
786	FL	Miami/Dade County, Keys
787	PR	Puerto Rico

801	UT	Salt Lake City
802	VT	entire state
803	SC	central
804	VA	Richmond, Petersburg
805	CA	San Luis Obispo, Oxnard
806	TX	Amarillo, Lubbock
807	ON	Ontario; Thunder Bay, northwestern
808	HI	entire state
809	—	Dominican Republic
810	MI	Flint, Port Huron, northeastern Detroit suburbs
812	IN	Evansville, southern
813	FL	Tampa
814	PA	Altoona, Erie
815	IL	Joliet, northern
816	MO	Kansas City, St. Joseph
817	TX	Fort Worth
818	CA	San Fernando
819	QC	Quebec; central and northern
828	NC	western
830	TX	Del Rio, New Braunfels
831	CA	Monterey, Santa Cruz
832	TX	Houston and suburbs
843	SC	coastal
845	NY	northern suburbs of New York City
847	IL	Chicago/northern suburbs
848	NJ	northern shore region
850	FL	panhandle
856	NJ	southwestern
857	MA	Boston
858	CA	San Diego/north suburbs
859	KY	Lexington, Covington
860	CT	northern and eastern
862	NJ	Newark, Paterson
863	FL	central
864	SC	northwestern
865	TN	eastern

867	YT, NT, NU	Yukon, Northwest Territories & Nunavut
868	—	Trinidad & Tobago
869	—	St. Kitts & Nevis
870	AR	eastern, north central
876	—	Jamaica
878	PA	Pittsburgh, New Castle
901	TN	Memphis
902	NS	Nova Scotia; entire province
902	PE	Prince Edward Island; entire province
903	TX	northeastern
904	FL	Jacksonville
905	ON	Ontario; Hamilton, Mississauga, Toronto suburbs
906	MI	Sault Ste. Marie, upper peninsula
907	AK	entire state
908	NJ	Elizabeth
909	CA	San Bernardino, Ontario, Riverside
910	NC	Wilmington, Fayetteville, southeastern
912	GA	southeastern
913	KS	Kansas City, Leavenworth
914	NY	White Plains, Westchester County
915	TX	El Paso, western
916	CA	Sacramento
917	NY	New York City (mostly cellular phones/pagers/ voicemail)
918	OK	northeastern
919	NC	Raleigh
920	WI	northeastern
925	CA	Concord, Walnut Creek, Livermore, inland East Bay
928	AZ	most of the state
931	TN	central except Nashville
936	TX	central
937	OH	Dayton
939	PR	Puerto Rico
940	TX	north central

941	FL	Sarasota
947	MI	Pontiac, Troy, northwestern Detroit suburbs
949	CA	Irvine, Newport Beach, southern Orange County
952	MN	Bloomington, southwest suburbs of Minneapolis
954	FL	Ft. Lauderdale, Broward County
956	TX	southeast
970	CO	Grand Junction, northern and western
971	OR	Portland, Salem, not Tillamook or Astoria
972	TX	Dallas and suburbs
973	NJ	Newark, Paterson
978	MA	northeastern
979	TX	Bryan/College Station, Wharton
980	NC	Charlotte
985	LA	southeastern except New Orleans & Baton Rouge
989	MI	Saginaw

AREA CODE LIST BY LOCATION

264	—	Anguilla
268	—	Antigua, Barbuda
242	—	Bahamas
246	—	Barbados
441	—	Bermuda
284	—	British Virgin Islands
345	—	Cayman Islands
767	—	Dominica
809	—	Dominican Republic
473	—	Grenada
876	—	Jamaica
664	—	Montserrat
869	—	St. Kitts & Nevis
758	—	St. Lucia
784	—	St. Vincent & the Grenadines
868	—	Trinidad & Tobago
649	—	Turks & Caicos Islands

780	AB	Alberta; Edmonton, northern
403	AB	Alberta; southern
907	AK	entire state
205	AL	Birmingham, Tuscaloosa
256	AL	Huntsville, northern and east central
251	AL	Mobile, southwestern
334	AL	Montgomery, southeastern
870	AR	eastern, north central
479	AR	Ft. Smith, Fayetteville, northwestern
501	AR	Little Rock, central
623	AZ	Glendale, Sun City, west Phoenix area
928	AZ	most of the state
602	AZ	Phoenix
480	AZ	Scottsdale, Mesa, Tempe, east Phoenix area
520	AZ	Tucson, Nogales, southeastern
604	BC	British Columbia; Vancouver
778	BC	British Columbia; Vancouver
250	BC	British Columbia; Victoria, Kamloops
714	CA	Anaheim
661	CA	Bakersfield, northern Los Angeles County
925	CA	Concord, Walnut Creek, Livermore, inland East Bay
559	CA	Fresno
949	CA	Irvine, Newport Beach, southern Orange County
562	CA	Long Beach (part of Los Angeles County)
213	CA	Los Angeles/inner downtown
323	CA	Los Angeles/outer downtown
209	CA	Modesto, Stockton
831	CA	Monterey, Santa Cruz
707	CA	north coast
341	CA	Oakland
510	CA	Oakland
626	CA	Pasadena, San Gabriel Valley
530	CA	Redding, Lake Tahoe, northeastern
916	CA	Sacramento
909	CA	San Bernardino, Ontario, Riverside

858	CA	San Diego/north surburbs
619	CA	San Diego, Coronado
818	CA	San Fernando
415	CA	San Francisco, Marin County
408	CA	San Jose
805	CA	San Luis Obispo, Oxnard
650	CA	San Mateo, San Francisco Airport
310	CA	Santa Monica, part of Los Angeles County
760	CA	southeastern
303	CO	Denver
720	CO	Denver
970	CO	Grand Junction, northern and western
719	CO	southeastern
203	CT	New Haven, Stamford, southwestern
860	CT	northern and eastern
202	DC	Washington, DC
302	DE	entire state
863	FL	central
772	FL	central east coast
754	FL	Ft. Lauderdale, Broward County
954	FL	Ft. Lauderdale, Broward County
239	FL	Ft. Myers, Naples
352	FL	Gainesville, Ocala
904	FL	Jacksonville
786	FL	Miami/Dade County, Keys
305	FL	Miami, Key West
386	FL	northeast
407	FL	Orlando
321	FL	Orlando, Melbourne
561	FL	Palm Beach County
850	FL	panhandle
941	FL	Sarasota
727	FL	St. Petersburg
813	FL	Tampa
229	GA	Albany, Valdosta, southwestern
706	GA	Athens, Columbus, Augusta, Rome

404	GA	Atlanta
470	GA	Atlanta
478	GA	Macon, central
770	GA	Marietta, Atlanta suburbs
912	GA	southeastern
671	GU	Guam/U.S. Pacific
808	HI	entire state
319	IA	Cedar Rapids, Waterloo, Burlington
641	IA	central
563	IA	Davenport/Quad Cities, Dubuque, Decorah
515	IA	Des Moines, Ames, Ft. Dodge
712	IA	Sioux City, Council Bluffs, western
208	ID	entire state
309	IL	Bloomington, Peoria, Moline/Quad Cities
217	IL	Champaign, Springfield
312	IL	Chicago/downtown
224	IL	Chicago/northern suburbs
847	IL	Chicago/northern suburbs
773	IL	Chicago/outside downtown
618	IL	East St. Louis, southern
331	IL	Hinsdale, some western suburbs of Chicago
630	IL	Hinsdale, some western suburbs of Chicago
815	IL	Joliet, northern
708	IL	La Grange, southwestern suburbs of Chicago
812	IN	Evansville, southern
260	IN	Ft. Wayne, northeastern
219	IN	Gary, northwestern
317	IN	Indianapolis
765	IN	Lafayette, Muncie
574	IN	South Bend, north central
913	KS	Kansas City, Leavenworth
785	KS	north central, northwestern
620	KS	southern except metro Wichita
316	KS	Wichita
606	KY	Ashland, Pikeville, Somerset, eastern
270	KY	Bowling Green, western

859	KY	Lexington, Covington
502	KY	Louisville, Frankfort
225	LA	Baton Rouge
337	LA	Lafayette, Lake Charles, southwestern
504	LA	New Orleans
318	LA	Shreveport, Alexandria, northern
985	LA	southeastern except New Orleans & Baton Rouge
617	MA	Boston
857	MA	Boston
339	MA	Lexington, Boston suburbs
781	MA	Lexington, Boston suburbs
351	MA	Lowell, Leominster, northeastern
978	MA	northeastern
774	MA	southeastern
413	MA	Springfield, western
508	MA	Worcester, Cape Cod, southeastern
204	MB	Manitoba; entire province
410	MD	Baltimore, northern and eastern
443	MD	Baltimore, northern and eastern
240	MD	Silver Spring; Washington, DC, suburbs; Frederick
301	MD	Washington, DC, suburbs
207	ME	entire state
734	MI	Ann Arbor, Monroe
313	MI	Detroit
810	MI	Flint, Port Huron, northeastern Detroit suburbs
616	MI	Grand Rapids
269	MI	Kalamazoo, southwestern
517	MI	Lansing
231	MI	northwest Lower Peninsula, Muskegon, Traverse City
248	MI	Pontiac, Troy, northwestern Detroit
947	MI	Pontiac, Troy, northwestern Detroit suburbs
989	MI	Saginaw
906	MI	Sault Ste. Marie, upper peninsula
586	MI	Warren, Macomb County
952	MN	Bloomington, southwest suburbs of Minneapolis

763	MN	Brooklyn Park
218	MN	Duluth, northern
612	MN	Minneapolis
507	MN	Rochester, southern
320	MN	St. Cloud, central and west central
651	MN	St. Paul area
573	MO	central and western except St. Louis
816	MO	Kansas City, St. Joseph
660	MO	Sedalia, north central
417	MO	Springfield, Joplin, southwestern
636	MO	St. Charles
314	MO	St. Louis
228	MS	Biloxi, Gulfport, extreme southern tip
662	MS	Greenville, Tupelo, northern half
601	MS	Jackson, Hattiesburg
406	MT	entire state
506	NB	New Brunswick; entire province
704	NC	Charlotte
980	NC	Charlotte
336	NC	Greensboro, High Point, Winston-Salem
252	NC	Greenville, Rocky Mount, Cape Hatteras
919	NC	Raleigh
828	NC	western
910	NC	Wilmington, Fayetteville, southeastern
701	ND	entire state
402	NE	eastern
308	NE	North Platte, Grand Island, western
603	NH	entire state
609	NJ	Atlantic City, Trenton, southeast, central west
908	NJ	Elizabeth
201	NJ	Hackensack, Jersey City
551	NJ	Hackensack, Jersey City
862	NJ	Newark, Paterson
973	NJ	Newark, Paterson
732	NJ	northern Jersey Shore region
848	NJ	northern shore region

856	NJ	southwestern
709	NL	Newfoundland & Labrador; entire province
505	NM	entire state
902	NS	Nova Scotia; entire province
775	NV	all except Las Vegas
702	NV	Las Vegas
518	NY	Albany, northeastern
607	NY	Binghamton, south central
716	NY	Buffalo
516	NY	Nassau County (western Long Island)
347	NY	New York City/Bronx, Brooklyn, Queens, Staten Island
718	NY	New York City/Bronx, Brooklyn, Queens, Staten Island
212	NY	New York City/Manhattan
646	NY	New York City/Manhattan
917	NY	New York City (mostly cellular phones/pagers/ voicemail)
845	NY	northern suburbs of New York City
585	NY	Rochester
631	NY	Suffolk County (eastern Long Island)
315	NY	Syracuse, Utica, north central
914	NY	White Plains, Westchester County
234	OH	Akron, Canton, Youngstown
330	OH	Akron, Canton, Youngstown
513	OH	Cincinnati
216	OH	Cleveland
614	OH	Columbus
937	OH	Dayton
440	OH	Lorain, Ashtabula, surrounds Cleveland on three sides
740	OH	southeastern
567	OH	Toledo, northwest
419	OH	Toledo, northwestern
580	OK	Lawton, panhandle, southern, western, north central
918	OK	northeastern

405	OK	Oklahoma City, central
289	ON	Ontario; Hamilton, Mississauga, Toronto suburbs
905	ON	Ontario; Hamilton, Mississauga, Toronto suburbs
519	ON	Ontario; London, Windsor
613	ON	Ontario; Ottawa
705	ON	Ontario; Sudbury, north central
807	ON	Ontario; Thunder Bay, northwestern
416	ON	Ontario; Toronto
647	ON	Ontario; Toronto
541	OR	Eugene, all but northwestern corner (Portland, Salem)
503	OR	Portland, Salem, Astoria, northwestern
971	OR	Portland, Salem, not Tillamook or Astoria
484	PA	Allentown, King of Prussia, outer Philadelphia suburbs
610	PA	Allentown, King of Prussia, outer Philadelphia suburbs
814	PA	Altoona, Erie
717	PA	Harrisburg
215	PA	Philadelphia
267	PA	Philadelphia
412	PA	Pittsburgh
878	PA	Pittsburgh, New Castle
570	PA	Scranton, Wilkes-Barre
724	PA	western
902	PE	Prince Edward Island; entire province
787	PR	Puerto Rico
939	PR	Puerto Rico
819	QC	Quebec; central and northern
450	QC	Quebec; Laval, Longueuil, suburbs of Montréal
514	QC	Quebec; Montréal, Île-Perrot
418	QC	Quebec; Quebec City, Gaspé, southeastern
401	RI	entire state
803	SC	central
843	SC	coastal
864	SC	northwestern

605	SD	entire state
306	SK	Saskatchewan; entire province
931	TN	central except Nashville
423	TN	Chattanooga, Johnson City, southeastern, northeastern
865	TN	eastern
901	TN	Memphis
615	TN	Nashville
731	TN	western except Memphis
325	TX	Abilene, San Angelo
806	TX	Amarillo, Lubbock
512	TX	Austin
979	TX	Bryan/College Station, Wharton
936	TX	central
361	TX	Corpus Christi
214	TX	Dallas and suburbs
469	TX	Dallas and suburbs
972	TX	Dallas and suburbs
830	TX	Del Rio, New Braunfels
915	TX	El Paso, western
682	TX	Fort Worth
817	TX	Fort Worth
409	TX	Galveston, Beaumont
281	TX	Houston
713	TX	Houston
832	TX	Houston and suburbs
432	TX	Midland, Odessa
940	TX	north central
903	TX	northeastern
210	TX	San Antonio
956	TX	southeast
430	TX	Tyler, Texarkana, northeastern
254	TX	Waco
435	UT	entire state except urban area
801	UT	Salt Lake City
571	VA	Arlington; suburbs of Washington, DC

703	VA	Arlington; suburbs of Washington, DC
276	VA	Bristol, southwestern
434	VA	Charlottesville, Lynchburg, Emporia
540	VA	Harrisonburg, Roanoke, northwestern, central western
757	VA	Norfolk
804	VA	Richmond, Petersburg
340	VI	U.S. Virgin Islands
802	VT	entire state
360	WA	Bellingham, Olympia, Vancouver
425	WA	Everett, Redmond, northern/eastern Seattle area
206	WA	Seattle
509	WA	Spokane, eastern
253	WA	Tacoma, southern Seattle area
715	WI	Eau Claire
262	WI	Kenosha, Racine, Waukesha, West Bend
608	WI	Madison, southwestern
414	WI	Milwaukee
920	WI	northeastern
304	WV	entire state
307	WY	entire state
867	YT, NT, NU	Yukon, Northwest Territories & Nunavut

THE TELEPHONE INQUIRY

When you see an interesting airplane listed in printed advertising or on a bulletin board, you will need to contact the seller (usually the owner). Generally, a telephone call will suffice. It sounds simple enough, but first consider what you are going to ask. Don't pretend to be an expert, but don't be a tire-kicker either. The owner wants to sell his airplane, not relive past experiences. Use the telephone and ask questions, making notes as you get answers.

Is this person the owner of the airplane? If not, ask for the owner's name, address, and telephone number. Owner means the individual owner or dealer owner (as recorded by FAA records). If it is not the owner, but someone selling the airplane for another party, it is most likely an airplane broker. Ask the year, make, and model of the plane. You would be surprised at the number of mistakes in airplane ads that appear in the trade magazines and papers. This way you are sure you are both talking about the same airplane. Sometimes an owner will have more than one airplane for sale. Inquire about:

- General appearance and condition of the plane
- Total hours on the airframe
- Make and model of engine
- Hours on the engine since new
- Hours since the last overhaul
- Type of overhaul
- Damage history
- Asking price
- Where it is located
- Color
- The N-number

The last three questions will sometimes elicit evasive answers when a freelance airplane broker is involved. Often freelance brokers are reluctant to give out particulars about a specific airplane, as they have no real control over the plane. The plane's owner has not listed it exclusively with the freelance broker, so the broker is afraid of losing his commission if you deal directly with the owner (which you are free to do).

A broker differs from a dealer. The dealer owns the aircraft being sold and the broker does not. Airplane brokers attempt to match buyers with known planes for sale and can be very useful when you are looking for a used airplane. A bonafide used airplane broker can do the leg work and ad reading on your behalf, saving you time. For their efforts, brokers receive a sales commission—sometimes from the seller, other times from the buyer, and often from both.

Sometimes an owner who is hot to sell (motivated) will fly the airplane to you for inspection. Don't ask someone to do this unless you are serious and have the money to make the purchase.

PREPURCHASE INSPECTION

The object of the prepurchase inspection of a used airplane is to preclude the purchase of a less-than-worthy airplane. No one wants to buy someone else's troubles. The prepurchase inspection must be completed in an orderly, well-planned manner. Take your time during this inspection, for these few minutes could well save you thousands of dollars and a lot of grief.

The very first item of inspection is the most-asked question of anyone selling anything: "Why are you selling it?" Fortunately, most people answer honestly. Often the current owner is moving up to a larger airplane and will start to tell you all about the new prospective purchase. Listen carefully because you can learn a lot about the owner from what is said, gaining insight into flying habits and how the plane you are considering purchasing was treated. Perhaps there are other commitments (the spouse saying, "Me or the plane") or maybe there are pressing money concerns and the airplane is no longer affordable. Both of these reasons are common and can be to your advantage, as they create a motivated seller.

Warning (again): If the seller is having financial difficulties, consider the quality of maintenance done on the airplane. Ask the seller if there are any known problems or defects with the airplane. Again, honesty usually prevails, but there is always the unknown. Remember: Buyer beware! It's your money and your safety. Now here are some more definitions, related to conducting a prepurchase inspection:

Airworthiness: The airplane must conform to the original type certificate or the STCs (supplemental type certificates) issued for this particular airplane (by serial number). In addition, the airplane must be in safe operating condition relative to wear and deterioration.

Annual inspection: All small airplanes must be inspected annually by an FAA-certified airframe and power-plant (A&P) mechanic who holds an IA (inspection authorization), by an FAA-certified repair station or the airplane's manufacturer. This is a complete inspection of the airframe, power plant, and all subassemblies.

100-hour inspection: An inspection made every 100 hours of the same scope as an annual inspection. It is required on all commercially operated small airplanes (rental, training, air taxi, and so forth). This inspection can be done by an A&P without an IA rating. An annual inspection will fulfill the 100-hour inspection requirement, but the reverse is not true.

Preflight inspection: A thorough inspection, by the pilot, of an aircraft before flight. The purpose is to spot obvious discrepancies by inspection of the exterior, interior, and engine area of the airplane.

Preventive maintenance: FAR Part 43 lists a number of airplane maintenance operations that are preventive in nature and can be done by a certificated pilot, provided the airplane is not flown in commercial service (described in Chap. 6).

Repairs and alterations: The two types of repairs and alterations are *major* and *minor*. Major repairs and alterations must be approved for return to service by an A&P mechanic holding an IA authorization, a repair station, or the FAA. Minor repairs and alterations can be returned to service by an A&P mechanic.

The prepurchase inspection consists of a walk-around inspection, test flight, and mechanic's inspection. The purpose of the inspection is to determine what is deteriorating on the air-plane from the effects of weather, stress, heat, vibration, friction, and age:

- Weather can affect the airplane with an arsenal of weapons, including: heat, sunlight, rain, snow, wind, temperature, humidity, and ultra-violet light.
- Stress includes overloading of the structure, causing deformation or outright structural failure from excessive landing weights, high-speed maneuvers, turbulence, and hard landings.
- Heat other than heat caused by weather is engine-associated and usually causes degrading of rubber parts such as hoses and plastic fittings. It can also be caused by a defective exhaust system.
- Vibration is caused by the engine or by the fluttering of control surfaces. Damage from vibration is in the form of cracks in the aircraft structure or skin.
- Friction can affect any moving part of the airplane. You can reduce frictional damage by using lubricants, but they can be fixed only with a complete repair.
- Age is chronological by years and the number of hours the plane has accumulated.

It is common for a prospective purchaser to desire a "low-time" airplane and engine, but there are times when this is detrimental, for example, a 20-year-old airplane with a mere 800 hours on the engine and airframe. This low-time airplane could harbor considerable deterioration due to lack of use. Like human bodies, an airplane requires exercise to stay healthy. Engines and airframes, as well as most moving parts, suffer when not used on a regular basis. This low-time example could require an engine overhaul, considerable airframe parts replacement and lubrication, and cosmetic work.

The prepurchase walk-around inspection is very similar to a preflight inspection, only more thorough. It's divided into four parts, exploring each section of the airplane: cabin, airframe, engine, and paperwork.

Cabin inspection

1. Starting in the cabin, ensure that all required paperwork is with the airplane:
   - Airworthiness certificate
   - Aircraft registration certificate
   - FCC radio station license (if required)
   - Flight manual or operating limitations
   - Logbooks (airframe, engine, and propeller). Note: Logbooks are not required to be kept in the airplane, but they must be available.
   - Equipment list and weight and balance chart
2. While inside the airplane looking for the paperwork, notice the general condition of the interior. Does it appear clean, or has it recently

been scrubbed after a long period of inattention? Look in the corners, just like you would when buying a used car. Care given the interior is a good indication of what care was given to the remainder of the airplane. By the way, the interior of the typical four-place airplane can cost upward of $3,500 to replace, some even more than that.

3. Look at the instrument panel. Does it have what you want and need? Are the instruments in good condition? Are any knobs missing or glass faces broken? Is the equipment all original or updated? If updates were made, are they neat in appearance and workable? Avionics updating is often done haphazardly with results that are neither pleasing to the eye nor workable for the pilot (Figs. 4-4 through 4-6).

4. Look out the windows. Are they clear, unyellowed, and uncrazed? Side windows are not expensive to replace and you can often do the job yourself. Windshields, however, are another story and another price (typically several hundred dollars) and you cannot do the work yourself.

5. Check the operation of the doors. They should close and lock with little effort and the seals should be tight, allowing no outside light to be visible around door edges.

6. Check the seats for freedom of movement and adjustability. Check the seat tracks and the adjustment locks for damage.

Fig. 4-4. *You won't go far with this panel, unless you are interested only in simple sport flying, and there is nothing wrong with that. Just be sure you get enough panel to do what you want.*

Fig. 4-5. *Although not state of the art, this is a good panel with everything you would normally need.*

7. Check the carpet for wetness or moldy condition, suggesting long-term exposure to water, possibly inducing corrosion.

8. Open the rear fuselage compartment behind the seats and use a flashlight to look back towards the tail. You are looking for corrosion, dirt, debris, damage, frayed cables, loose fittings, and previously repaired areas.

9. Be on the lookout for mouse houses, usually fluffy piles of insulation, fabric, or paper materials. The biggest problem with mouse houses is that mice have bad bathroom manners; they urinate in their beds. Mouse urine is one of the most caustic liquids known to aircraft mechanics because it eats through aluminum like battery acid. Watch for signs of it.

10. Finally, write down the times displayed on the tachometer and Hobbs meters, then reference them when checking the logbooks.

Airframe inspection

1. Stand behind the airplane and look at all surfaces, comparing one to another. Are they positioned as they should be or does one look out of place compared with the others? This could mean past damage

Fig. 4-6. *A modern digital panel found in a light twin.*

and/or a rigging problem. Repeat this procedure from each side and from the front of the airplane.

2. Sight along each flying and control surface, looking for dents, bends, or other signs of damage. Dents, wrinkles, or tears of the metal skin may indicate prior damage or just careless handling.

3. An additional indicator of past damage is the cherry rivet. The cherry rivet is commonly used during repairs, particularly in locations where bucking from the back would be impossible (inside wings and control surfaces). You can recognize the cherry rivet by a small hole in its center, where it is pulled into itself during tightening. The hole does not appear on regular rivets. Note that not all cherry rivets indicate damage, as some are used during original manufacture.

4. Examine each discrepancy very carefully. A total of all dings and dents will reveal if the airplane has had an easy or a rough life. Wrinkles in the skin are usually caused by hidden structural damage and should be checked by a mechanic (Figs. 4-7 and 4-8). Damaged belly skins on retractables usually mean a wheels-up landing.

5. Is the paint in good condition, or is some of it flaking onto the ground underneath the airplane (Fig. 4-9)? Paint jobs are expensive, yet necessary for protecting metal surfaces from corrosive elements.

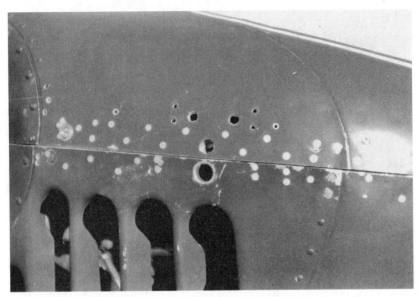

Fig. 4-7. *Avoid planes with metal parts in poor condition, such as this cowl.*

6. Corrosion or rust on surfaces or control systems is cause for alarm. Corrosion is to aluminum what rust is to iron: a form of oxidation that is very destructive. Corrosion is generally indicated by blistered paint (similar to a rusted car) and/or a white powdery coating of bare metal. Any signs of corrosion or rust must be brought to the attention of a mechanic.

7. Check landing gear for evidence of being sprung. Check the tires for signs of unusual wear that may indicate structural damage. Look at the oleo strut(s) for signs of fluid leakage and proper extension.

8. Move all control surfaces and check each for damage. They should be free and smooth in movement, not loose or subject to rattle, and they should not exhibit sideward movement (they should move only in their designed plane). When the controls are centered, the surfaces should also be centered. If they are not centered, a rigging problem exists.

9. Remove the wing root fairings and check the condition of the wiring and hoses inside. Look for mouse houses and check the wing attachment bolts.

10. Check the antennas for proper mounting and any obvious damage.

Engine inspection

1. Look for signs of oil leakage when checking the engine. Do this by closely examining the engine, the area inside the cowl, and the fire

Fig. 4-8. *Repaired minor damage, such as this split wheel pant, will be added to the total valuation.*

Fig. 4-9. *Peeling paint can break the bank.*

Fig. 4-10. *Removing the engine cowling makes inspection easy.*

wall. Oil will be dripping to the ground or onto the nose wheel if the leaks are bad enough. Assume that the seller has cleaned all the old drips away, but remember that oil leaves stains (Fig. 4-10).

2. Check all hoses and lines for signs of deterioration or chafing and assure that all the connections are tight and that there are no indications of leakage. Examine the baffles for proper shape, alignment, and tightness. Baffles control the airflow for engine cooling and if they are improperly positioned, the engine will not be properly cooled. The long-term result of poor baffling is damage to the engine.

3. Check control linkages and cables for obvious damage and easy movement.

4. Check the battery box and battery for corrosion.

5. Examine the exhaust pipes for rigidity, then reach inside and rub your finger along the inside wall. If your finger comes back perfectly clean, you can assume that someone washed the inside of the pipe(s) to remove any telltale deposits that had accumulated there (Fig. 4-11). Oily or sooty deposits can be caused by a carburetor in need of adjustment or a large amount of oil blow-by. The latter can indicate an engine in need of major expenditures for overhaul. A light gray dusty coating means proper engine operation.

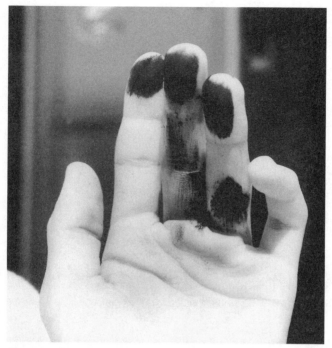

Fig. 4-11. *Goo on fingers like this means a mechanic should check the engine.*

6. Check for exhaust stains on the belly of the plane to the rear of the stacks. This area has probably been washed, but look anyway. If you find black oily goo, then see a mechanic (Fig. 4-12).

7. Inspect the propeller for damage, such as nicks, cracks, or gouges. These (often small) defects can cause stress areas on the propeller and require checking by a mechanic. Also observe any movement that could suggest propeller looseness at the hub.

8. Look for oil change or other maintenance stickers. Compare the times written on these stickers with those you wrote down from the tachometer and Hobbs meter. Note the additional sticker information for later referral to logbooks.

Paperwork check

If you are satisfied with what you've seen up to this point, you are ready to chase the paperwork trail of the airplane.

Checking the airplane's paperwork involves the logbooks and the owner's file of repair bills. The logbooks are the legal records of airplane maintenance. Unfortunately, logbooks are not always as detailed as they should be and can

Fig. 4-12. *A dark trail from the exhaust stack calls for a mechanic to check the engine.*

be somewhat cryptic to read when it comes to mechanic's entries. Repair bills
for mechanical work, however, will contain detailed information about parts
and labor. After all, the mechanic will leave no stone unturned when extract-
ing money from an airplane owner. Compare the repair bills for all major work
to logbook entries.

1. Pull out the logbooks and start reading them. Be sure that you're
   looking at the proper logs for this particular aircraft, determine that
   they are the original logs, and verify serial numbers of the airframe,
   engine, and equipment listed for the airplane.

2. Sometimes logbooks get lost and must be replaced by new ones
   because of carelessness or theft. The latter is the reason that many
   owners keep photocopies of their logs in the aircraft and the
   originals in a safe place. Replacement logs often lack very important
   information or could be outright frauds. Fraud is always a distinct
   possibility, so be on guard if the original logs are not available. In
   the event of missing original logbooks, a complete check of the airplane's FAA records will provide most of the missing information.

3. Look in the back of the airframe logbook for Airworthiness Directive
   (AD) compliance. Check that it's up to date and that any required
   periodic inspections have been done. Then return to the most recent
   entry, usually an annual or 100-hour inspection. The annual inspection will be a statement like the following:

March 27, 1997

Total Time: 2,815 hrs.

I certify that this aircraft has been inspected in accordance with an annual inspection and was determined to be in airworthy condition.

[signed]

[IA # 0000000]

4. Return to the first entry in the logbook, looking for similar entries, always keeping track of the total time for continuity purposes and to show the regularity of usage (number of hours flown between inspections). Also, look for statements about major repairs and modifications. The latter will be flagged by the phrase "Form 337 filed." A copy of Form 337 should be with the logs and will explain what work was completed; the work may also be described in the logbook (Fig. 4-13). Form 337 is filed with the FAA and copies become a part of the official record of each airplane, so they are retrievable from the FAA (contact the Civil Aviation Registry at (405) 954-3116 for further information). This is a good point to remember if 337s are missing from the logbook. See Appendix E for AC 43-9.1E about Form 337.

5. Review the current weight and balance sheet with the logbook and with equipment you see (radios, etc.). The engine logbook will be similar to the airframe logbook and will contain information from the annual or 100-hour inspections. Total time will be noted and possibly an indication of time since any overhaul work, although you may have to do some math here. It's quite possible that this log and engine are not the originals for the aircraft. As long as the facts are well-documented in the logs, there is no cause for alarm. After all, this would be the case if the original engine was replaced with a rebuilt engine.

6. Check the engine's ADs for compliance and the appropriate entries in the log.

7. Pay particular attention to numbers that show the results of a differential compression check. These numbers can say a lot about the overall health of the engine. Each cylinder checked is represented by a fraction, with the bottom number always 80. The 80 indicates the air pressure (pounds per square inch) used for the check. It is the industry standard. The top number is the air pressure that the combustion chamber maintained while tested; 80 would be perfect, but is unattainable and therefore the top figure will always be lower. A lower number is caused by air pressure loss resulting from

MAJOR REPAIR AND ALTERATION
(Airframe, Powerplant, Propeller, or Appliance)

U.S. Department
of Transportation
Federal Aviation
Administration

For FAA Use Only
Office Identification

INSTRUCTIONS: Print or type all entries. See FAR 43.9, FAR 43 Appendix B, and AC 43.9-1 (or subsequent revision thereof) for instructions and disposition of this form. This report is required by law (49 U.S.C. 1421). Failure to report can result in a civil penalty not to exceed $1,000 for each such violation (Section 901 Federal Aviation Act of 1958).

| 1. Aircraft | Make | | Model | |
| | Serial No. | | Nationality and Registration Mark | |

| 2. Owner | Name (As shown on registration certificate) | Address (As shown on registration certificate) |

3. For FAA Use Only

4. Unit Identification				5. Type	
Unit	Make	Model	Serial No.	Repair	Alteration
AIRFRAME	~~~~~~~~~~~~~~~~~~~ (As described in Item 1 above) ~~~~~~~~~~~~~~~~~~~				
POWERPLANT					
PROPELLER					
APPLIANCE	Type				
	Manufacturer				

6. Conformity Statement

A. Agency's Name and Address	B. Kind of Agency	C. Certificate No.
	U.S. Certificated Mechanic	
	Foreign Certificated Mechanic	
	Certificated Repair Station	
	Manufacturer	

D. I certify that the repair and/or alteration made to the unit(s) identified in item 4 above and described on the reverse or attachments hereto have been made in accordance with the requirements of Part 43 of the U.S. Federal Aviation Regulations and that the information furnished herein is true and correct to the best of my knowledge.

| Date | Signature of Authorized Individual |

7. Approval for Return To Service

Pursuant to the authority given persons specified below, the unit identified in item 4 was inspected in the manner prescribed by the Administrator of the Federal Aviation Administration and is ☐ APPROVED ☐ REJECTED

BY	FAA Flt. Standards Inspector	Manufacturer	Inspection Authorization	Other (Specify)
	FAA Designee	Repair Station	Person Approved by Transport Canada Airworthiness Group	
Date of Approval or Rejection	Certificate or Designation No.	Signature of Authorized Individual	

FAA Form 337 (12-88)

Fig. 4-13. FAA Form 337.

loose, worn, or broken rings, scored or cracked cylinder walls, or burned, stuck, or poorly seated valves. A mechanic can determine which is the cause of the lack of compression and, of course, repair the damage. Normal readings are in the 70s, with a minimum of 70/80, and should be uniform (within two or three pounds) for all cylinders. A discrepancy between cylinders could indicate the need for a top overhaul of one or more cylinders. The FAA says that a loss in excess of 25 percent is cause for further investigation. That would be a reading of 60/80, indicating a very tired engine in need of much work and much money. If the engine had a top or major overhaul (or any repairs), there will be a description of the work done, a date, and the total time on the engine when the work was completed.

8. Read the information from the last oil change, which may contain a statement about debris found on the oil screen or in the oil filter. Oil changes, however, are often done by owners, and may not be recorded in the log (all preventive maintenance is required to be logged). How often was the oil changed? Every 25 hours shows excellent care, but every 50 hours is acceptable. Oil is cheap insurance for a long engine life. If there is a record of oil analysis, ask for it. Oil analysis can indicate internal engine wear problems.

9. Get your notebook out and cross-check those meter readings and the information from the engine maintenance stickers.

LOGBOOK COMMENTS

As time goes on there are more and more airplanes with high hours on them. It isn't uncommon to see Cessna 150, 152s, or 172s or Piper PA-28 models with more than 5000 hours on the airframe. And, rare is the airplane with fewer than a couple of thousand hours. It is sometimes difficult to sell a single-engine airplane today if it has more than 5000 hours on the airframe. Damage history can also make an airplane difficult to sell.

It is a given that even properly repaired structural damage will reduce the value of the aircraft to some extent, depending on the type of damage and the make/model of airplane. It follows that the more hours on the airplane the more likelihood that the airplane has a damage history. As a result, unscrupulous sellers sometimes find it easy to lose some or all of the logbooks for an airplane, making it difficult (but not impossible) to determine the real use time and damage history.

As an additional screening tool, check the aircraft against the FAA Aviation Maintenance Alerts, which show service difficulty reports (SDRs) for all U.S. registered aircraft. The web site for the alerts is http://afs600.faa.gov/srchFolder.asp?Category = alerts.

To check if a specific airplane has been involved in a reported accident, use the query form provided by the National Transportation Safety Board at www.ntsb.gov/ntsb/query.asp#query_start. For even further investigation, you can look for Incident Reports (damage to aircraft of a minor nature not involving an NTSB report/investigation) starting at www.nasdac.faa.gov.

As one final check on an airplane, use the FAA's Request for Copies of Aircraft Records Entry Screen to request a copy of an aircraft's complete file. This should include such information as Form 337 (used for repairs and modifications) and STC information. The query page is located at http://162.58.35.241/e.gov/ND/airrecordsND.asp.

If you find information that the seller has failed to tell you about, or lied about, confrontation is going to be necessary if you wish to continue looking at this particular airplane. Further, it would make all information related by the seller very suspect. In the defense of airplane owners, it is possible for an owner to be unaware of a plane's true history. A sad fact.

Test flight

The test flight is a short flight to determine if the airplane feels right to you. It is not meant to be a "rip-snortin' slam-bang shake-down ride." The owner or a competent flight instructor should accompany you on the test flight to eliminate problems of currency and ratings with the FAA and the owner's insurance company. (The plane is insured, isn't it?) It will also foster better relations with the owner. The flight's duration should be a minimum of one hour, two preferred.

Start the engine and pay particular attention to the gauges. Do they jump to life or are they sluggish? Watch the oil pressure gauge in particular. Did the pressure rise within a few seconds of start? Scrutinize the other gauges. Is everything "in the green"? Watch them during the ground run-up, then again during the takeoff and climb-out. Do the numbers match those specified in the operational manual?

Operate all avionics. This may require flying to an airport that is equipped for IFR operations. A short cross-country jaunt would be an excellent chance to become familiar with the plane. Observe the gyro instruments for stability. Check the ventilation and heating system for proper operation. Make a few turns and some stalls, and fly level. Does the airplane perform as expected? Can it be trimmed for hands-off flight?

Return to the airport for a couple of landings. Watch for improper brake operation and nosewheel shimmy.

Park the plane, shut the engine down, and open the engine compartment to again look for oil leaks. Look along the belly for signs of oil leakage and blow-by. A one- or two-hour flight should be enough to dirty things up again.

This would also be a very good time to get an oil sample for analysis. Before flight would not be advisable, as it could have been freshly changed and thus would contain no contaminants.

Mechanic's inspection

If you are still satisfied with the airplane and want to pursue the matter further, have it inspected by an A&P or IA mechanic familiar with the make and model of airplane you are considering. The inspection will not be free, but it could save you thousands of dollars. The average for a prepurchase inspection is three to four hours of labor, at shop rates, for a simple airplane such as a fixed-gear Cessna or Piper. Retractables and twins will be higher as they take longer, because of their complexity.

The mechanic will do a search of ADs and factory service bulletins (SBs), make a full review of the logs for AD compliance, match installed equipment with logbook entries, check that copies of 337s match logbook entries, and make an overall analysis of the airplane. For the engine, a borescope examination and a compression check must be done. A borescope is the best means of determining the real condition of the engine combustion chambers. It is an optical device for looking into each cylinder and viewing the top of the piston, the valves, and the cylinder walls. A compression check reports how well the combustion chamber seals itself (rings and valve action).

Be sure the static system and transponder are inspected and operating, as required for VFR and IFR flight.

Ask the mechanic to make a complete estimate of the extent of work needed and what it will cost. This may become a determining factor in finalizing the purchase, or in setting the final purchase offer.

Avionics check

Depending upon the complexity of the avionics system on the airplane and your intended use for the plane, you may be well advised to take the plane to an avionics shop to have the full radio system checked.

Such a check of the installed avionics should disclose any malfunctioning or improperly installed equipment. It should also check for equipment not allowed by the FCC (not type accepted or out of date).

As with the mechanic's estimate of work needed, have the avionics shop provide you with a list of discrepancies and estimated costs for repair.

POINTS OF ADVICE

Always use your personal mechanic for the prepurchase inspection, someone you are paying to watch out for your interests. Do not hire someone who may have an interest in the sale of the plane, such as an employee of the seller. Get the plane checked even if the annual was just done, unless you know and trust the mechanic who did the annual inspection. You may be able to make a deal with the owner regarding the cost of the mechanic's inspection, particularly if an annual is due.

It is not uncommon to see airplanes listed for sale with the phrase "annual on date of sale." Always be leery of this, because who knows how complete this annual will be? The FARs are very explicit in their inspection requirements, but not all mechanics do what is required or do equal work. To make matters worse, this annual is part of the airplane deal and is done by the seller as part of the sale. Who is looking out for your interests?

If airplane sellers refuse anything that has been mentioned in this chapter, then thank them for their time, walk away, and look elsewhere. Never let a seller control the situation. Your money, your safety, and possibly your very life are at stake. Airplanes are not hot sellers, and there is rarely a line forming to make a purchase. You are the buyer and have the final word.

PROFESSIONAL SALES

For one reason or another you may find that the task of finding a suitable used airplane to be overwhelming or perhaps too time-consuming. Perhaps you fear your inexperience, which is not at all uncommon and not at all unwise either. Want help? Turn to a professional.

Brokers

Airplane brokers are involved with all facets of airplane sales, from sales only to freelancing to purchasing brokers. For purchasing assistance, you could contract with an airplane purchasing broker to do all the leg work for you. A purchasing broker works for you, the purchaser, and will no doubt be a great time saver.

The broker's job is to locate and screen potential aircraft for you to consider for purchase. For the work done, you will pay your broker a fee called a *commission*, which can range from 6 to 10 percent of the purchase price.

A good broker can save you loads of time by weeding out the chaff from the grain. Many offer full service, from airplane location to paperwork assistance to training in your new airplane.

Before contracting with any broker, contact the AOPA (Airplane Owners and Pilots Association). They maintain a complaint file on brokers and can provide you with information that may prevent you from working with a less than reputable broker.

Dealers

Airplane dealers are not like car dealers. They seldom have new airplanes to choose from, but will have a selection of used airplanes. Many specialize in specific makes and models. Most large dealers advertise in the printed press.

An advantage to purchasing a used airplane from a dealer is integrity. Large dealers are more willing to back a purchased airplane than will a private individual. After all, they have a reputation to keep and have a vested interest in your return business. Airplanes offered by dealers have generally been well inspected and serviced as needed.

Of course nothing is free, so plan on spending top dollar when purchasing from a dealer. But remember you are getting some protection from financial disaster for that price.

Canadian airplanes

You may see airplanes registered in Canada among the advertisements you read. Don't discount these airplanes just because they are located in Canada. You can successfully import an airplane from Canada. As a side note, Canadian airplane maintenance procedures are much stricter than those of the United States.

In the United States, FAR Part 91 governs the maintenance requirements of all personal and business use airplanes. The more strict Part 135 governs airplanes used for hire, charter, or scheduled air transportation. In Canada, maintenance requirements for all civilian personal and business use airplanes are very similar to the stricter U.S. Part 135 requirements. As a result, Canadian airplanes are generally maintained to a higher standard than U.S.-registered airplanes.

BASICS ABOUT CANADIAN REGISTRATION

Airplanes manufactured in the United States were given an N-number and a set of logbooks. Hence, it has an FAA record that is searchable. When registered in Canada, an airplane gets a new set of Canadian logbooks (airframe, engine, propeller, and avionics). Also, a journal log is started. The journal log fulfills the Canadian requirement of logging every flight.

In most instances, the U.S. logbooks will be destroyed when the airplane is registered in Canada. Whether the airplane is new or used at the time of Canadian registry, the airplane's history may be tossed away.

LOAN GLITCH

Most airplane financing companies, and banks that specialize in airplane financing, require an airplane to be registered in the United States prior to releasing money on a loan. This makes for a "catch 22" situation. The Canadian airplane owner doesn't want the airplane to leave Canada and lose its Canadian registration before getting paid. The U.S. purchaser, using the airplane as collateral, can't get the money to pay the Canadian owner until the airplane is registered in the United States. There is generally no direct way around this problem, unless the airplane is not being used as collateral.

GETTING A U.S. REGISTRATION

To register the airplane in the United States, it must first be unregistered in Canada. The Canadian owner should be able to handle this through the Canadian Department of Transport, which will then send confirmation of deregistration to the FAA.

After the FAA receives confirmation that the Canadian registration has been canceled, the airplane is registered in the United States and receives an N-number. The newly U.S.-registered airplane must then be inspected before flight by an FAA inspector (employee of the FAA) and the N-numbers placed on the airplane.

5

As You Purchase

IF YOU HAVE COMPLETELY INSPECTED the prospective purchase and found the airplane acceptable at an agreeable price, you should formulate a sales contract. A sales contract is a legal document, signed by the seller and the purchaser. It should state the sales price, terms of sale, and any warranties provided.

The sales contract must include a description of the airplane and the selling price at a minimum. Contingencies, if any, should be listed along with a list of airplane inspection discrepancies and notation of the party responsible for repairs. It is also recommended to have a statement specifically saying that the airplane is airworthy, has a current inspection, and that all Airworthiness Directives and factory service bulletins have been complied with.

As a result of the complexities and diversities of the various state laws involving commercial agreements, such as sales contracts, you should hire a competent attorney to assist you.

If all this legal discussion and recommendation sounds like overkill, remember that, after your home, the purchase of an airplane will probably represent one of the largest investments you will ever make. Protect yourself!

After the sales contract is signed by the seller and the purchaser, you're ready to sit down and complete the paperwork that will lead to airplane ownership.

THE TITLE SEARCH

The first step in the paperwork of purchasing an airplane is to ensure that the airplane has a clear title and that it can be legally sold by the seller. This is called a title search, performed by checking the plane's individual records at the Mike Monroney Aeronautical Center in Oklahoma City, Oklahoma. These records include title information, chain of ownership, major repair and alteration (Form 337) information, and other data pertinent to a particular airplane. The FAA files this information by N-number in individual folders for each airplane. It is currently not an automated system, but hopefully the FAA will soon move into the modern computer era with the rest of the world.

The object of a title search is to make sure that there are no liens or other hidden encumbrances against clear ownership of the airplane. You, your attorney, or another personal representative can conduct the search. Most prospective airplane purchasers find it inconvenient to travel to Oklahoma City to research the FAA's files themselves, so they contract with a third party specializing in title search service. Contact:

AIC Title Services, Inc.
4400 Will Rogers Parkway, Suite 201
Oklahoma City, OK 73108
(800) 288-2519, (405) 948-1811, fax (405) 948-1869
www.aictitleservice.com

AOPA
www.aopa.org

King Aircraft Title, Inc.
14801 S.W. 65th Street
Mustang, OK 73064
(800) 688-1832, (405) 376-5055, fax (405) 376-0717

As part of the title search, ask the title company to obtain copies of any FAA Form 337s that have been filed for that particular airplane. Mechanics are required by regulation to fill out and submit this form every time they make any *major repairs or alterations* to any certified airplane.

A copy of each completed Form 337 is normally kept with the maintenance logs, however they often become misplaced (by accident or otherwise). It is possible that work entered on a Form 337 won't appear in the logbook, as some mechanics do not always complete the required paperwork.

If the title search indicates a Form 337 revealing damage or other work you were not told about, you should confront the airplane's owner to determine if you are being purposefully misled or if the current owner was not as careful at record keeping or purchasing as you.

Title insurance

A title search provides a service, but is not all inclusive. A title search is current only up to the moment it is performed; it is no guarantee against lien filings that were in the process of being recorded or those never filed.

To protect the purchaser against unrecorded liens, FAA recording mistakes, late filings, or other *clouds* on the title, it is recommended you purchase title insurance. The title insurance is generally available from title search service providers for a fee.

IRS liens

The Internal Revenue Service may hold liens on aircraft and not record such with the FAA. This is very important to know, as an airplane's title could appear perfectly clean, yet the airplane could be seized a few months later, or even years later, by the IRS, based on its lien.

Traditional title searches will not disclose IRS liens. Section 49.17 of the FARs states, "A notice of Federal tax lien is not recordable under this part, since it is required to be filed elsewhere by the Internal Revenue Code (26 U.S.C. 6321, 6323; 26 CFR 301.6321-1, 301.6323-1)." Title insurance will not protect you either.

IRS liens are filed as "personal property tax liens" at the local level. In most instances, this would be a local county court. The recommended means of searching for such liens usually entails working through an attorney. The attorney, possibly acting as a middleman by contracting with a third party, completes the records search at the court house.

Using this suggested method provides two levels of protection, the first by completing a search and the second by having a responsible party in case of an incomplete or incorrect search.

LAST LOOK

Before taking possession of the airplane, make one last inspection before signing any documents. This will confirm that no new damage has occurred and that any items questioned during the prepurchase inspection have been repaired. There should be paperwork documenting any and all repair or service work completed.

Documents

The following documents must be presented with the airplane at the time of sale:

- Airworthiness certificate (Fig. 5-1)
- Airframe logbook
- Engine logbook(s)
- Propeller logbook(s)
- Equipment list (including weight and balance data)
- Flight manual

Assistance

Purchasing an airplane requires that many forms be completed and, although they are not complicated, you may want help in completing them. Check

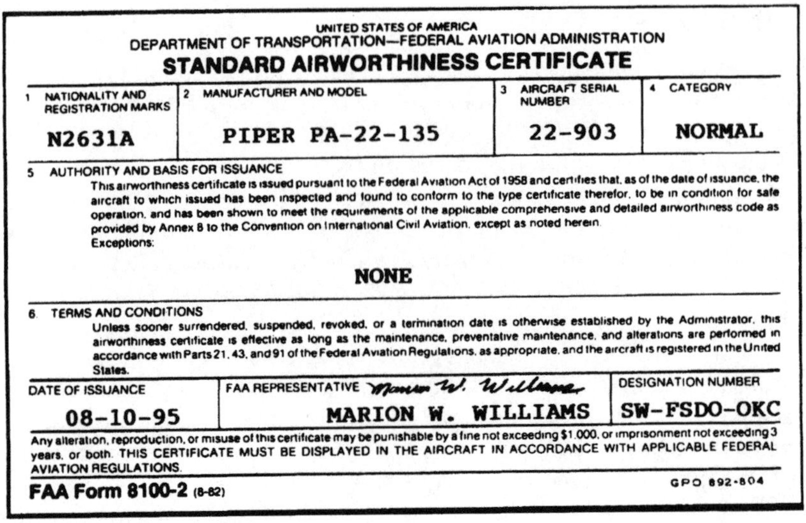

UNITED STATES OF AMERICA
DEPARTMENT OF TRANSPORTATION—FEDERAL AVIATION ADMINISTRATION
STANDARD AIRWORTHINESS CERTIFICATE

1 NATIONALITY AND REGISTRATION MARKS	2 MANUFACTURER AND MODEL	3 AIRCRAFT SERIAL NUMBER	4 CATEGORY
N2631A	PIPER PA-22-135	22-903	NORMAL

5 AUTHORITY AND BASIS FOR ISSUANCE
This airworthiness certificate is issued pursuant to the Federal Aviation Act of 1958 and certifies that, as of the date of issuance, the aircraft to which issued has been inspected and found to conform to the type certificate therefor, to be in condition for safe operation, and has been shown to meet the requirements of the applicable comprehensive and detailed airworthiness code as provided by Annex 8 to the Convention on International Civil Aviation, except as noted herein.
Exceptions:

NONE

6 TERMS AND CONDITIONS
Unless sooner surrendered, suspended, revoked, or a termination date is otherwise established by the Administrator, this airworthiness certificate is effective as long as the maintenance, preventative maintenance, and alterations are performed in accordance with Parts 21, 43, and 91 of the Federal Aviation Regulations, as appropriate, and the aircraft is registered in the United States.

DATE OF ISSUANCE	FAA REPRESENTATIVE	DESIGNATION NUMBER
08-10-95	MARION W. WILLIAMS	SW-FSDO-OKC

Any alteration, reproduction, or misuse of this certificate may be punishable by a fine not exceeding $1,000, or imprisonment not exceeding 3 years, or both. THIS CERTIFICATE MUST BE DISPLAYED IN THE AIRCRAFT IN ACCORDANCE WITH APPLICABLE FEDERAL AVIATION REGULATIONS.

FAA Form 8100-2 (8-82) GPO 892-804

Fig. 5-1. *Standard Airworthiness Certificate example.*

with your FBO or contract with a professional closing service to assist you.

King Aircraft Title, Inc., for a small fee, will provide closing services via telephone, then prepare and file the necessary forms to complete the transaction. This is particularly convenient if the parties involved in the transaction are in different geographical locations, which would be the case when purchasing an airplane long distance. Contact King Aircraft Title, Inc., at (800) 688-1832.

Another source of help in completing the necessary paperwork is a bank, particularly if that bank has a vested interest in your airplane (if they hold the lien). Many banks will not accept a title search unless they requested it. The search may well be completed by the same organization you contracted with, but don't worry; you'll get to pay for this one, too. One way or another, banks charge the customer for everything. The forms you will be completing include:

AC Form 8050-2, Bill of Sale is the standard means of recording the transfer of ownership (Figs. 5-2 and 5-3).

AC Form 8050-1, Aircraft Registration Application is filed with the bill of sale. If you are purchasing the airplane under a contract of conditional sale, then that contract must accompany the registration application in lieu of the AC Form 8050-2. Retain the pink copy of the registration application and keep it in the airplane until the new registration is issued by the FAA (see Figs. 5-4 through 5-6).

AC 8050-41, Release of Lien must be filed by a seller who still owes money on the airplane.

Paperwork Reduction Act Statement: The information collected is used to register an aircraft or hold an aircraft in trust. The information is required to register and prove ownership of an aircraft. We estimate that it will take .5 hour to complete. Use of this form is optional. Please note that an agency may not conduct or sponsor, and a person is not required to respond to, a collection of information unless it displays a currently valid OMB control number. The OMB control number associated with this collection is 2120-0042.

U.S. Department of Transportation
Federal Aviation Administration
Flight Standards Service
FAA Aircraft Registration Branch
P.O. Box 25504
Oklahoma City, OK 73125-0504

AIRCRAFT BILL OF SALE INFORMATION

PREPARATION: Prepare this form in duplicate. Except for signatures, all data should be typewritten or printed. Signatures must be in ink. The name of the purchaser must be identical to the name of the applicant shown on the application for aircraft registration.

When a trade name is shown as the purchaser or seller, the name of the individual owner or co-owners must be shown along with the trade name.

If the aircraft was not purchased from the last registered owner, conveyances must be submitted completing the chain of ownership from the last registered owner, through all intervening owners, to the applicant.

REGISTRATION AND RECORDING FEES: The fee for issuing a certificate of aircraft registration is $5.00. An additional fee of $5.00 is required when a conditional sales contract is submitted in lieu of bill of sale as evidence of ownership along with the application for aircraft registration ($5.00 for the issuance of the certificate, and $5.00 for recording the lien evidenced by the contract). The fee for recording a conveyance is $5.00 for each aircraft listed thereon. (There is no fee for issuing a certificate of aircraft registration to a governmental unit or for recording a bill of sale that accompanies an application for aircraft registration and the proper registration fee.)

MAILING INSTRUCTIONS:

If this form is used, please mail the original or copy which has been signed in ink to FAA Aircraft Registration, P.O. Box 25504, Oklahoma City, OK   73125-0504.

AC Form 8050-2(9/92)(NSN 0052-00-629-0003) Supersedes Previous Edition

Fig. 5-2. *Instructions for AC Form 8050-2, Bill of Sale.*

AC 8050-64, Assignment of Special Registration Number is issued upon written request. All N-numbers consist of the prefix N followed by one to five numbers, one to four numbers, and a single-letter suffix, or one to three numbers and a two-letter suffix. A special registration number is similar to a personalized license plate for an automobile, sometimes called a *vanity plate* (Fig. 5-7).

UNITED STATES OF AMERICA	FORM APPROVED

UNITED STATES OF AMERICA
U.S. DEPARTMENT OF TRANSPORTATION FEDERAL AVIATION ADMINISTRATION

FORM APPROVED
OMB NO. 2120-0042

AIRCRAFT BILL OF SALE

FOR AND IN CONSIDERATION OF $ THE
UNDERSIGNED OWNER(S) OF THE FULL LEGAL AND
BENEFICIAL TITLE OF THE AIRCRAFT DES-CRIBED AS
FOLLOWS:

UNITED STATES
REGISTRATION NUMBER   **N**

AIRCRAFT MANUFACTURER & MODEL

AIRCRAFT SERIAL No.

DOES THIS   DAY OF   ,
HEREBY SELL, GRANT, TRANSFER AND
DELIVER ALL RIGHTS, TITLE, AND INTERESTS
IN AND TO SUCH AIRCRAFT UNTO:

Do Not Write In This Block
FOR FAA USE ONLY

PURCHASER

NAME AND ADDRESS
(IF INDIVIDUAL(S), GIVE LAST NAME, FIRST NAME, AND MIDDLE INITIAL.)

DEALER CERTIFICATE NUMBER

AND TO   EXECUTORS, ADMINISTRATORS, AND ASSIGNS TO HAVE AND TO HOLD
SINGULARLY THE SAID AIRCRAFT FOREVER, AND WARRANTS THE TITLE THEREOF:

IN TESTIMONY WHEREOF   HAVE SET   HAND AND SEAL THIS   DAY OF

SELLER

NAME(S) OF SELLER (TYPED OR PRINTED)	SIGNATURE(S) (IN INK) (IF EXECUTED FOR CO-OWNERSHIP, ALL MUSTSIGN.	TITLE (TYPED OR PRINTED)

ACKNOWLEDGMENT (NOT REQUIRED FOR PURPOSES OF FAA RECORDING: HOWEVER, MAY BE REQUIRED BY LOCAL LAW FOR
VALIDITY OF THE INSTRUMENT.)

ORIGINAL: TO FAA:
AC Form 8050-2 (9/92) (NSN 0052-00-629-0003) Supersedes Previous Edition

Fig. 5-3. *AC Form 8050-2, Bill of Sale.*

FCC Form 605, Quick-Form Application must be completed if radio
transmitting (COM, GPS/COM, or NAV/COM) equipment is on
board and if you plan to fly outside of the United States, such as
into Canada or Mexico. No license is required for noncommercial
domestic operations (Figs. 5-8 and 5-9).

You must be a licensed radio operator to use the radio transmitting equipment, except for domestic noncommercial flights. Use Form 605 Schedule E

FORM APPROVED
OMB No. 2120-0042

Paperwork Reduction Act Statement: The information collected on this form is necessary to ensure applicant eligibility. The information is used to determine that the applicant meets the necessary qualifications as owner of an amateur built aircraft. We estimate that it will take approximately 30 minutes to complete the form. The information collection is required to obtain a benefit. The information collected becomes part of the aircraft registration system. Please note that an agency may not conduct or sponsor, and a person is not required to respond to, a collection of information unless it displays a currently valid OMB control number. OMB 2120-0042

U.S. Department of Transportation
Federal Aviation Administration
Mike Monroney Aeronautical Center
FAA Aircraft Registry
P.O. Box 25504
Oklahoma City, OK 73125

UNITED STATES OF AMERICA-DEPARTMENT OF TRANSPORTATION

FEDERAL AVIATION ADMINISTRATION-MIKE MONRONEY AERONAUTICAL CENTER

AIRCRAFT REGISTRATION INFORMATION

PREPARATION: Prepare this form in triplicate. Except for signatures, all data should be typewritten or printed. Signatures must be in ink. The name of the applicant should be identical to the name of the purchaser shown on the applicant's evidence of ownership.

EVIDENCE OF OWNERSHIP: The applicant for registration of an aircraft must submit evidence of ownership that meets the requirements prescribed in Part 47 of the Federal Aviation Regulations. AC Form 8050-2, Aircraft Bill of Sale, or its equivalent may be used as evidence of ownership. If the applicant did not purchase the aircraft from the last registered owner, the applicant must submit conveyances completing the chain of ownership from the registered owner to the applicant.

The purchaser under a CONTRACT OF CONDITIONAL SALE is considered the owner for the purpose of registration and the contract of conditional sale must be submitted as evidence of ownership.

A corporation which does not meet citizenship requirements must submit a certified copy of its certificate of incorporation.

REGISTRATION AND RECORDING FEES: The fee for issuing a certificate of aircraft registration is $5; therefore, a $5 fee should accompany this application. An additional $5 recording fee is required when a conditional sales contract is submitted as evidence of ownership. There is no recording fee for a bill of sale submitted with the application.

MAILING INSTRUCTIONS: Please send the WHITE original and GREEN copy of this application to the Federal Aviation Administration Aircraft Registry, Mike Monroney Aeronautical Center, P.O. Box 25504, Oklahoma City, Oklahoma 73125. Retain the pink copy after the original application, fee, and evidence of ownership have been mailed or delivered to the Registry. When carried in the aircraft with an appropriate current airworthiness certificate or a special flight permit, this pink copy is temporary authority to operate the aircraft.

CHANGE OF ADDRESS: An aircraft owner must notify the FAA Aircraft Registry of any change in permanent address. This form may be used to submit a new address.

AC Form 8050-1 (5/03) (NSN 0052-00-628-9007)

Fig. 5-4. *Instructions for AC Form 8050-1, Aircraft Registration.*

UNITED STATES OF AMERICA DEPARTMENT OF TRANSPORTATION
FEDERAL AVIATION ADMINISTRATION–MIKE MONRONEY AERONAUTICAL CENTER
AIRCRAFT REGISTRATION APPLICATION

CERT. ISSUE DATE

UNITED STATES
REGISTRATION NUMBER **N**

AIRCRAFT MANUFACTURER & MODEL

AIRCRAFT SERIAL No.

FOR FAA USE ONLY

TYPE OF REGISTRATION (Check one box)

☐ 1. Individual ☐ 2. Partnership ☐ 3. Corporation ☐ 4. Co-owner ☐ 5. Gov't. ☐ 8. Non-Citizen Corporation

NAME OF APPLICANT (Person(s) shown on evidence of ownership. If individual, give last name, first name, and middle initial.)

TELEPHONE NUMBER: (   )
ADDRESS (Permanent mailing address for first applicant listed.) (If P.O. BOX is used, physical address must also be shown.)

Number and street: _____

Rural Route: _____   P.O. Box:

CITY	STATE	ZIP CODE

☐ **CHECK HERE IF YOU ARE ONLY REPORTING A CHANGE OF ADDRESS**
ATTENTION! Read the following statement before signing this application.
This portion MUST be completed.

A false or dishonest answer to any question in this application may be grounds for punishment by fine and / or imprisonment (U.S. Code, Title 18, Sec. 1001).

CERTIFICATION

I/WE CERTIFY:

(1) That the above aircraft is owned by the undersigned applicant, who is a citizen (including corporations) of the United States.

(For voting trust, give name of trustee: _____), or:

CHECK ONE AS APPROPRIATE:

a. ☐ A resident alien, with alien registration (Form 1-151 or Form 1-551) No. _____

b. ☐ A non-citizen corporation organized and doing business under the laws of (state) _____ and said aircraft is based and primarily used in the United States. Records or flight hours are available for inspection at _____

(2) That the aircraft is not registered under the laws of any foreign country; and

(3) That legal evidence of ownership is attached or has been filed with the Federal Aviation Administration.

NOTE: If executed for co-ownership all applicants must sign. Use reverse side if necessary.

TYPE OR PRINT NAME BELOW SIGNATURE

EACH PART OF THIS APPLICATION MUST BE SIGNED IN INK.	SIGNATURE	TITLE	DATE
	SIGNATURE	TITLE	DATE
	SIGNATURE	TITLE	DATE

NOTE   Pending receipt of the Certificate of Aircraft Registration, the aircraft may be operated for a period not in excess of 90 days, during which time the PINK copy of this application must be carried in the aircraft.

AC Form 8050-1 (5/03) (0052-00-628-9007)

Fig. 5-5. *AC Form 8050-1 (keep the pink copy in the plane).*

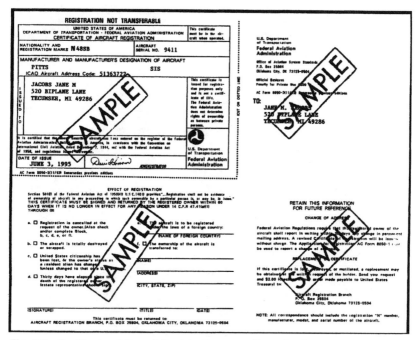

Fig. 5-6. *Certificate of Aircraft Registration you will receive from the FAA.*

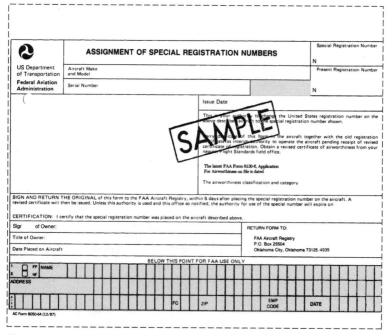

Fig. 5-7. *AC Form 8050-64, Assignment of Special Registration Numbers.*

FCC 605
Main Form

Quick-Form Application for Authorization in the Ship, Aircraft, Amateur, Restricted and Commercial Operator, and General Mobile Radio Services

Approved by OMB
3060 - 0850
See instructions for
public burden estimate

---

1) Radio Service Code:

Application Purpose (Select only one) (　　)

2) **NE** – New
MD – Modification
AM – Amendment

RO – Renewal Only
RM – Renewal / Modification
CA – Cancellation of License

WD – Withdrawal of Application
DU – Duplicate License
AU – Administrative Update

---

3) If this request if for Developmental License or STA (Special Temporary Authorization) enter the appropriate code and attach the required exhibit as described in the instructions. Otherwise enter 'N' (Not Applicable). | (　) **D** **S** N/A

4) If this request is for an Amendment or Withdrawal of Application, enter the file number of the pending application currently on file with the FCC. | File Number

5) If this request is for a Modification, Renewal Only, Renewal / Modification, Cancellation of License, Duplicate License, or Administrative Update, enter the call sign (serial number for Commercial Operator) of the existing FCC license. If this is a request for consolidation of DO & DM Operator Licenses, enter serial number of DO. Also, if filing for a ship exemption, you must provide call sign. | Call Sign/Serial #

6) If this request is for a New, Amendment, Renewal Only, or Renewal Modification, enter the requested expiration date of the authorization (this item is optional). | MM　　DD

7) Does this filing request a Waiver of the Commission's rules? If 'Y', attach the required showing as described in the instructions. | (　) **Y**es **N**o

8) Are attachments (other than associated schedules) being filed with this application? | (　) **Y**es **N**o

---

Applicant/Licensee Information

9) FCC Registration Number (FRN):

10) Applicant /Licensee is a(n): (　) Individual / Corporation | Unincorporated Association / Limited Liability Company | Trust / Partnership | Government Entity / Consortium | Joint Venture

11) First Name (if individual): | MI: | Last Name: | Suffix:

12) Entity Name (if other than individual):

13) Attention To:

14) P.O. Box: | And/Or | 15) Street Address:

16) City: | 17) State: | 18) Zip Code/Postal Code: | 19) Country:

20) Telephone Number: | 21) FAX Number:

22) E-Mail Address:

Ship Applicants/Licensees Only

23) Enter new name of vessel:_____

Aircraft Applicants/Licensees Only

24) Enter the new FAA Registration Number (the N-number):_____
　　NOTE: Do not enter the leading "N".

FCC 605 – Main Form
January 2005 - Page 1

Fig. 5-8. *FCC Form 605, Quick-Form Application (page 1).*

to apply for a Restricted Radiotelephone Permit. Note that FCC Form 605 may be completed on line at www.fcc.gov/Forms/Form605/605.html.

Who really owns the airplane

Fraudulent airplane ownership is a rising problem. Although not as widespread as auto theft, each instance results in aggravation, embarrassment, and expense.

25) Is the applicant/licensee exempt from FCC application Fees?	( ) Yes No
26) Is the applicant/licensee exempt from FCC regulatory Fees?	( ) Yes No

General Certification Statements

1) The applicant/licensee waives any claim to the use of any particular frequency or of the electromagnetic spectrum as against the regulatory power of the United States because of the previous use of the same, whether by license or otherwise, and requests an authorization in accordance with this application.

2) The applicant/licensee certifies that all statements made in this application and in the exhibits, attachments, or documents incorporated by reference are material, are part of this application, and are true, complete, correct, and made in good faith.

3) Neither the applicant/licensee nor any member thereof is a foreign government or a representative thereof.

4) The applicant/licensee certifies that neither the applicant/licensee nor any other party to the application is subject to a denial of Federal benefits pursuant to Section 5301 of the Anti-Drug Abuse Act of 1988, 21 U.S.C. § 862, because of a conviction for possession or distribution of a controlled substance. **This certification does not apply to applications filed in services exempted under Section 1.2002(c) of the rules, 47 CFR § 1.2002(c).** See Section 1.2002(b) of the rules, 47 CFR § 1.2002(b), for the definition of "party to the application" as used in this certification.

5) Amateur or GMRS applicant/licensee certifies that the construction of the station would NOT be an action which is likely to have a significant environmental effect (see the Commission's rules 47 CFR Sections 1.1301-1.1319 and Section 97.13(a) rules (available at web site http://wireless.fcc.gov/rules.html).

6) Amateur applicant/licensee certifies that they have READ and WILL COMPLY WITH Section 97.13(c) of the Commission's rules (available at web site http://wireless.fcc.gov/rules.html) regarding RADIOFREQUENCY (RF) RADIATION SAFETY and the amateur service section of OST/OET Bulletin Number 65 (available at web site http://www.fcc.gov/oet/info/documents/bulletins/).

Certification Statements For GMRS Applicants/Licensees

1) Applicant/Licensee certifies that he or she is claiming eligibility under Rule Section 95.5 of the Commission's rules.

2) Applicant/Licensee certifies that he or she is at least 18 years of age.

3) Applicant/Licensee certifies that he or she will comply with the requirement that use of frequencies 462.650, 467.650, 462.700 and 467.700 MHz is not permitted near the Canadian border North of Line A and East of Line C. These frequencies are used throughout Canada and harmful interference is anticipated.

4) Non-Individual applicants/licensees certify that they have NOT changed frequency or channel pairs, type of emission, antenna height, location of fixed transmitters, number of mobile units, area of mobile operation, or increase in power.

Certification Statements for Ship Applicants/Licensees (Including Ship Exemptions)

1) Applicant/Licensee certifies that they are the owner or operator of the vessel, a subsidiary communications corporation of the owner or operator of the vessel, a state or local government subdivision, or an agency of the US Government subject to Section 301 of the Communications Act.

2) This application is filed with the understanding that any action by the Commission thereon shall be limited to the voyage(s) described herein, and that apart from the provisions of the specific law from which the applicant/licensee requests an exemption, the vessel is in full compliance with all applicable statues, international agreements and regulations.

Signature
27) Typed or Printed Name of Party Authorized to Sign

First Name:	MI:	Last Name:	Suffix:
28) Title:			
Signature:		29) Date:	

Failure to Sign This Application May Result in Dismissal Of The Application And Forfeiture Of Any Fees Paid

WILLFUL FALSE STATEMENTS MADE ON THIS FORM OR ANY ATTACHMENTS ARE PUNISHABLE BY FINE AND/OR IMPRISONMENT (U.S. Code, Title 18, Section 1001) AND / OR REVOCATION OF ANY STATION LICENSE OR CONSTRUCTION PERMIT (U.S. Code, Title 47, Section 312(a)(1)), AND / OR FORFEITURE (U.S. Code, Title 47, Section 503).

FCC 605 – Main Form
January 2005 - Page 2

Fig. 5-9. *FCC Form 605, Quick-Form Application (page 2).*

TITLE FRAUD

An example of a simple form of airplane theft is title fraud. That is when you pay for an airplane and another person's name appears on the title, unbeknownst to you. The story goes like this:

A student pilot finds an airplane and makes a purchase deal with the owner. Funds change hands and a local flight instructor is asked to complete the paperwork. Fraud enters the picture when the instructor's name is entered on the FAA forms as the owner, rather than the purchaser. The instructor then takes the airplane and is never seen again.

This old story has been flying around the hangar circuit for years, but it is something to think about. Other methods of fraud exist, both by omission and commission.

LINKS OF OWNERSHIP

The chain of ownership follows the paperwork filed with the FAA; no other paperwork is acknowledged as proof of ownership. This means that if ownership is not changed with the FAA, then technically the ownership never changed. For example:

Mr. Smith sells his Piper Cub to Mr. Adams in 1983. Mr. Adams dies in 1988, leaving the airplane as part of his estate. The airplane is subsequently advertised for sale, and Mr. Kent decides to purchase it. Mr. Kent runs a title check on the Cub and finds that the airplane is owned by Mr. Smith, not the estate of Mr. Adams.

How could this happen? Simple! Mr. Adams never completed or filed the paperwork necessary to make the change with the FAA. Ownership will have to be cleared before Mr. Kent can purchase the airplane. Most likely he will need legal assistance to sort this mess out.

FIXING RESPONSIBILITY

A new specter in ownership fraud is fixing financial responsibility in the event of injury (property damage or personal injury). For example, imagine you are upgrading to a larger airplane and you sell your trusty Ercoupe to Mr. Benson. The necessary papers are completed and presented with the airplane at the time of sale. The next time you hear about the Ercoupe is when you are served papers in a lawsuit. It seems the airplane crashed and several people were injured on the ground. Mr. Benson never filed the paperwork with the FAA and you are still listed as the legal owner. The injured parties are seeking redress for their losses.

Make copies of all papers related to the sale (and purchase) of an aircraft and save them for future reference. With proper copies as proof of sale and an attorney's help, you should survive this type of legal action.

OUTRIGHT THEFT

The registration numbers and general appearance of stolen aircraft are often altered, making simple detection difficult. The stolen airplane is then listed for sale through a broker with a place of business and good reputation. The broker, through no fault of its own, is ignorant of the theft and sells the airplane to an unsuspecting buyer.

For example, Mr. Cody selects Piper N1234A from an airplane broker's inventory. He gets a title check and finds the title to be clear. All appears in good order and Mr. Cody makes the purchase. Four months later the FBI shows up and interviews Mr. Cody. They explain that he is the victim of fraud; the real N1234A is in the Midwest. The airplane Mr. Cody purchased was stolen from Southern California, renumbered, and sold.

The Piper is returned to its rightful owner, leaving Mr. Cody with no airplane and without a considerable sum of money. He may have recourse against the broker, but it will be a costly battle.

TOO GOOD

Some things are just too good to be true and, being human, we often cannot see this and jump at such offers. So goes another story: You are seated in the airport lounge one afternoon when a sharp-looking Cessna 210 lands. The pilot enters and asks if anyone is interested in purchasing the airplane. The price is unusually low and the plane looks good. You check the plane, fly it, and wind up purchasing it. The pilot claims to be the owner listed on the aircraft's papers. However, the real owner is in Florida for the winter and the airplane was stolen by the pilot from a small airport in the Northeast.

It is likely that the fraud won't be exposed until the airplane is reported stolen in the spring when the owner returns from Florida. The FBI will visit and take the airplane, leaving you with a few memories and an empty wallet.

ALL THOSE LITTLE PIECES

Some used airplanes are *rebuilds*, constructed from many used and new parts around an aircraft data plate (the metal information plate containing the make, model, and serial number). This happens when an unscrupulous person purchases or otherwise obtains the logbooks and data plate for a plane that was wrecked or scrapped.

The logbooks often will not reflect the full extent of the repairs made. Unfortunately, some planes reconstructed in this fashion will never fly properly, while others are downright dangerous to take into the air. The average purchaser can easily be fooled by a reconstruction job, which is why a mechanic should inspect any airplane being considered for purchase.

Self-defense

How do you protect yourself? Where can you turn for assistance in making sure everything is as represented? Actually, there is no single easy answer. The following suggestions may, however, help you avoid a problematic situation:

- Purchase from a known individual or dealer. Ask for positive ID.
- Check serial numbers of the airframe, engine, propeller, and avionics against the logs, paperwork, and title search information.

- If you are not dealing directly with the owner, contact the owner on the telephone or in person and determine the true ownership of the airplane.
- Do a title check and purchase title insurance.
- If the deal appears too good to be true, it probably is!
- When selling an airplane, do a title check 90 days after the sale to ensure title transfer to the new owner.

If anyone gets upset about your being cautious, so what! You are protecting your interests. If something is wrong, back out of the deal. You may even want to call the authorities and report any suspicious activity.

INSURANCE

Insure your airplane from the moment you sign on the dotted line. No one can afford to take risks. For the purposes of airplane ownership, you will be concerned with two types of insurance: *liability* and *hull*.

Liability insurance protects you or your heirs in instances of claims for bodily injury or property damage losses resulting from your operation of an airplane. It is generally true that if someone is injured or killed as a result of your flying, you will be sued, even if you are not at fault. Fault and litigation have very little to do with one another. Hull insurance protects your investment from loss caused by nature, fire, theft, vandalism, or operation. Lending institutions require hull insurance for their protection. There are three types of hull insurance:

All Risk (not in motion): Coverage while stored, tied down, parked, or being manually moved.

All Risk (not in flight): Same coverage as "not in motion," and also while moving on the ground under power.

All Risk: Coverage at all times.

Required insurance

Only a few states require liability insurance on noncommercial airplanes:

- California
- Maryland
- Minnesota
- New Hampshire
- South Carolina
- Virginia

No doubt other states are considering similar requirements. State insurance department telephone numbers are listed in App. C.

A couple of insurance pits to avoid

Beware of insurance policies with exclusions, which create areas of no protection for the insured. Insurance companies use exclusions to avoid payoff in the event of a loss, for instance, by requiring that all installed equipment be functioning properly during operation.

For example, a survivable accident occurs after an engine malfunction. During the postcrash investigation, the insurance company discovers the ADF wasn't working properly. They refuse to pay for the loss because they require all installed equipment to be functioning properly during flight operations. The fact that the nonworking ADF had nothing to do with the cause of the accident is, in their eyes, irrelevant.

Another exclusion example, not as common as it once was, is that no payment will be made for any covered losses if any FARs have been violated. It is safe to say that when an airplane is involved in an accident, some FAR has been violated. Hence, no coverage!

Avoid policies that have part replacement limitations or exclusions, or other specific rules, that set maximum predetermined values for replacement airframe parts. Sometimes called a *component parts* schedule, such policies limit the amount paid for replacement parts and can leave you holding the bag for the difference of the sum paid by the insurance company and the actual cost of repair parts.

Limits

Beware of a policy that provides a $1 million total coverage with per-seat limits of $100,000. If you own a two-place airplane, your $1 million passenger liability is effectively reduced to only $100,000. After all, you carry only one other person in a two-placer. A combined single limit is the recommended coverage; then you don't have to deal with per-seat limits and know your total coverage from the very start.

Always read the policy carefully and understand the exclusions. Ask questions about the policy and demand changes, even refuse the policy. Do whatever is necessary to get the coverage you desire. Carefully selecting insurance coverage can save you money, but the savings must be based on solidly informed decisions. Discuss your specific needs with an experienced aviation insurance agent. During your discussion about coverage, include questions about in-motion, not-in-motion, in-flight, not-in-flight, and ground-risk-only coverage. Understand your insurance coverage completely.

Check various aviation publications for telephone numbers of aviation underwriters; many list toll-free telephone numbers. Consult with more than one company because services, coverage, and rates differ. Ask other pilots and owners about their insurance and how well they have been served by the agents and companies from whom they purchased coverage. If your bank has an aviation department, check with them about the reputations of insurance companies.

A word of advice: Whenever you add something new to your airplane, such as avionics, you have added to the overall value of your investment. Be sure that the hull coverage limits you choose stay aligned with the actual value of your airplane.

Other insurance

Examine your personal health and life insurance policies. Make sure you are covered as a passenger in or while operating a small airplane. It is not uncommon to find exclusions in personal policies that will leave you (and your family) financially uncovered in the event of injury or death involving a small airplane.

GETTING THE AIRPLANE HOME

In most cases you, the new owner, will fly the airplane home to where it will be based. However, in some instances the plane will be flown by the past owner or a dealer. Whatever the case is, be sure the pilot is qualified to handle the airplane, legally and technically.

Legally: Means rated and current for the particular airplane, possessing a current medical, and covered by insurance.

Technically: Means that an individual can actually handle the airplane in a safe and proper manner.

Special flight permits

In some instances the purchased aircraft may not be legal to fly because of damage or an outdated annual inspection. If that is the case, contact the FAA office nearest where the aircraft is located. They can arrange for a Special Flight Permit, generally called a *ferry permit*, which allows an otherwise airworthy aircraft to be flown to a specified destination one time, one way. Before flying an aircraft on a ferry permit, check with your insurance company to ensure that you and your investment will be covered if there is a loss.

LOCAL REGULATIONS

Many states and local jurisdictions register and tax aircraft. Sales taxes may apply to commercial sales, whether made in your home state or not. They may apply when sales are noncommercial between two private parties, or only if the sale is over a certain dollar limit. Use taxes and personal property taxes may also apply.

At the time of this writing, only Alaska, Delaware, Montana, New Hampshire, and Oregon have no sales or use taxes. You can be sure that, in these times of fiscal shortfalls, the tax collector is working full time and will find you should you neglect to pay your fair share.

States requiring registration generally do so for the purpose of collecting a personal property tax. Just because a state does not require registration, it may still impose a tax. Personal property tax rates generally are under 1 percent of the actual cash value of the airplane. Note, however, that some states leave the rate up to local jurisdictions and may be quite aggressive with collections.

Ignorance of the law does not excuse you from compliance. Appendix C contains a list of various state aviation agencies and notations about state aircraft registration requirements, personal property tax, and tax information sources. Contact your respective state agency for information regarding additional rules and regulations.

6

After You Purchase

AN AIRPLANE OWNER MUST RECOGNIZE that properly caring for the airplane directly influences the well-being of both the financial investment and the safety aspects of airplane ownership. Care includes proper maintenance, storage, and cleaning. Fortunately, most steps taken for financial reasons also aid the safety aspect; the opposite is also true.

MAINTENANCE

All airplanes need maintenance, repairs, and inspections. In general, service work is expensive, but there are ways to save money and learn about your airplane in the process.

Finding a good FBO

When considering an FBO (fixed-base operator), check with other airplane owners and see what they think and who they like (Fig. 6-1). In selecting an FBO, ask the following questions:

- How long has the FBO been in business?
- What is the quality of work done?
- Are the FBO's prices reasonable?
- Is timely service available?
- Can you discuss problems with the FBO?
- Is the FBO an authorized dealer for your make of airplane?
- Is flight instruction available from the FBO?
- Are avionics services available from the FBO?

These are important points to consider, but the easiest and quickest means of sizing up an FBO is to examine the facility. Do the buildings look

Fig. 6-1. *This nice general aviation airport is located in Frederick, Md. and offers all services, including a place to eat.*

clean and well-kept and does the general area offer proper physical security for an airplane?

Most operations that appear clean are well run, but not all! Check with the local Better Business Bureau for any recent complaints about the FBO. Call the local FAA office and ask them for their comments. Check with the agency or government entity responsible for the airport and leasing airport space.

Choose an FBO with a maintenance shop that encourages owner-assisted annuals and maintenance-involved owners. *Walk* from a shop that resists your desire to participate in detailed maintenance decisions about your aircraft. *Run* from a shop that prohibits customers from entering the shop floor (usually for *insurance reasons*).

If you are not allowed to be involved in the hands-on end of maintenance and are prohibited from seeing what the mechanic is doing you will have no idea what or how, if anything, is being done to your airplane.

Know your airplane

After you buy your airplane, learn how it is put together. A used airplane has the potential for many minor problems. Minor in cost, maybe, but not in terms of disaster potential.

Ask yourself, have you ever looked into your engine with a borescope, seen the wing spars, examined any of your control cables, or closely checked your strut forks and attach points? Have you personally gone over the AD compliance list, confirmed it, and done a visual inspection of the airplane to reconfirm it?

There are no unimportant screws, nuts, bolts, or anything else on an airplane and it all needs to be correct—all the time. If you don't understand your airplane, don't fly it until you figure it out.

As pilot-in-command (PIC) you are responsible for determining the airworthiness of the aircraft. FAR 91.163(a) says "The owner or operator is primarily responsible for maintaining that aircraft in an airworthy condition. . . ." Simply stated, your mechanic is not responsible for the mechanical condition of your airplane; you, the pilot/owner, are responsible.

One way to gain understanding of your airplane is to take your airplane apart and put it back together again. Work with your A&P or IA and learn about your airplane. If your A&P or IA doesn't want to help you, or bother with you, then you need a new mechanic.

As a last point, after any maintenance is performed, inspect the logs, then check the airplane and make sure the airplane and the logs agree. If, during this process you see something you can't explain or don't like, ask your mechanic (or someone else knowledgeable of your particular make/model airplane). You should also check the bill for the work.

Servicing your own plane

In accordance with regulations, preventive maintenance is simple or minor preservation operations and the replacement of small standard parts not involving complex assembly operations.

FARs (Federal Aviation Regulations) specify that preventive maintenance may be done by the pilot or owner of an airplane not used in commercial service. All preventive maintenance work must be done in such a manner with materials of such quality that the airframe, engine, propeller, or assembly worked on will be at least equal to its original condition according to regulations.

FAR Part 43, Appendix A lists the allowable preventive maintenance items. Note that only those operations listed are considered preventive maintenance. As listed in Part 43, Appendix A:

Preventive maintenance is limited to the following work, provided it does not involve complex assembly operations:

(1) Removal, installation, and repair of landing gear tires.

(2) Replacing elastic shock absorber cords on landing gear.

(3) Servicing landing gear shock struts by adding oil, air, or both.

(4) Servicing landing gear wheel bearings, such as cleaning and greasing.

(5) Replacing defective safety wiring or cotter keys.

(6) Lubrication not requiring disassembly other than removal of nonstructural items such as cover plates, cowlings, and fairings.

(7) Making simple fabric patches not requiring rib stitching or the removal of structural parts or control surfaces.

(8) Replenishing hydraulic fluid in the hydraulic reservoir.

(9) Refinishing decorative coating of fuselage, balloon baskets, wings tail group surfaces (excluding balanced control surfaces), fairings, cowlings, landing gear, cabin, or cockpit interior when removal or disassembly of any primary structure or operating system is not required.

(10) Applying preservative or protective material to components where no disassembly of any primary structure or operating system is involved and where such coating is not prohibited or is not contrary to good practices.

(11) Repairing upholstery and decorative furnishings of the cabin and cockpit, when the repairing does not require disassembly of any primary structure or operating system or interfere with an operating system or affect the primary structure of the aircraft.

(12) Making small simple repairs to fairings, nonstructural cover plates, cowlings, and small patches and reinforcements not changing the contour so as to interfere with proper air flow.

(13) Replacing side windows where that work does not interfere with the structure or any operating system such as controls, electrical equipment, etc.

(14) Replacing safety belts.

(15) Replacing seats or seat parts with replacement parts approved for the aircraft, not involving disassembly of any primary structure or operating system.

(16) Troubleshooting and repairing broken circuits in landing light wiring circuits.

(17) Replacing bulbs, reflectors, and lenses of position and landing lights.

(18) Replacing wheels and skis where no weight and balance computation is involved.

(19) Replacing any cowling not requiring removal of the propeller or disconnection of flight controls.

(20) Replacing or cleaning spark plugs and setting of spark plug gap clearance.

(21) Replacing any hose connection except hydraulic connections.

(22) Replacing prefabricated fuel lines.

(23) Cleaning or replacing fuel and oil strainers or filter elements.

(24) Replacing and servicing batteries.

(25) Not applicable to airplanes.

(26) Replacement or adjustment of nonstructural standard fasteners incidental to operations.

(27) Not applicable to airplanes.

(28) The installation of antimisfueling devices to reduce the diameter of fuel tank filler openings provided the specific device has been made a part of the aircraft type certificate data by the aircraft manufacturer, the aircraft manufacturer has provided FAA-approved instructions for installation of the specific device, and installation does not involve the disassembly of the existing tank filler opening.

(29) Removing, checking, and replacing magnetic chip detectors.

(30) The inspection and maintenance tasks prescribed and specifically identified as preventive maintenance in a primary category aircraft type certificate or supplemental type certificate holder's approved special inspection and preventive maintenance program when accomplished on a primary category aircraft provided:

(i) They are performed by the holder of at least a private pilot certificate issued under part 61 who is the registered owner (including co-owners) of the affected aircraft and who holds a certificate of competency for the affected aircraft (1) issued by a school approved under Sec. 147.21(f) of this chapter; (2) issued by the holder of the production certificate for that primary category aircraft that has a special training program approved under Sec. 21.24 of this subchapter, or (3) issued by another entity that has a course approved by the Administrator, and

(ii) The inspections and maintenance tasks are performed in accordance with instructions contained by the special inspection and preventive maintenance program approved as part of the aircraft's type design or supplemental type design.

BEFORE YOU TRY IT

Before you undertake any of the allowable preventive maintenance procedures, discuss your planned work with a licensed mechanic. The advice you receive will help you avoid costly mistakes. You may have to pay a consultation fee, but it will be money well spent.

REQUIRED LOGBOOK ENTRIES

Entries must be made in the appropriate logbook whenever preventive maintenance is done. The aircraft cannot legally fly without the logbook entry, which must include a description of work, the date completed, the name of the person doing the work, and approval for return to service (signature and certificate number) by the pilot/owner approving the work.

Be aware that the FAA frowns on inaccurate logbook entries, as stated in Sec. 43.12 of the FARs:

Maintenance records: falsification, reproduction, or alteration.

(a) No person may make or cause to be made:

(1) Any fraudulent or intentionally false entry in any record or report that is required to be made, kept, or used to show compliance with any requirement under this part;

(2) Any reproduction, for fraudulent purpose, of any record or report under this part; or

(3) Any alteration, for fraudulent purpose, of any record or report under this part.

(b) The commission by any person of an act prohibited under paragraph (a) of this section is a basis for suspending or revoking the applicable airman, operator, or production certificate, Technical Standard Order Authorization, FAA-Parts Manufacturer Approval, or Product and Process Specification issued by the Administrator and held by that person.

See App. E for additional information regarding airplane records.

TOOLS

The following minimum number of quality tools should allow you to perform preventive maintenance operations on your airplane:

- Multipurpose Swiss Army knife
- $\frac{1}{4}$- and $\frac{3}{8}$-inch ratchet drive with a flex head as an option
- 2-, 4-, or 6-inch drive extensions
- Sockets from $\frac{5}{16}$ to $\frac{3}{4}$ inch in $\frac{1}{16}$-inch increments
- 6-inch crescent wrench
- 10-inch monkey wrench
- 6- or 12-point closed (box) wrenches from $\frac{3}{8}$ to $\frac{3}{4}$ inch
- Set of open-end wrenches from $\frac{3}{8}$ to $\frac{3}{4}$ inch
- Set of ignition wrenches (open and closed end)
- Pair of channel lock pliers (medium size)
- Phillips screwdriver set in the three common sizes
- Flathead screwdriver set short (2-inch) to long (8-inch) sizes
- Plastic electrician's tape
- Container of assorted approved nuts and bolts

- Spare set of spark plugs
- A bag or box to keep tools in order and protected

MANUALS

Before attempting any preventive maintenance, you will need the proper manuals for your make and model of airplane. These may be available for purchase from the manufacturer (if still in business), your FBO, or from a specialized source, such as:

Essco, Inc.
378 S. Van Buren Avenue
Barberton, OH 44203
(230) 644-7724
www.esscoaircraft.com

or

The Airport Shoppe
2635 Cunningham Avenue
San Jose, CA 95148
(800) 634-4744
www.airportshoppe.com

or, for all AD compliance information,

Aerotech Publications, Inc.
P.O. Box 1359
Southold, NY 11971
(800) 235-6444, (631) 765-9375
www.adlog.com

or the type club supporting your particular airplane. Address and telephone numbers for type clubs appear in App. D.

OFFICIAL ENCOURAGEMENT

The FAA encourages pilots and owners to carefully maintain their airplanes and recognizes that proper preventive maintenance provides the pilot or owner with a better understanding of the airplane, saves money, and offers a great sense of accomplishment. If an FBO or airport manager voices concerns about your doing preventive maintenance on your own airplane, Advisory Circular 150/5190-2A may help you. It states, in part:

Restrictions on Self-Service: Any unreasonable restriction imposed on the owners and operators of aircraft regarding the servicing of their own aircraft and equipment may be considered a violation of agency policy. The owner of an aircraft should be permitted to fuel, wash, repair, paint, and otherwise take care of his [or her] own aircraft, provided there is no attempt to perform such services for others. Restrictions which have the effect of diverting activity of this type to a commercial enterprise amount to an exclusive right contrary to law.

With these words the FAA has allowed, as a right, owners of aircraft to save their hard-earned dollars and become familiar with their airplanes, which no doubt contributes to safety.

RESPONSIBILITY

One last point of information that many pilots and owners fail to understand is the maintenance statement found in FAR 91.163(a): "The owner or operator of an aircraft is primarily responsible for maintaining that aircraft in an airworthy condition . . . ." Put very simply, your mechanic is not primarily responsible for the mechanical condition of your airplane; you, as the pilot and/or owner, are responsible.

STORING THE AIRPLANE

Proper aircraft storage is more than mere parking in a hangar or at a tiedown. Unfortunately, too few pilots pay heed to the requirements of properly storing their airplanes during periods of nonuse. Sad but true, the majority of small airplanes are stored outdoors because hangar space is limited at most airports. It is not uncommon to find a two-to-five-year waiting list for hangar space. When found, hangar space is very expensive.

Tiedown

Basic airplane storage is a tiedown. Most pilots should be familiar with the basics of proper tiedown. After all, most trainer airplanes are tied down, and the preflight and postflight parts of each lesson include untying and tying the plane down (Fig. 6-2).

An airplane should be parked facing into the wind, if possible. This is not always feasible because many airports have engineered tiedown systems

Fig. 6-2. *If you don't tie your airplane down, this can happen.*

with predetermined aircraft placement. A proper tiedown must include secure anchors, such as concrete piers with metal loops on their top surface. Each anchor must provide a minimum of 3000 pounds of holding power (4000 pounds for light twins). Three anchors are used for each airplane.

Some airports use an anchor and cable system for tiedowns. This consists of stout steel cables connecting properly installed anchors in two long parallel lines. The lines are approximately one airplane length apart. Airplanes are parked perpendicular to the cables, allowing the wings to be tied to one cable and the tail to the other cable. All planes are parked side by side, oftentimes facing in alternate directions (Fig. 6-3).

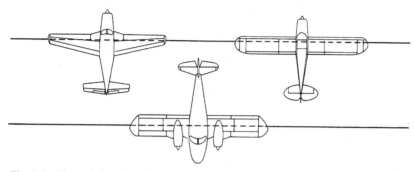

Fig. 6-3. *Typical airport anchor and cable system.* Courtesy of FAA AC20-35C

Tiedown ropes must have a minimum 3000 pounds of tensile strength (4000 pounds for twins). The use of nylon or Dacron ropes is recommended over natural fiber ropes, such as manila, which tend to shrink when wet and are prone to rot and mildew. Chains or steel rope can be used if you take care to prevent damage at the tiedown points on the airplane. Fasten tiedowns only to those points on the aircraft designed for the purpose. This means use the tiedown rings, not just any handy surface (Fig. 6-4).

Cover the pitot tube and fuel vents to keep insects and debris out, plug or cover nacelles or the cowling to discourage birds from building nests in the engine compartment, and lock control surfaces with an internal control lock or fasten gust locks on the control surfaces to prevent movement during windy conditions. Mark all external control surface locks with colorful ribbons as a reminder to remove them before flight.

A FEW DON'TS

- Don't depend on wooden stakes driven into the ground for tiedown anchors.
- Don't use the wing struts for tiedown points.
- Don't use cheap or lightweight rope for tiedowns.

Fig. 6-4.
Use the proper tie points.
Courtesy of FAA AC20-35C

- Don't leave an airplane parked and not tied down.
- Don't forget to lock the controls.

HEAT AND SUNLIGHT PROTECTION

Research shows that the interior temperature of a parked aircraft can reach as much as 185°F. This heat buildup will not only affect the avionics, but it will cause problems with instrument panels, upholstery, and a variety of other plastic items. It is for this reason that many aircraft tied down outside have a cover over the windshield, either inside or outside.

An inside cover's reflective surface protects the plane's interior and reduces heat by reflecting the sun's rays. The shield (cover) attaches to the interior of the aircraft with Velcro fasteners (Fig. 6-5).

Fig. 6-5. *Interior window covers reflect the sunlight and help to reduce interior temperatures.*

Fig. 6-6. *An exterior aircraft cover keeps out sunlight and protects window surfaces from airborne debris.*

Exterior covers provide similar protection for the cabin interior, and provide additional exterior protection for refueling caps and fresh-air vents. They also protect expensive window surfaces from blowing debris (Fig. 6-6).

Hangars

The two types of indoor storage are a large hangar and an individual hangar. An individual hangar is generally called a *tee-hangar* because of its shape. Many owners prefer tee-hangars because they offer more security for the airplane. Security means one airplane stored in the hangar, so no one has access to the airplane without your knowledge and there is less exposure to airplane thieves and vandals.

Also, you generally avoid the dreaded *hangar rash* disease with private hangar storage. Hangar rash is associated with large hangars and occurs when planes are carelessly handled by noncaring ground personnel. In short, they get banged into one another, walls, and equipment. The result is numerous small dings, dents, and scratches (Fig. 6-7).

An often forgotten point of indoor storage is the hours of airplane accessibility. The hangar is locked when the FBO is closed, requiring prior arrangements for access during unusual hours. Of course, the plane stored in an individual hangar is accessible at any time.

NONUSE OF THE AIRPLANE

Nothing is worse for an airplane than sitting on the ground unused. Of course, this is true of all mechanical things, but particularly so for airplanes.

Fig. 6-7. *Hangar rash comes from careless handling in tight quarters.*

All too often airplanes are parked for months and months without ever moving, particularly over the winter. Then one warm spring day, the pilot arrives at the airport, sticks a finger into the wind, and says, "Let's go flying." A quick preflight swing around the plane, assuring that both wings are still attached, then the pilot unties it, jumps in, and starts it up—madly rushing off into the wild blue yonder, all without a care as to what has happened to the airplane while it sat unattended all winter.

Engines in aircraft that are flown only occasionally tend to exhibit more cylinder wall corrosion than engines in aircraft that are flown on a regular basis. The recommended method of preventing cylinder wall corrosion and other corrosion is to fly the aircraft at least weekly. While flying, the engine warms to operating temperature, which vaporizes moisture and other by-products of combustion that can cause engine damage.

Proper storage and preservation techniques are a must, but they are not done by many owners. As you shall see, these procedures take time and effort. Some people don't, for one reason or another, want to extend themselves. Yet, without proper care, the airplane value will be reduced and flying safety impaired. Aircraft storage recommendations are broken down into three categories of storage:

- Flyable storage (used infrequently)
- Temporary storage (up to 90 days of nonuse)
- Indefinite storage (over 90 days of nonuse)

Flyable storage

Most modern aircraft are built of corrosion-resistant Alclad aluminum that will last indefinitely under normal conditions, if kept clean. However, Alclad aluminum is subject to oxidation (corrosion), the first indication of which is the formation of white deposits or spots on metal surfaces. The corrosion can resemble white dust or take on a linty look on bare metal, and cause discoloration or blistering of painted surfaces. Storage in a dry hangar is essential for good preservation.

Minimum care of the engine calls for a weekly propeller pull-through and, at a minimum, a flight once every 30 days. Rotate the propeller by hand without starting the engine. Rotate the propeller six revolutions on four- and six-cylinder nongeared engines, then stop the propeller 45 to 90 degrees from the original position. For six-cylinder geared engines, rotate the propeller four revolutions and stop the propeller 30 to 60 degrees from the original position.

The monthly flight must be at least 30 minutes long to ensure that the engine has reached normal oil and cylinder temperatures. Ground run-up is not an acceptable substitute for flight.

Temporary storage

When an airplane will remain inoperative for a period not exceeding 90 days, it becomes necessary to take specific measures to protect it from the elements. The following steps apply:

1. Fill the fuel tanks with the correct grade of fuel.
2. Clean and wax the aircraft.
3. Clean any grease and oil from the tires and coat them with a tire preservative.
4. Cover the nosewheel to protect it from oil drips.
5. Block up the fuselage to remove the weight from the tires.
6. Cover all airframe openings to keep vermin, insects, and birds out.
7. Remove the top spark plug and spray preservative oil at room temperature through the upper spark plug hole of each cylinder with the piston roughly in bottom dead center position.
8. Rotate the crankshaft as each pair of opposite cylinders is sprayed. Stop the crankshaft with no piston at top dead center. Reinstall the spark plugs.
9. Apply preservative to the engine interior by spraying approximately two ounces of preservative oil through the oil filler tube.
10. Seal all engine openings exposed to the atmosphere using suitable plugs or moisture-resistant tape. Attach red streamers at each sealed location.

Indefinite storage

An aircraft unused in excess of 90 days must be completely preserved. The following steps outline the basics of such preservation, which represents loads of work at considerable expense (before beginning, consult with a mechanic):

1. Drain the engine oil and refill with MIL-C-6259 type-II oil. The aircraft should then be flown for 30 minutes, reaching but not exceeding normal oil and cylinder temperatures. Allow the engine to cool to ambient temperature.

2. Perform all the steps outlined in the previous section, "Temporary Storage," except replacing the spark plugs.

3. Install dehydrator plugs in each of the top spark plug holes and make sure that each plug is blue in color when installed. Protect and support the spark plug leads.

4. If the engine is equipped with a pressure-type carburetor, drain the carburetor by removing the drain and vapor vent plugs from the regulator and fuel control unit. With the mixture control set to Rich, inject lubricating oil into the fuel inlet, at a pressure not to exceed 10 lb/in^2, until oil flows from the vapor vent opening. Allow excess oil to drain, plug the inlet, and tighten the drain and vapor vent plugs. Wire the throttle in the open position, place bags of desiccant in the intake, and seal the opening with moisture-resistant paper and tape or a cover plate.

5. Place a bag of desiccant in the exhaust pipes and seal the openings with moisture-resistant tape.

6. Seal the cold-air inlet to the heater muff with moisture-resistant tape to exclude moisture and foreign objects.

7. Seal the engine breather by inserting a dehydrator plug in the breather hose and clamping it in place.

8. Lubricate all airframe items.

9. Remove the battery and store it in a cool, dry place.

10. Place covers over the windshield and rear windows.

Engines with installed propellers that are preserved for storage in accordance with this section should have a tag affixed to the propeller in a conspicuous place, with the following notation on the tag:

DO NOT TURN PROPELLER: ENGINE PRESERVED.
PRESERVATION DATE (written in)

Periodic examinations are necessary for aircraft placed in indefinite storage. Engines should have the cylinder dehydrator plugs visually inspected every 15 days and replaced as soon as the color indicates unsafe conditions for storage. If the dehydrator plugs have changed color on half or more than half of the cylinders, all desiccant material on the engine should be replaced.

The cylinder bores of all engines prepared for indefinite storage should be resprayed with a corrosion-preventive mixture every six months, more frequently if a bore inspection shows that corrosion has started. Before spraying, check the engine for corrosion by inspecting the interior of at least one cylinder on each engine through a spark plug hole. If the cylinder shows rust, spray the cylinder with corrosion preventive oil and turn the prop over six times, then respray all cylinders. Remove at least one rocker box cover from each engine and inspect the valve mechanism. Replace all desiccant packets and dehydrator plugs at this time.

Return to service

Preparing the airplane for a return to service requires an inspection by a mechanic. You must also:

1. Remove the aircraft from the blocks and check the tires for proper inflation.
2. Check the nose strut for proper inflation.
3. Remove all covers and plugs, and inspect the interior of the airframe for debris and foreign matter.
4. Clean and inspect the exterior of the aircraft.
5. Remove the cylinder dehydrator plugs, tape, and desiccant bags used to preserve the engine.
6. Drain the corrosive preventive mixture and fill with recommended lubricating oil.
7. If the carburetor has been preserved with oil, drain it by removing the drain and vapor vent plugs from the regulator and fuel control unit. With the mixture control set to Rich, inject service-type gasoline into the fuel inlet, at a pressure not to exceed 10 lb/in^2, until all the oil is flushed from the carburetor. Reinstall the carburetor plugs and attach the fuel line.
8. With the bottom plugs removed, rotate the propeller to clear excess preservative oil from the cylinders. Reinstall the spark plugs and rotate the propeller by hand through compression strokes of all cylinders to check for possible liquid lock.
9. Install the battery.
10. Thoroughly clean and visually inspect the airplane.
11. Start the engine in the normal manner.
12. Perform a test flight per the airframe manufacturer's instructions.

These procedures are generic recommendations, applicable to many general aviation aircraft. For particulars on a specific airplane, check the service manual. Also note that some of the aforementioned procedures required for storage and return to service are legally beyond the scope of pilot and owner preventive maintenance.

Lots of work

As you can see, nonuse of an airplane can be complicated and expensive. If you do not plan to use a plane for periods in excess of three months, you may want to consider selling it. You are not getting your money's worth from the airplane, and the lack of use will cost you additional maintenance dollars in the long run.

CLEANING THE AIRPLANE

Normal upkeep of any airplane requires cleaning, which means everything from washing and waxing the exterior to properly caring for the interior. The results will be a sharp-looking airplane that you can be proud of.

Cleaning the exterior

Exterior care of an airplane is important not only for the plane's appearance and value, but for safety as well. While cleaning the airplane, you will notice small imperfections and minor damage, which can be corrected before they become big problems. Completely washing with automotive-type cleaners will produce acceptable results and be less expensive than using specialized aircraft cleaners. Automotive polishes provide adequate protection for painted or unpainted surfaces. The term *polish* in this context means new space-age silicone preparations, also called *sealers*.

Inside cleaning

The interior of an airplane is seen by everyone, including the pilot and passengers. It is also used by everyone and therefore is a problem to keep clean. My recommendation is a very thorough cleaning with standard automobile cleaners.

Cleaning aids

Purchasing specialty cleaning products with the word *airplane* as part of the product name or description can be very expensive. As an alternative, the following items are available at most grocery stores or automotive supply houses, are reasonable in cost, and work well:

- A heavy-duty spray cleaner for the hard-to-remove stuff. Keep it away from the windshield, instruments, and painted surfaces. Do *not* use oven cleaner; it is a corrosive.

- An engine cleaner to degrease the engine area and oleo struts. It dissolves grease and can be washed off with water, under pressure. When using it in the engine compartment, cover the magnetos and alternator with plastic bags to keep the cleaner and rinse water out. Engine cleaner is also good for cleaning the plane's belly and is reasonably safe on painted surfaces if rinsed per instructions.

- A biodegradable degreaser product based on citric acid, which is one of the newer cleaning products available. It smells like orange juice and is available at most large automotive parts stores. Use it sparingly around aluminum and rinse well.

- Silicon-based spray lubricant to stop squeaks and ease movement. It is also good on cables, controls, seat runners, and doors. Keep it off the windows.

- Rubbing compound to clean away exhaust stains. Use it carefully or you could remove more than just stains.

- Touch-up paint, which may be difficult to find since many airplane colors are unavailable in small quantities for touch-up use. However, don't despair because any good automotive supply house (and some department stores) have inexpensive cans of spray paint to match almost any automotive color. Try to find one that closely matches your airplane's paint color. Remember that a touch-up is just that, not a complete repaint job, and don't expect it to be more. Touching up small blemishes and chips protects the airframe and improves the appearance slightly.

- Spray furniture wax, which is invaluable for use on hard plastic surfaces such as instrument panels and other vinyl-covered objects.

- Liquid resin rubber/vinyl cleaners, which are excellent for noncloth upholstery and other flexible surfaces. The surfaces will look and smell like new but could also be slippery.

- Vacuum cleaners, which are essential for cleaning an airplane. Sweeping will not do the job.

Note that I didn't mention window cleaners. This is one area where you should use only products designed for cleaning airplane windows. Automotive and household products contain chemicals that are harmful to the plastics used in airplane windows.

AIRPLANE THEFT

Airplane and airplane equipment theft is something that owners must guard against at all times. Although airplane thefts are higher in some areas of the South and Southwest because of drug-running, airplane theft is considered a nationwide problem.

Theft prevention

Airplane owners can do several things to protect their investment from thieves. Among these are using various devices designed to prevent the operation of the ignition switch, throttle, and/or controls. On the exterior of the airplane, special wheel locks or propeller locks can be installed.

No device will provide complete protection from a determined thief, who will counter any measures you take to protect your airplane and steal it anyway. The idea is to make the airplane undesirable to thieves because it will take too much time to steal.

Equipment theft

Theft of equipment, such as avionics, has been on the upswing for several years and will no doubt continue. This is a problem that generally occurs more often near large urban areas than at small country airports. Contributing factors could be that the more sophisticated and better equipped aircraft are concentrated in a smaller area (one airport) and that there are more thieves in metropolitan areas. Both factors probably go hand in hand. Just because you are in a rural area, however, you aren't exempt from becoming a victim.

A fairly recent type of thievery involves stealing radio equipment and replacing it with similar stolen equipment. You could go for months or years and never notice that the swap has been made. This delay can make tracing equipment ownership nearly impossible, to say nothing of what happens to the airplane's equipment list and logbook entries. Certain preventive measures will aid in the prevention of equipment theft:

- Mark installed equipment (avionics) with a driver's or pilot's license number.
- Park in well-lighted areas. The dark corner of a deserted airport affords little protection for an airplane.
- Use covers inside the windows to keep prying eyes out.
- Use a radio alert burglar alarm. These systems have been installed at a few airports with great success. Ask the FBO if such a protection system is available at your airport.
- Photograph the instrument panel. This will show what equipment was installed, preventing any quibbling with the insurance company in the event of a loss.

- Keep your logbooks in a secure place. They aren't required to be kept in the airplane, only made available for inspection on request. Many owners keep photocopies in the plane.
- Maintain a list of serial numbers and other data pertaining to your plane's equipment.

Although some of these recommendations may sound redundant with FAA requirements for logbooks and the like, they will serve you well in the event of a loss.

A sad problem that many airplane owners face today is the outright destruction of equipment that is protected by antitheft devices or markings. If thieves can't have it, they don't want you to have it either. This, of course, applies to cars, boats, and nearly everything people work hard to buy and maintain.

How to report airplane theft

In the event your airplane is stolen, you must report the loss to the proper authorities as soon as possible. The less delay when reporting, the greater the chances of prompt recovery. Take the following steps immediately:

1. Report the theft to the local law enforcement agency with jurisdiction over the airport. This will cause the aircraft's registration number to be entered into the NCIC (National Crime Information Center) computer database. It will also cause a notice of the theft (called a "lookout") to be flashed among local jurisdictions.

2. Request that the officer taking the report to notify the nearest FAA flight service station of the theft. This will alert the air traffic control system of the theft, and the information will be sent to all airports and centers. An owner's report will not be accepted; it must come from an official source.

3. After you have notified the police or other law enforcement agency, call the Aviation Crime Prevention Institute at (800) 969-5473 and report the same information.

4. Notify your insurance company of the loss.

Note: The Aviation Crime Prevention Institute (ACPI) of Hagerstown, Maryland, is a nonprofit organization dedicated to the elimination of aviation related crime. For further information about ACPI contact it at (301) 791-9791 or www.acpi.org.

Part 2

The Used Airplane Fleet

This part contains the history, photographs, and specifications for most of the used airplanes you will find on today's market. There are, however, a few makes and models that have not been included because of their scarcity. You will notice that most of the airplanes included have an average value of under $100,000.

Specifications and performance information are as accurate as possible, having come from the FAA, manufacturers, and owner organizations. However, keep in mind that each airplane is an individual example of the art, and might not meet the exact data presented. Additionally, many older airplanes have been modified to perform well beyond original published performance data.

Complete performance data are not presented for every make and model because of the typically small changes in the specifications and performance data from year to year; significant changes are noted.

7

Two-Place Airplanes

MOST PILOTS ARE QUITE FAMILIAR WITH TWO-PLACE AIRPLANES. After all, the majority of us learned to fly in a two-place trainer. Two-placers come in two basic styles: *tandem seating* (one person sitting behind the other) and *side by side* (both occupants sitting beside each other, often quite snugly).

For most of these airplanes, the key word is *simplicity*. They have small, efficient engines, few moving parts, and minimal required maintenance— all totaling up to lower purchase and ownership costs.

Some of the older two-placers, because of their age and type of construction (tube and fabric), are considered to be representatives of a bygone era. Examples are the Piper J3, Aeronca 7AC, and Taylorcraft BC-12D. But today's manufacturers are again producing airplanes of that type. In fact, both the Christen Husky, a newly designed utility airplane first introduced in 1987, and the American Champion series, based on the Aeronca 7 series, are in current production. As you can see, some older airplanes are not so old after all and can represent good, solid flying.

The latest addition to the two-place airplane market is Diamond Aircraft Industries' Katana, a modern composite airplane first produced in 1995.

Most modern two-place airplanes are of all-metal construction, provide reasonable speed and performance, and represent loads of affordable, simple flying fun. Some models are aerobatic-approved. A few two-place airplanes are quite advanced, with many of the complexities found on larger airplanes: retractable landing gear, variable-pitch props, large engines, and more. The all-metal Swift from the 1940s is such an example.

You will no doubt notice that many of these airplanes have the nosewheel "on the tail," which is called *conventional landing gear*. There was a time when all airplanes had conventional landing gear; now most have tricycle gear. Arguments rage about which type of landing gear is better, and the question will never be settled. The point is that anyone can be competent with either. So never forsake an airplane just because it has conventional gear. You can learn to handle it. Furthermore, conventional geared airplanes are better suited for some purposes than those equipped with tri-gear.

AERONCA

Aeronca airplanes were among the most popular trainers of the postwar period. Although not many are used today as trainers, they do make inexpensive sport planes.

All Aeronca two-place airplanes are of tube-and-fabric design. The airframe is made of a welded steel tube structure that is covered with fabric. The wings are also covered with fabric. Coverings were originally grade-A cotton cloth; today, most Aeroncas are covered with one of the new synthetic products.

Aeroncas are fun and easy to fly and are among the least expensive airplanes to operate. Although rather slow by today's speed standards, you can see and enjoy what you're flying over. All Aeroncas (with one exception) have conventional landing gear, but don't let that scare you because they're honest little airplanes that display few bad habits.

Planes in the 7 Champ series were based on the World War II TA Defender and the L3 Liaison airplanes. Most 7ACs were manufactured with a 65-hp A-65 Continental engine. Although some were built with Franklin or Lycoming engines, most now have the Continental. The 7ACs, often referred to as *airknockers*, were produced from 1945 through 1948. All were painted yellow with red trim. According to historical records, Aeronca once produced 56 Champs in a single day. The usual time required to build one was 291 hours. The original sticker price for a 7AC was $2999.

As an improvement to the series, the 7CCM (powered with a 90-hp Continental engine) was introduced in 1948. It had a slightly larger fin than the 7AC and a few minor structural changes. The military configuration of this plane was the L-16B.

The 7EC was the last try for Aeronca. Major changes for this model included a Continental C-90-12F engine, an electrical system, and a metal propeller.

All told, more than 7200 Champs were built before production was halted. There are many still around and rare is the small airport where you will not see at least one.

During the same time period, Aeronca also built the 11 Chief series. It had a wider body than the Champ and seated two, side by side. Like the Champ, it was powered by a Continental A-65 engine. An updated version, the 11CC Super Chief (powered by a Continental C-85-8F with a metal propeller) was introduced in 1947.

As with so many of the postwar-era airplanes, a good design was hard to kill. Aeronca was a good design and in 1954 the Champion Aircraft Company of Osceola, Wisconsin was formed, reintroducing the Aeronca 7EC model as the Champion Traveler. The new 7EC was upholstered and carpeted, and had a propeller spinner.

A tricycle-gear version of the airplane, called the 7FC Tri-Traveler, was built for a short period of time. Unfortunately for the manufacturer, the tri-gear version was never as popular as the Piper or Cessna competition.

A new Champ series started with the 7GC in 1959. These planes were produced with various engines. All display good short field capabilities:

7GC (1959): 140-hp Lycoming

7GCB (1960): 150-hp Lycoming

7GCBC (1967): Long-wing version of 7GCB (with flaps)

The Citabria series of aerobatic airplanes was introduced in 1964 (the name Citabria is "airbatic" spelled backwards). They, like their predecessors, have been very popular planes. Citabria model numbers change with the engines:

7ECA (1964–1965): 100-hp Continental

7ECA (1966–1971): 115-hp Lycoming

7GCAA (1965–1977): 150-hp Lycoming

7KCAB (1967–1977): 150-hp, fuel-injected Lycoming

Bellanca, another old name in airplane manufacturing, merged with Champion in 1970. They continued producing the Champion line, then attempted to introduce a very low-cost airplane, the 7ACA. It was built to sell for $4995 and was powered by a two-cylinder, 60-hp Franklin engine. The 7ACA was never a popular model and those existing today have had their engines changed to either Continental or Lycoming.

The Decathlon 8KCAB series, with a newly designed wing and a more powerful engine, was introduced in 1971. The new wing featured near-symmetrical airfoil, a shorter span, and a wider cord than the Citabria wing. The wing changes, coupled with the 6G positive and 5G negative flight load limits and an inverted engine system, made the Decathlon perfect for aerobatics. It was initially available with a 150-hp engine and in 1977 came optionally powered with a 180-hp engine.

The last entry from Bellanca was the 8GCBC Scout. It is a strong-hearted workhorse built for pipeline/powerline patrol and ranching; several fish and game departments also fly them. Bellanca produced their last Champion-type airplane in 1980.

Production records indicate that the following numbers of aircraft were produced:

- 440 100-hp 7ECAs
- 910 115-hp 7ECAs
- 396 7GCAAs
- 1214 7GCBCs

- 618 7KCABs

- 638 8KCABs

Today, American Champion Aircraft produces the line. Since acquiring the line's production rights in 1990, they have made many improvements to the airplanes, including a new metal spar wing. Models currently produced are 7ECA, 7GCBC, 8GCBC, and 8KCAB.

In regards to flying safety, the 7/8 series of airplanes display a high incidence rate of groundloop accidents: nearly 28 percent of all accidents. Look for these airplanes under the names Aeronca, American Champion, Champion, and Bellanca (Figs. 7-1 through 7-7).

Make: Aeronca
Model: 7AC Champ
Year: 1945–1948
Engine
    Make: Continental (Lycoming and Franklin alternates)
    Model: A-65
    Horsepower: 65
    TBO: 1800 hours
Speeds
    Maximum: 95 mph
    Cruise: 86 mph
    Stall: 38 mph
Transitions
    Takeoff over 50-foot obstacle: NA
    Ground run: 630 feet

Fig. 7-1. *Aeronca 7AC Champ.*

Landing over 50-foot obstacle: NA
Ground roll: 880 feet
Weights
Gross: 1220 pounds
Empty: 740 pounds
Dimensions
Length: 21 feet, 6 inches
Height: 7 feet
Span: 35 feet
Other
Fuel capacity: 13 gallons
Rate of climb: 370 fpm
Seats: Two, tandem

Make: Aeronca
Model: 7CCM Champ
Year: 1947–1950
Engine
Make: Continental
Model: C-90
Horsepower: 90
TBO: 1800 hours
Speeds
Maximum: 103 mph
Cruise: 90 mph
Stall: 42 mph
Transitions
Takeoff over 50-foot obstacle: NA
Ground run: 475 feet
Landing over 50-foot obstacle: NA
Ground roll: 850 feet
Weights
Gross: 1300 pounds
Empty: 810 pounds
Dimensions
Length: 21 feet, 6 inches
Height: 7 feet
Span: 35 feet
Other
Fuel capacity: 19 gallons
Rate of climb: 650 fpm
Seats: Two, tandem

Make: Aeronca
Model: 11AC Chief
Year: 1946–1947

Fig. 7-2. *Aeronca 11AC Chief.*

Engine
    Make: Continental
    Model: A-65
    Horsepower: 65
    TBO: 1800 hours
Speeds
    Maximum: 90 mph
    Cruise: 83 mph
    Stall: 38 mph
Transitions
    Takeoff over 50-foot obstacle: NA
    Ground run: 580 feet
    Landing over 50-foot obstacle: NA
    Ground roll: 880 feet
Weights
    Gross: 1250 pounds
    Empty: 786 pounds
Dimensions
    Length: 20 feet, 4 inches
    Height: 7 feet
    Span: 36 feet, 1 inch
Other
    Fuel capacity: 15 gallons
    Rate of climb: 360 fpm
Seats: Two, side by side

Make: Aeronca
Model: 11CC Super Chief
Year: 1947–1949
Engine
    Make: Continental
    Model: C-85
    Horsepower: 85

TBO: 1800 hours
Speeds
    Maximum: 102 mph
    Cruise: 95 mph
    Stall: 40 mph
Transitions
    Takeoff over 50-foot obstacle: NA
    Ground run: 720 feet
    Landing over 50-foot obstacle: NA
    Ground roll: 880 feet
Weights
    Gross: 1350 pounds
    Empty: 820 pounds
Dimensions
    Length: 20 feet, 7 inches
    Height: 7 feet
    Span: 36 feet, 1 inch
Other
    Fuel capacity: 15 gallons
    Rate of climb: 600 fpm
Seats: Two, side by side

Make: Champion
Model: 7-EC/FC (90) Traveler/Tri-Traveler
Year: 1955–1962
Engine
    Make: Continental
    Model: C-90
    Horsepower: 90

Courtesy of James Koepnick, EAA

Fig. 7-3. *Champion Tri-Traveler.*

TBO: 1800 hours
Speeds
    Maximum: 135 mph
    Cruise: 105 mph
    Stall: 44 mph
Transitions
    Takeoff over 50-foot obstacle: 980 feet
    Ground run: 630 feet
    Landing over 50-foot obstacle: 755 feet
    Ground roll: 400 feet
Weights
    Gross: 1450 pounds
    Empty: 860 pounds
Dimensions
    Length: 21 feet, 6 inches
    Height: 7 feet, 2 inches
    Span: 35 feet, 2 inches
Other
    Fuel capacity: 24 gallons
    Rate of climb: 700 fpm
Seats: Two, tandem

Make: Champion
Model: 7-EC/FC (115) Traveler/Tri-Traveler
Year: 1961–1962
Engine
    Make: Lycoming
    Model: O-235-C1
    Horsepower: 115
    TBO: 2000 hours
Speeds
    Maximum: 135 mph
    Cruise: 125 mph
    Stall: 44 mph
Transitions
    Takeoff over 50-foot obstacle: 630 feet
    Ground run: 375 feet
    Landing over 50-foot obstacle: 755 feet
    Ground roll: 400 feet
Weights
    Gross: 1500 pounds
    Empty: 968 pounds
Dimensions
    Length: 21 feet, 6 inches
    Height: 7 feet, 2 inches

Span: 35 feet, 2 inches
Other
    Fuel capacity: 24 gallons
    Rate of climb: 900 fpm
    Seats: Two, tandem

Make: Champion
Model: 7-ECA (100) Citabria
Year: 1964–1965
Engine
    Make: Continental
    Model: O-200
    Horsepower: 100
    TBO: 1800 hours
Speeds
    Maximum: 117 mph
    Cruise: 112 mph
    Stall: 51 mph
Transitions
    Takeoff over 50-foot obstacle: 890 feet
    Ground run: 480 feet
    Landing over 50-foot obstacle: 755 feet
    Ground roll: 400 feet
Weights
    Gross: 1650 pounds
    Empty: 980 pounds
Dimensions
    Length: 22 feet, 7 inches
    Height: 6 feet, 7 inches
    Span: 33 feet, 5 inches
Other
    Fuel capacity: 35 gallons
    Rate of climb: 650 fpm
Seats: Two, tandem

Make: Champion
Model: 7-ECA (115) Citabria
Year: 1966–1980
Engine
    Make: Lycoming
    Model: O-235
    Horsepower: 115
    TBO: 2000 hours
Speeds
    Maximum: 119 mph
    Cruise: 114 mph

Fig. 7-4. *Champion 7-ECA.*

Stall: 51 mph
Transitions
Takeoff over 50-foot obstacle: 716 feet
Ground run: 450 feet
Landing over 50-foot obstacle: 775 feet
Ground roll: 400 feet
Weights
Gross: 1650 pounds
Empty: 1060 pounds
Dimensions
Length: 22 feet, 7 inches
Height: 6 feet, 7 inches
Span: 33 feet, 5 inches
Other
Fuel capacity: 35 gallons
Rate of climb: 725 fpm
Seats: Two, tandem

Make: American Champion
Model: 7-ECA (118) Aurora
Year: 1995–
Engine
Make: Lycoming
Model: O-235-K2C
Horsepower: 118
TBO: 2000 hours
Speeds
Maximum: 121 mph
Cruise: 116 mph
Stall: 50 mph
Transitions
Takeoff over 50-foot obstacle: 716 feet
Ground run: 450 feet

Landing over 50-foot obstacle: 775 feet
Ground roll: 400 feet
Weights
Gross: 1650 pounds
Empty: 1150 pounds
Dimensions
Length: 22 feet, 7 inches
Height: 7 feet, 7 inches
Span: 33 feet, 5 inches
Other
Fuel capacity: 35 gallons
Rate of climb: 725 fpm
Seats: Two, tandem

Make: Champion
Model: 7GC series
Year: 1959–1980
Engine
Make: Lycoming
Model: O-320 (O-290-A2B 1959 only)
Horsepower: 150 (140 1959 only)
TBO: 2000 hours (1500 1959 only)
Speeds
Maximum: 130 mph
Cruise: 125 mph
Stall: 49 mph
Transitions
Takeoff over 50-foot obstacle: 535 feet
Ground run: 375 feet
Landing over 50-foot obstacle: 755 feet
Ground roll: 450 feet
Weights
Gross: 1650 pounds
Empty: 1140 pounds
Dimensions
Length: 22 feet, 8 inches
Height: 6 feet, 7 inches
Span: 33 feet, 6 inches
Other
Fuel capacity: 39 gallons
Rate of climb: 1120 fpm
Seats: Two, tandem

Make: Champion
Model: 7GCBC
Year: 1967–1980

Engine
    Make: Lycoming
    Model: O-320-A2D
    Horsepower: 150
    TBO: 2000 hours
Speeds
    Maximum: 130 mph
    Cruise: 128 mph
    Stall: 45 mph
Transitions
    Takeoff over 50-foot obstacle: 475 feet
    Ground run: 375 feet
    Landing over 50-foot obstacle: 755 feet
    Ground roll: 450 feet
Weights
    Gross: 1650 pounds
    Empty: 1140 pounds
Dimensions
    Length: 22 feet, 8 inches
    Height: 6 feet, 7 inches
    Span: 33 feet, 6 inches
Other
    Fuel capacity: 35 gallons
    Rate of climb: 1145 fpm
Seats: Two, tandem

Make: American Champion
Model: 7GCBC Explorer
Year: 1994–
Engine
    Make: Lycoming
    Model: O-320-B2B
    Horsepower: 160
    TBO: 2000 hours
Speeds
    Maximum: 135 mph
    Cruise: 131 mph
    Stall: 44 mph
Transitions
    Takeoff over 50-foot obstacle: 650 feet
    Ground run: 310 feet
    Landing over 50-foot obstacle: 690 feet
    Ground roll: 310 feet
Weights
    Gross: 1800 pounds

Fig. 7-5. *American Champion 7GCBC Explorer.*

Empty: 1200 pounds
Dimensions
    Length: 22 feet, 8 inches
    Height: 7 feet, 7 inches
    Span: 34 feet, 3 inches
Other
    Fuel capacity: 35 gallons
    Rate of climb: 1345 fpm
Seats: Two, tandem

Make: Champion
Model: 7KCAB Citabria
Year: 1967–1977
Engine
    Make: Lycoming
    Model: IO-320-E2A
    Horsepower: 150
    TBO: 2000 hours
Speeds
    Maximum: 133 mph
    Cruise: 125 mph
    Stall: 50 mph
Transitions
    Takeoff over 50-foot obstacle: 535 feet
    Ground run: 375 feet
    Landing over 50-foot obstacle: 755 feet
    Ground roll: 400 feet
Weights
    Gross: 1650 pounds
    Empty: 1060 pounds

Dimensions
    Length: 22 feet, 8 inches
    Height: 6 feet, 7 inches
    Span: 33 feet, 5 inches
Other
    Fuel capacity: 39 gallons
    Rate of climb: 1120 fpm
Seats: Two, tandem

Make: Bellanca
Model: 7-ACA Champ
Year: 1971–1972
Engine
    Make: Franklin
    Model: 2A-120-B
    Horsepower: 60
    TBO: 1500 hours
Speeds
    Maximum: 98 mph
    Cruise: 83 mph
    Stall: 39 mph
Transitions
    Takeoff over 50-foot obstacle: 900 feet
    Ground run: 525 feet
    Landing over 50-foot obstacle: NA
    Ground roll: 300 feet
Weights
    Gross: 1220 pounds
    Empty: 750 pounds
Dimensions
    Length: 21 feet, 9 inches
    Height: 7 feet
    Span: 35 feet, 1 inch
Other
    Fuel capacity: 13 gallons
    Rate of climb: 400 fpm
Seats: Two, tandem

Make: Bellanca/American Champion
Model: 8-KCAB (150) Decathlon
Year: 1971–1994
Engine
    Make: Lycoming
    Model: IO-320 (optional C/S propeller)
    Horsepower: 150
    TBO: 2000 hours

Fig. 7-6. *Bellanca 8-KCAB Decathlon.*

Speeds
    Maximum: 147 mph
    Cruise: 137 mph
    Stall: 54 mph
Transitions
    Takeoff over 50-foot obstacle: 1450 feet
    Ground run: 840 feet
    Landing over 50-foot obstacle: 1462 feet
    Ground roll: 668 feet
Weights
    Gross: 1800 pounds
    Empty: 1260 pounds
Dimensions
    Length: 22 feet, 11 inches
    Height: 7 feet, 8 inches
    Span: 32 feet
Other
    Fuel capacity: 40 gallons
    Rate of climb: 1000 fpm
Seats: Two, tandem

Make: Bellanca/American Champion
Model: 8-KCAB (180) Decathlon
Year: 1977–
Engine
    Make: Lycoming
    Model: AEIO-360-H1A
    Horsepower: 180
    TBO: 1400 hours

Speeds
    Maximum: 158 mph
    Cruise: 150 mph
    Stall: 54 mph
Transitions
    Takeoff over 50-foot obstacle: 1310 feet
    Ground run: 710 feet
    Landing over 50-foot obstacle: 1462 feet
    Ground roll: 668 feet
Weights
    Gross: 1800 pounds
    Empty: 1315 pounds
Dimensions
    Length: 22 feet, 11 inches
    Height: 7 feet, 8 inches
    Span: 32 feet
Other
    Fuel capacity: 40 gallons
    Rate of climb: 1230 fpm
Seats: Two, tandem

Make: Bellanca/American Champion
Model: 8-GCBC Scout
Year: 1974–
Engine
    Make: Lycoming
    Model: O-360-C2E (optional C/S propeller)
    Horsepower: 180
    TBO: 2000 hours

Courtesy of American Champion

Fig. 7-7. *American Champion 8-GCBC Scout.*

Speeds
    Maximum: 135 mph
    Cruise: 122 mph
    Stall: 52 mph
Transitions
    Takeoff over 50-foot obstacle: 1090 feet
    Ground run: 510 feet
    Landing over 50-foot obstacle: 1245 feet
    Ground roll: 400 feet
Weights
    Gross: 2150 pounds
    Empty: 1315 pounds
Dimensions
    Length: 23 feet, 10 inches
    Height: 8 feet, 8 inches
    Span: 36 feet, 2 inches
Other
    Fuel capacity: 35 gallons
    Rate of climb: 1080 fpm
Seats: Two, tandem

AVIAT

Christen Industries introduced a "blank sheet of paper" airplane in 1987 called the Husky, a two-place airplane resembling a Piper Super Cub. (The new Husky was designed to fill the gap that opened when Piper stopped building Super Cubs.)

Airplanes such as the Husky are used extensively by the United States Border Patrol, U.S. Forest Service, various state fish and game commissions, law enforcement agencies, pipeline and powerline patrols, and ranchers.

It is doubtful that a large number of Husky airplanes will appear on the used market for a long time, because of their newness and low production numbers. However, they represent a cost-effective alternative to purchasing an aging Super Cub.

The Husky is known to be a stout airplane with a good accident record. On March 19, 1991, Aviat, Inc. purchased the rights to produce the Husky A-1. The 1991 base price was $65,625, in 1995 it was $78,785, and now the plane lists for over $140,000 (Fig. 7-8).

Sub-models of A-1A and A-1B have minimal differences from the original A-1. Both use the Lycoming O-360-A1P engine; however, the B model has an increase of 200 pounds gross.

Make: Aviat
Model: Husky A-1

Fig. 7-8. *Aviat Husky.*

Year: 1987–
Engine
    Make: Lycoming
    Model: O-360-C1G
    Horsepower: 180
    TBO: 2000 hours
Speeds
    Maximum: 145 mph
    Cruise: 140 mph
    Stall: 42 mph
Transitions
    Takeoff over 50-foot obstacle: 625
    Ground run: 200 feet
    Landing over 50-foot obstacle: 1400
    Ground roll: 350 feet
Weights
    Gross: 1800 pounds
    Empty: 1190 pounds
Dimensions
    Length: 22 feet, 7 inches
    Height: 6 feet, 7 inches
    Span: 35 feet, 6 inches
Other
    Fuel capacity: 52 gallons
    Rate of climb: 1500 fpm
Seats: Two, tandem

BEECHCRAFT/RAYTHEON

Beech Aircraft Corporation, although not well known for its two-seat airplanes, has produced two important models. Both are modern, all-metal aircraft.

The most recent two-place Beechcraft is the Skipper, a low-wing, all-metal, tricycle-gear airplane. Naturally, it's quite up to date in design and appearance, having been introduced in 1979. Its wings are unusually sleek, owing to their honeycomb ribs and bonded skin construction. No rivets are used on the wing surfaces. The Skipper is often confused with the Piper Tomahawk (PA-38), because of a similarity in the general shape of the planes.

Declining general aviation sales in the early 1980s essentially forced the Skipper out of production. None were built in 1981 and a complete production halt soon followed. A total of 312 Skippers were built.

The remaining member of the Beech two-place family is the T-34 Mentor (Beech model 45). This tandem-seat, low-wing, all-metal airplane saw extensive service with the military as a trainer. Many pilots consider them war birds, because of their prolonged military history. They were produced from 1948 until 1958.

Until a few years ago, most civilian T-34s were flying with the Civil Air Patrol or military-sponsored flight clubs, with only a few finding their way into the civilian market. Although Mentors are now readily available on the used market, they command a premium price.

No longer supported by Beech for parts and service, T-34s are expensive to own, pretty to look at, and amazing to fly. Late models are powered with the 285-hp Continental IO-520 engine.

You will normally see good examples of T-34s at airshows. Just look for the military-trainer yellow color that many are painted (Figs. 7-9 and 7-10).

Make: Beechcraft
Model: 77 Skipper
Year: 1979–1981
Engine
    Make: Lycoming
    Model: O-235-L2C
    Horsepower: 115
    TBO: 2000 hours
Speeds
    Maximum: 122 mph
    Cruise: 112 mph
    Stall: 54 mph
Transitions
    Takeoff over 50-foot obstacle: 1280 feet
    Ground run: 780 feet
    Landing over 50-foot obstacle: 1313 feet

Fig. 7-9. *Beechcraft 77 Skipper.*

Ground roll: 670 feet
Weights
    Gross: 1675 pounds
    Empty: 1103 pounds
Dimensions
    Length: 24 feet
    Height: 6 feet, 11 inches
    Span: 30 feet
Other
    Fuel capacity: 29 gallons
    Rate of climb: 720 fpm
Seats: Two, side by side

Make: Beech
Model: 225-hp T-34 Mentor
Engine
    Make: Continental
    Model: O-470
    Horsepower: 225
    TBO: 1500 hours
Speeds
    Maximum: 189 mph
    Cruise: 173 mph
    Stall: 54 mph
Transitions
    Takeoff over 50-foot obstacle: 1200 feet
    Ground run: NA
    Landing over 50-foot obstacle: 960 feet
    Ground roll: NA

Fig. 7-10. *Beechcraft T-34 Mentor in military colors.*

Weights
    Gross: 2975 pounds
    Empty: 2246 pounds
Dimensions
    Length: 25 feet, 11 inches
    Height: 9 feet, 7 inches
    Span: 32 feet, 10 inches
Other
    Fuel capacity: NA
    Rate of climb: 1120 fpm
Seats: Two, tandem

Make: Beech
Model: 285-hp T-34 Mentor
Engine
    Make: Continental
    Model: IO-520
    Horsepower: 285
    TBO: 1700 hours
Speeds
    Maximum: 202 mph
    Cruise: 184 mph
    Stall: 54 mph
Transitions
    Takeoff over 50-foot obstacle: 1044 feet
    Ground run: 820 feet
    Landing over 50-foot obstacle: 735 feet
    Ground roll: 420 feet
Weights
    Gross: 3200 pounds
    Empty: 2170 pounds

Dimensions
    Length: 25 feet, 11 inches
    Height: 9 feet, 6 inches
    Span: 32 feet, 10 inches
Other
    Fuel capacity: 80 gallons
    Rate of climb: 1130 fpm
Seats: Two, tandem

CESSNA

The first Cessna two-placer appeared in 1946 as the model 120. The 120 was a metal-fuselage airplane with fabric-covered metal structure wings. Naturally, it was a taildragger. The seating, as in all the Cessna two-placers, was side by side. Control wheels were used rather than sticks for control.

The model 140 that followed was a deluxe version of the 120 with an electrical system, flaps, and a plushier cabin. Many 120s have been updated to look like 140s with the addition of extra side windows and electrical systems.

The 140A was the last Cessna two-place airplane produced for almost a decade. It was all-metal and had a Continental C-90 engine. It's interesting to note that the 140A airplane sold new for $3695 and now commands more than four times that price. Production ceased in 1950 after more than 7000 120s, 140s, and 140As were manufactured.

The 120/140 series fair well in accident numbers when compared to other similar airplanes (taildraggers).

Cessna introduced a new two-place trainer airplane in 1959. Starting life as a tricycle-gear version of the 140A, the model 150 was destined to become the most popular training aircraft ever made. Always being improved, the 150 underwent numerous changes between 1959 and 1977, including the following:

- Omni-vision (rear windows) in 1964
- Swept tail in 1966
- Aerobat model in 1970

The Aerobat 150A, stressed to 6Gs positive and 3Gs negative, has been economically popular for aerobatic training.

Cessna rolled out the 152 in 1978. The 152 is nearly identical to the 150, the biggest change being the Lycoming O-235 engine, which burns 100 low-lead aviation fuel.

A used Cessna 150 or 152 is possibly the best buy in today's two-place used airplane market. If you select it carefully and care for it, you should not lose money when selling the airplane. Nor should you have problems selling it. Few pilots have never flown a Cessna 150 or 152 (Figs. 7-11 through 7-15).

Fig. 7-11. *Cessna 140.*

Make: Cessna
Model: 120 and 140
Year: 1946–1950
Engine
    Make: Continental
    Model: C-90
    Horsepower: 90
    TBO: 1800 hours
Speeds
    Maximum: 125 mph
    Cruise: 115 mph
    Stall: 45 mph
Transitions
    Takeoff over 50-foot obstacle: 1850 feet
    Ground run: 650 feet
    Landing over 50-foot obstacle: 1530 feet
    Ground roll: 460 feet
Weights
    Gross: 1500 pounds
    Empty: 850 pounds
Dimensions
    Length: 20 feet, 9 inches
    Height: 6 feet, 3 inches
    Span: 32 feet, 8 inches
Other
    Fuel capacity: 25 gallons
    Rate of climb: 680 fpm
Seats: Two, side by side

Make: Cessna
Model: 150
Year: 1959–1977

Fig. 7-12. *Cessna 1959 150.*

Fig. 7-13. *Cessna 1972 150.*

Engine
    Make: Continental
    Model: O-200
    Horsepower: 100
    TBO: 1800 hours
Speeds
    Maximum: 125 mph
    Cruise: 122 mph
    Stall: 48 mph
Transitions
    Takeoff over 50-foot obstacle: 1385 feet
    Ground run: 735 feet
    Landing over 50-foot obstacle: 1075 feet
    Ground roll: 445 feet

Fig. 7-14. *Cessna 150 Aerobat.*

Weights
    Gross: 1600 pounds
    Empty: 1060 pounds
Dimensions
    Length: 23 feet, 9 inches
    Height: 8 feet, 9 inches
    Span: 32 feet, 8 inches
Other
    Fuel capacity: 26 gallons
    Rate of climb: 670 fpm
Seats: Two, side by side

Make: Cessna
Model: 152
Year: 1978–1985
Engine
    Make: Lycoming
    Model: O-235-L2C
    Horsepower: 110
    TBO: 2000 hours
Speeds
    Maximum: 127 mph
    Cruise: 123 mph
    Stall: 50 mph
Transitions
    Takeoff over 50-foot obstacle: 1340 feet
    Ground run: 725 feet

Fig. 7-15. *Cessna 1983 152.*

Landing over 50-foot obstacle: 1200 feet
Ground roll: 475 feet
Weights
Gross: 1670 pounds
Empty: 1129 pounds
Dimensions
Length: 24 feet, 1 inch
Height: 8 feet, 6 inches
Span: 33 feet, 2 inches
Other
Fuel capacity: 26 gallons
Rate of climb: 715 fpm
Seats: Two, side by side

COMMONWEALTH

The Commonwealth Skyranger was first introduced just prior to World War II. The first versions were built by the Rearwin Aircraft and Engine Company. After the war they were produced for a short time by the Commonwealth Aircraft Company, which went out of business when the postwar boom went bust. Although a good airplane, it was never produced again.

The instrument panel somewhat resembles the dashboard in an automobile of a similar age. The central engine cluster is actually a single unit displaying RPM, oil pressure and temperature, cylinder head temperature, and battery voltage.

These little airplanes are tube-and-fabric design and, being tail draggers, require some handling skill. They can be flown inexpensively (for about four gallons of fuel per hour), but they are expensive to recover.

The Skyrangers have not been produced for many years and there is no product support. Parts could be a problem, as this is really an orphan airplane (Fig. 7-16).

Fig. 7-16. *Commonwealth Skyranger.*

Make: Commonwealth
Model: Skyranger
Year: 1945–1946
Engine
    Make: Continental
    Model: C-85-12
    Horsepower: 85
    TBO: 1800 hours
Speeds
    Maximum: 145 mph
    Cruise: 115 mph
    Stall: 52 mph
Transitions
    Takeoff over 50-foot obstacle: NA
    Ground run: 800
    Landing over 50-foot obstacle: NA
    Ground roll: NA
Weights
    Gross: 1450 pounds
    Empty: 965 pounds
Dimensions
    Length: 21 feet, 9 inches
    Height: 6 feet, 5 inches
    Span: 34 feet
Other
    Fuel capacity: 24 gallons
    Rate of climb: 725 fpm
Seats: Two, side by side

DIAMOND

Diamond Aircraft Industries operations in Canada were established in June of 1992 to develop, certify, and manufacture a light, all-composite, two-place aircraft, based on the already popular Austrian DV20 design, to serve the particular needs of the North American flight training market. The company is the largest general aviation manufacturer of single engine aircraft in Canada and the third largest in North America.

Among the 40 changes made to the initial Austrian design, called the Katana DA20, was a wider cabin, electric trim, doubling of the electrical system capacity, and the use of North American standard airframe hardware. In the first year of production the Katana received the prestigious *Flying Magazine* Eagle Award for Best Light Airplane.

The Katana DA20-A1 airframe is constructed of glass and carbon fiber/epoxy composites, making it light, durable, and corrosion resistant. The composite construction allows for the easy use of complex curves providing the excellent handling characteristics of the Katana.

The Katana DA20-A1 is powered by an 81-hp Rotax 912-F3 engine. The engine incorporates modern technologies including liquid cooling, electronic ignition, and self-adjusting fuel mixture. The engine meets the certification requirements of FAR 33.

In 1998 Diamond introduced the Katana DA20-C1, powered by the 125-hp Teledyne Continental IO-240-B3B engine. Aircraft performance enhancements include an increased rate of climb, shorter takeoff distance, faster cruise speed, improved density altitude capabilities, and increased engine TBO (Fig. 7-17).

Make: Diamond
Model: Katana DA20-A1
Year: 1995–1998
Engine
    Make: Rotax
    Model: 912-F3
    Horsepower: 81
    TBO: 1200 hours
Speeds
    Maximum: 185 mph
    Cruise: 134 mph
    Stall: 43 mph
Transitions
    Takeoff over 50-foot obstacle: 1560 feet
    Ground run: 1120 feet
    Landing over 50-foot obstacle: 1490 feet
    Ground roll: 748 feet
Weights
    Gross: 1609 pounds
    Empty: 1095 pounds
Dimensions
    Length: 23 feet, 6 inches

Fig. 7-17. *Diamond Katana DA20-C1.*

Height: 6 feet, 11 inches
Span: 35 feet, 7 inches
Other
Fuel capacity: 20 gallons
Rate of climb: 680 fpm
Seats: Two, side by side

Make: Diamond
Model: Katana DA20-C1
Year: 1998–
Engine
    Make: Teledyne Continental
    Model: IO-240B
    Horsepower: 125
    TBO: 2000 hours
Speeds
    Maximum: 188 mph
    Cruise: 152 mph
    Stall: 39 mph
Transitions
    Takeoff over 50-foot obstacle: 1263 feet
    Ground run: 952 feet
    Landing over 50-foot obstacle: 1235 feet
    Ground roll: 550 feet
Weights
    Gross: 1654 pounds
    Empty: 1166 pounds
Dimensions
    Length: 23 feet, 6 inches
    Height: 7 feet, 2 inches
    Span: 35 feet, 8 inches
Other
    Fuel capacity: 25 gallons
    Rate of climb: 1105 fpm
Seats: Two, side by side

ERCOUPE

Ercoupes were designed to be the most foolproof airplanes ever built. Originally they had only a control wheel, which was used for all directional maneuvering (on the ground and in the air). The control wheel operated the ailerons and rudder via control interconnections, and steered the nosewheel. As a result of the control interconnections, no coordination was required to make turns.

The single control wheel design had a limited amount of travel, or authority, to avoid entering into a stall. Being unable to enter a stall made the original Ercoupe "spinproof," which led to the issuance of limited class pilot licenses in the late 1940s when spin training was still required. Many Ercoupes have been modified to include rudder pedals, and in later production years rudder pedals became a factory option.

The first Ercoupes had metal fuselages and fabric-covered wings; later models were all metal. All Ercoupes have tricycle landing gear. One popular feature is the fighterlike canopy that can be opened during flight.

Crosswind landings are unique. The trailing beam main gear takes the shock of the crabbed landing, then makes any directional correction. This is not as novel as you may think; the Boeing 707 lands the same way, unable to drop a wing during crosswind landings because of low engine-to-ground clearance. Early Ercoupe model numbers indicated engine horsepower:

- 65-hp 415C
- 75-hp 415D
- 85-hp 415G

Ercoupe was the original name for these airplanes, but several companies have been associated with the airplane's production. Forney Aircraft Company of Fort Collins, Colorado acquired the production rights in 1956. Forney produced the F1, which was powered by a Continental C-90-12 engine, until 1960. Alon Inc., of McPhearson, Kansas, purchased the Ercoupe production rights in the mid-1960s after Forney gave up. They manufactured the A-2 as a 90-hp aircraft, with optional rudder pedals and sliding canopy, that sold for $7825.

Alon gave up in 1967 and sold the production rights to Mooney Aircraft of Kerrville, Texas. Mooney produced the A-2 as the A-2A, then completely redesigned the Ercoupe in 1968 and called it the Mooney M-10 Cadet. It didn't sell well and was quickly discontinued.

Interestingly, the Ercoupe was designed to be safe. Stalls were very minimal and spins were impossible. A placard reads: "This Airplane Characteristically Incapable of Spinning." Yet, when Mooney produced the Cadet, stalls and spins were introduced.

The last remnants of the Ercoupe went the way of so many other good things in life when production ceased in 1970, probably for all time. More than 5000 Ercoupes were manufactured (including Forney, Alon, and Mooney). A large number of these fine little planes are still flying, no doubt owing to their small thirst for fuel, low used prices, and low pilot skill requirements.

A mechanic who is familiar with Ercoupes and their special problems should do the prepurchase inspection. Remember that some Ercoupes are 50 years old.

The Ercoupe production rights are owned by Univair Aircraft Corporation and new parts are available. You will find these airplanes under the names Ercoupe, Forney, Alon, or Mooney (Figs. 7-18 through 7-20).

Make: Ercoupe
Model: 415
Year: 1946–1949

Fig. 7-18. *Ercoupe 415D.*

Engine
    Make: Continental
    Model: C-75-12(F)
    Horsepower: 75
    TBO: 1800 hours
Speeds
    Maximum: 125 mph
    Cruise: 114 mph
    Stall: 56 mph
Transitions
    Takeoff over 50-foot obstacle: 2250 feet
    Ground run: 560 feet
    Landing over 50-foot obstacle: 1750 feet
    Ground roll: 350 feet
Weights
    Gross: 1400 pounds
    Empty: 815 pounds
Dimensions
    Length: 20 feet, 9 inches
    Height: 5 feet, 11 inches
    Span: 30 feet
Other
    Fuel capacity: 24 gallons
    Rate of climb: 550 fpm
Seats: Two, side by side

Make: Forney
Model: F1
Year: 1957–1960

Engine
  Make: Continental
  Model: C-90-12F
  Horsepower: 90
  TBO: 1800 hours
Speeds
  Maximum: 130 mph
  Cruise: 120 mph
  Stall: 56 mph
Transitions
  Takeoff over 50-foot obstacle: 2100 feet
  Ground run: 500 feet
  Landing over 50-foot obstacle: 1750 feet
  Ground roll: 600
Weights
  Gross: 1400 pounds
  Empty: 890 pounds
Dimensions
  Length: 20 feet, 9 inches
  Height: 5 feet, 11 inches
  Span: 30 feet
Other
  Fuel capacity: 24 gallons
  Rate of climb: 600 fpm
Seats: Two, side by side

Make: Alon
Model: A-2 and Mooney A2-A
Year: 1965–1968
Engine
  Make: Continental
  Model: C-90-16F
  Horsepower: 90
  TBO: 1800 hours

Courtesy of Air Pix

Fig. 7-19. *Alon A-2.*

Speeds
    Maximum: 128 mph
    Cruise: 124 mph
    Stall: 56 mph
Transitions
    Takeoff over 50-foot obstacle: 2100 feet
    Ground run: 540 feet
    Landing over 50-foot obstacle: 1750 feet
    Ground roll: 650 feet
Weights
    Gross: 1450 pounds
    Empty: 930 pounds
Dimensions
    Length: 20 feet, 2 inches
    Height: 5 feet, 11 inches
    Span: 30 feet
Other
    Fuel capacity: 24 gallons
    Rate of climb: 640 fpm
Seats: Two, side by side

Make: Mooney
Model: M-10 Cadet
Year: 1969–1970
Engine
    Make: Continental
    Model: C-90
    Horsepower: 90
    TBO: 1800 hours
Speeds
    Maximum: 118 mph
    Cruise: 110 mph

Courtesy of Ercoupe Pilot's Assn.

Fig. 7-20. *Mooney M-10 Cadet.*

Stall: 46 mph
Transitions
    Takeoff over 50-foot obstacle: 1953 feet
    Ground run: 534 feet
    Landing over 50-foot obstacle: 1015 feet
    Ground roll: 431 feet
Weights
    Gross: 1450 pounds
    Empty: 950 pounds
Dimensions
    Length: 20 feet, 8 inches
    Height: 7 feet, 8 inches
    Span: 30 feet
Other
    Fuel capacity: 24 gallons
    Rate of climb: 835 fpm
Seats: Two, side by side

GULFSTREAM

In 1969, American Aviation introduced the AA-1 as an all-metal, side-by-side, two-place airplane that was manufactured with advanced construction methods. The new model's wings were made by bonding the metal skins to the frame with special adhesives, heat, and pressure—but no rivets. The intention was to create a low-drag, sleek wing. Similar methods are now used throughout the aerospace industry.

Unfortunately, there have been some problems with the bonding process used on this model, so it is common to find AA-1s with rivets in the wings for extra skin attachment strength.

Two nice attributes of this series of airplanes is a canopy that can be opened during flight and excellent visibility.

The Gulfstream airplanes are known for their quick responsiveness to control pressure, and they will cruise considerably faster than the Cessna 150. They are referred to as a "pilot's airplane" and are often called "mini-fighters." They are also hot and unforgiving of pilot inattention, land fast, and require a lot of room for takeoff. Early poor safety records have improved with better pilot training. For the latter reason, it is highly recommended that owner and pilots complete the flight proficiency training available from the American Yankee Association (see App. D).

The AA-1, AA-1B, Trainer, and TR-2 have the same airframe and basic engine combination:

- 108-hp Lycoming O-235-C2C engine (1969–1976)
- 115-hp Lycoming O-235-L2C for AA-1C, T-Cat, and Lynx

Manufacturing records show the following totals built:

- 459 AA-1s
- 470 AA-1As
- 634 AA-1Bs
- 211 AA-1Cs

Because of confusion about the manufacturer's identity, you may see these airplanes advertised as Gulfstream American, Grumman American, or American Aviation (Fig. 7-21).

Courtesy of FletchAir, Inc./photo by G. Miller

Fig. 7-21. *Gulfstream American AA-1.*

Make: Gulfstream
Model: AA-1, AA-1B, TR-2 Trainer
Year: 1969–1976
Engine
    Make: Lycoming
    Model: O-235-C2C
    Horsepower: 108
    TBO: 2000 hours
Speeds
    Maximum: 138 mph
    Cruise: 124 mph
    Stall: 60 mph
Transitions
    Takeoff over 50-foot obstacle: 1590 feet
    Ground run: 890 feet
    Landing over 50-foot obstacle: 1100 feet
    Ground roll: 410 feet
Weights
    Gross: 1500 pounds
    Empty: 1000 pounds

Dimensions
    Length: 19 feet, 3 inches
    Height: 6 feet, 9 inches
    Span: 24 feet, 5 inches
Other
    Fuel capacity: 24 gallons
    Rate of climb: 710 fpm
Seats: Two, side by side

Make: Gulfstream
Model: AA-1C T-Cat/Lynx
Year: 1977–1978
Engine
    Make: Lycoming
    Model: O-235-L2C
    Horsepower: 115
    TBO: 2000 hours
Speeds
    Maximum: 145 mph
    Cruise: 135 mph
    Stall: 60 mph
Transitions
    Takeoff over 50-foot obstacle: 1590 feet
    Ground run: 890 feet
    Landing over 50-foot obstacle: 1125 feet
    Ground roll: 425 feet
Weights
    Gross: 1600 pounds
    Empty: 1066 pounds
Dimensions
    Length: 19 feet, 3 inches
    Height: 7 feet, 6 inches
    Span: 24 feet, 5 inches
Other
    Fuel capacity: 22 gallons
    Rate of climb: 700 fpm
Seats: Two, side by side

LUSCOMBE

Luscombe two-seat airplanes are noted as having one of the strongest airframe and wing structures ever manufactured. They also have a reputation for some of the poorest ground-handling of any small plane.

The first point is true and the second is no doubt spread by those who don't really know better. Luscombes can be touchy on landings because of the narrow-track landing gear, but if the pilot is "tailwheel" proficient and stays alert, there should be no problems (the same applies to all conventional-gear airplanes).

Luscombes were produced from 1946 through 1949 by the Luscombe Airplane Corporation of Dallas, Texas. In the early 1950s, Texas Engineering and Manufacturing Company (TEMCO) built the Luscombe 8F version. In 1955, Silvaire Aircraft Corporation was formed in Fort Collins, Colorado and produced only the 8F. All production stopped in 1960 after a total of 6057 Series-8 Luscombes were manufactured.

Like most other makes of airplanes, the various models increase the engine horsepower:

- 8A (65-hp Continental engine)
- 8B (65-hp Lycoming engine)
- 8C/D (75-hp Continental engine)
- 8E (85-hp Continental engine)
- 8F (90-hp Continental engine)

Because of superior handling qualities, a Luscombe purchaser should carefully inspect for damage caused by overstress from aerobatics (Figs. 7-22 and 7-23).

Make: Luscombe
Model: 8A
Year: 1946–1949
Engine
    Make: Continental
    Model: A-65
    Horsepower: 65
    TBO: 1800 hours
Speeds
    Maximum: 112 mph
    Cruise: 102 mph
    Stall: 48 mph
Transitions
    Takeoff over 50-foot obstacle: 1950 feet
    Ground run: 1050 feet
    Landing over 50-foot obstacle: 1540 feet
    Ground roll: 450 feet
Weights
    Gross: 1260 pounds
    Empty: 665 pounds
Dimensions
    Length: 19 feet, 8 inches
    Height: 6 feet, 1 inch
    Span: 34 feet, 7 inches

Fig. 7-22. *Luscombe 8C.*

Other
    Fuel capacity: 14 gallons
    Rate of climb: 550 fpm
Seats: Two, side by side

Make: Luscombe
Model: 8E
Year: 1946–1947
Engine
    Make: Continental
    Model: C-85
    Horsepower: 85
    TBO: 1800 hours
Speeds
    Maximum: 122 mph
    Cruise: 112 mph
    Stall: 48 mph
Transitions
    Takeoff over 50-foot obstacle: 1875 feet
    Ground run: 650 feet
    Landing over 50-foot obstacle: 1540 feet
    Ground roll: 450 feet
Weights
    Gross: 1400 pounds

Empty: 810 pounds
Dimensions
    Length: 19 feet, 8 inches
    Height: 6 feet, 1 inch
    Span: 34 feet, 7 inches
Other
    Fuel capacity: 25 gallons
    Rate of climb: 800 fpm
Seats: Two, side by side

Make: Luscombe
Model: 8F
Year: 1948–1960
Engine
    Make: Continental
    Model: C-90
    Horsepower: 90
    TBO: 1800 hours
Speeds
    Maximum: 128 mph
    Cruise: 120 mph
    Stall: 48 mph
Transitions
    Takeoff over 50-foot obstacle: 1850 feet
    Ground run: 550 feet

Fig. 7-23. *Luscombe 8F.*

Landing over 50-foot obstacle: 1540 feet
Ground roll: 450 feet
Weights
    Gross: 1400 pounds
    Empty: 870 pounds
Dimensions
    Length: 20 feet
    Height: 6 feet, 3 inches
    Span: 35 feet
Other
    Fuel capacity: 25 gallons
    Rate of climb: 900 fpm
Seats: Two, side by side

PIPER

No other name is more often associated with small airplanes than Piper. Many people refer to every small airplane as a Piper Cub.

The most famous of the Cub series is the J3, first introduced in 1939 and manufactured until World War II, then again produced after the war. Production rates rose to the point of one airplane completed every 10 minutes. Production ended in 1947 after a total of 14,125 J3 Cubs were built.

The J4 Cub Coupe, built only before the war, used many J3 parts, which kept costs of design and production to a minimum.

The J5 Cub Cruiser, also introduced prior to the war, seated three people and, like the J4, shared many J3 parts. The pilot sat in a single bucket seat up front, with a bench seat for two in the rear. After the war, the J5-C and PA-12 appeared as outgrowths of the J5. All of this series are underpowered by today's standards and none will fly three adults on a hot day unless the original engine has been replaced with one with higher horsepower. Some 3760 PA-12s were manufactured, however, owing to World War II military contracts; the actual number of J-5Cs produced is unknown, but was far less. Many STCs are available for the PA-12, including engine power increases to 180 hp.

The letter "J" in the early Piper model numbers indicated Walter C. Jamouneau, Piper's chief engineer. The "PA" designator was later used to simply indicate "Piper Aircraft."

In 1947, the PA-11 Cub Special replaced the J3. The engine was completely enclosed by a cowling and the PA-11 could be soloed from the front seat, unlike the J3, which requires solo flight from the rear seat.

The PA-15 was introduced as the Vagabond in 1948 and was about as basic an airplane as you could get. It had a 65-hp Lycoming O-145 engine and seated two, side by side. The main landing gear was solid, with the only shock-absorbing action coming from the pilot's finesse during landings.

Soon the PA-17, also called the Vagabond, followed as an upgraded PA-15, with such niceties as bungee-type landing gear, floor mats, and the Continental A-65 engine, which is said to have considerably more pep than the Lycoming O-145 of the same horsepower rating.

The PA-18 Super Cub, although built with a completely redesigned airframe, shows its true heritage by outwardly resembling a J3. Production of the Super Cub started in 1949. More than 9000 were built, with a variety of engines from 90 to 150 hp, before the demise of the model.

The last two-place Piper airplane of tube-and-fabric design was the PA-22-108 Colt. It was really a two-seat, flapless version of the very popular four-place Tri-Pacer, powered with a 108-hp Lycoming engine. The Colt was originally marketed for use as a trainer, but most are now used for personal flying.

In 1978, Piper introduced a modern trainer, the PA-38 Tomahawk. It's an all-metal, low-wing, tricycle-gear airplane. Unfortunately, the model has been the victim of numerous ADs and, as a result of economic problems in the aviation market, was built for only a few years. A total of about 2500 PA-38s were produced.

Although Piper reorganized into The New Piper Aircraft, Inc., there are no current plans for the reintroduction of any two-seat models. In general, parts are not available from the factory for older models in this class.

All Piper two-seaters offer good value. They are easily maintained and parts are generally available. Their safety records are average, with a major improvement seen when the tricycle-geared airplanes entered the market (Figs. 7-24 through 7-31).

Fig. 7-24. *Piper J3 Cub.*

Make: Piper
Model: J3 Cub
Year: 1939–1947
Engine
    Make: Continental
    Model: A-65

Horsepower: 65
TBO: 1800 hours
Speeds
Maximum: 87 mph
Cruise: 75 mph
Stall: 38 mph
Transitions
Takeoff over 50-foot obstacle: 730 feet
Ground run: 370 feet
Landing over 50-foot obstacle: 470 feet
Ground roll: 290 feet
Weights
Gross: 1220 pounds
Empty: 680 pounds
Dimensions
Length: 22 feet, 4 inches
Height: 6 feet, 8 inches
Span: 35 feet, 2 inches
Other
Fuel capacity: 9 gallons
Rate of climb: 450 fpm
Seats: Two, tandem

Make: Piper
Model: J4 Cub Coupe
Year: 1939–1941
Engine
Make: Continental
Model: A-65
Horsepower: 65
TBO: 1800 hours

Courtesy of Piper

Fig. 7-25. *Piper J4 Cub.*

Speeds
    Maximum: 95 mph
    Cruise: 80 mph
    Stall: 42 mph
Transitions
    Takeoff over 50-foot obstacle: 750 feet
    Ground run: 370 feet
    Landing over 50-foot obstacle: 480 feet
    Ground roll: 300 feet
Weights
    Gross: 1301 pounds
    Empty: 650 pounds
Dimensions
    Length: 22 feet, 6 inches
    Height: 6 feet, 10 inches
    Span: 36 feet, 2 inches
Other
    Fuel capacity: 25 gallons
    Rate of climb: 450 fpm
Seats: Two, side by side

Make: Piper
Model: J5
Year: 1941
Engine
    Make: Continental
    Model: A-75
    Horsepower: 75
    TBO: 1800 hours
Speeds
    Maximum: 95 mph
    Cruise: 80 mph
    Stall: 43 mph
Transitions
    Takeoff over 50-foot obstacle: 1250 feet
    Ground run: 750 feet
    Landing over 50-foot obstacle: 900 feet
    Ground roll: 400 feet
Weights
    Gross: 1450 pounds
    Empty: 820 pounds
Dimensions
    Length: 22 feet, 6 inches
    Height: 6 feet, 10 inches
    Span: 35 feet, 6 inches

Other
    Fuel capacity: 25 gallons
    Rate of climb: 400 fpm
Seats: Three

Make: Piper
Model: J5C
Year: 1946
Engine
    Make: Lycoming
    Model: O-235
    Horsepower: 100
    TBO: 2000 hours
Speeds
    Maximum: 110 mph
    Cruise: 95 mph
    Stall: 45 mph
Transitions
    Takeoff over 50-foot obstacle: 1050 feet
    Ground run: 650 feet
    Landing over 50-foot obstacle: 950 feet
    Ground roll: 450 feet
Weights
    Gross: 1550 pounds
    Empty: 860 pounds
Dimensions
    Length: 22 feet, 6 inches
    Height: 6 feet, 10 inches
    Span: 35 feet, 6 inches
Other
    Fuel capacity: 20 gallons
    Rate of climb: 650 fpm
Seats: Three

Make: Piper
Model: PA-11 Cub Special (65)
Year: 1947–1949
Engine
    Make: Continental
    Model: A-65
    Horsepower: 65
    TBO: 1800 hours
Speeds
    Maximum: 100 mph
    Cruise: 87 mph
    Stall: 38 mph

Fig. 7-26. *Piper PA-11.*

Transitions
    Takeoff over 50-foot obstacle: 730 feet
    Ground run: 370 feet
    Landing over 50-foot obstacle: 470 feet
    Ground roll: 290 feet
Weights
    Gross: 1220 pounds
    Empty: 730 pounds
Dimensions
    Length: 22 feet, 4 inches
    Height: 6 feet, 8 inches
    Span: 35 feet, 2 inches
Other
    Fuel capacity: 18 gallons
    Rate of climb: 550 fpm
Seats: Two, tandem

Make: Piper
Model: PA-11 Cub Special (90)
Year: 1948–1949
Engine
    Make: Continental
    Model: C-90
    Horsepower: 90
    TBO: 1800 hours
Speeds
    Maximum: 112 mph
    Cruise: 100 mph
    Stall: 40 mph
Transitions
    Takeoff over 50-foot obstacle: 475 feet
    Ground run: 250 feet

Landing over 50-foot obstacle: 550 feet
    Ground roll: 290 feet
Weights
    Gross: 1220 pounds
    Empty: 750 pounds
Dimensions
    Length: 22 feet, 4 inches
    Height: 6 feet, 8 inches
    Span: 35 feet, 2 inches
Other
    Fuel capacity: 18 gallons
    Rate of climb: 900 fpm
Seats: Two, tandem

Fig. 7-27. *Piper PA-12 Cruiser.*

Make: Piper
Model: PA-12
Year: 1946–1947
Engine
    Make: Lycoming
    Model: O-235
    Horsepower: 100 (some have 108 hp)
    TBO: 2000 hours
Speeds
    Maximum: 115 mph
    Cruise: 105 mph
    Stall: 49 mph
Transitions
    Takeoff over 50-foot obstacle: 720 feet
    Ground run: 410 feet
    Landing over 50-foot obstacle: 470 feet
    Ground roll: 360 feet

Weights
    Gross: 1500 pounds
    Empty: 855 pounds
Dimensions
    Length: 23 feet, 1 inch
    Height: 6 feet, 9 inches
    Span: 35 feet, 6 inches
Other
    Fuel capacity: 30 gallons
    Rate of climb: 650 fpm
Seats: Three

Fig. 7-28. *Piper PA-15 Vagabond.*

Make: Piper
Model: PA-15/17 Vagabond
Year: 1948–1950
Engine
    Make: Continental (Lycoming on PA-15)
    Model: A-65
    Horsepower: 65
    TBO: 1800 hours
Speeds
    Maximum: 100 mph
    Cruise: 90 mph
    Stall: 45 mph
Transitions
    Takeoff over 50-foot obstacle: 1572 feet
    Ground run: 800 feet

Landing over 50-foot obstacle: 1280 feet
Ground roll: 450 feet
Weights
Gross: 1100 pounds
Empty: 620 pounds
Dimensions
Length: 18 feet, 7 inches
Height: 6 feet
Span: 29 feet, 3 inches
Other
Fuel capacity: 12 gallons
Rate of climb: 510 fpm
Seats: Two, side by side

Fig. 7-29. *Piper PA-18 Super Cub.*

Make: Piper
Model: PA-18 Super Cub (90)
Year: 1950–1961
Engine
Make: Continental
Model: C-90
Horsepower: 90
TBO: 1800 hours
Speeds
Maximum: 112 mph
Cruise: 100 mph
Stall: 42 mph
Transitions
Takeoff over 50-foot obstacle: 1150 feet
Ground run: 400 feet

Landing over 50-foot obstacle: 800 feet
Ground roll: 385 feet
Weights
Gross: 1500 pounds
Empty: 800 pounds
Dimensions
Length: 22 feet, 5 inches
Height: 6 feet, 6 inches
Span: 35 feet, 3 inches
Other
Fuel capacity: 18 gallons
Rate of climb: 700 fpm
Seats: Two, tandem

Make: Piper
Model: PA-18 Super Cub (125)
Year: 1951–1952
Engine
Make: Lycoming
Model: O-290-D
Horsepower: 125
TBO: 2000 hours
Speeds
Maximum: 125 mph
Cruise: 112 mph
Stall: 41 mph
Transitions
Takeoff over 50-foot obstacle: 650 feet
Ground run: 420 feet
Landing over 50-foot obstacle: 725 feet
Ground roll: 350 feet
Weights
Gross: 1500 pounds
Empty: 845 pounds
Dimensions
Length: 22 feet, 5 inches
Height: 6 feet, 6 inches
Span: 35 feet, 3 inches
Other
Fuel capacity: NA
Rate of climb: 940 fpm
Seats: Two, tandem

Make: Piper
Model: PA-18 Super Cub (150)
Year: 1955–1994

Engine
    Make: Lycoming
    Model: O-320-A2A
    Horsepower: 150
    TBO: 2000 hours
Speeds
    Maximum: 130 mph
    Cruise: 115 mph
    Stall: 43 mph
Transitions
    Takeoff over 50-foot obstacle: 500 feet
    Ground run: 200 feet
    Landing over 50-foot obstacle: 725 feet
    Ground roll: 350 feet
Weights
    Gross: 1750 pounds
    Empty: 930 pounds
Dimensions
    Length: 22 feet, 5 inches
    Height: 6 feet, 6 inches
    Span: 35 feet, 3 inches
Other
    Fuel capacity: 36 gallons
    Rate of climb: 960 fpm
    Seats: Two, tandem

Make: Piper
Model: PA-22-108 Colt
Year: 1961–1963
Engine
    Make: Lycoming
    Model: O-235-C1B
    Horsepower: 108
    TBO: 2000 hours
Speeds
    Maximum: 120 mph
    Cruise: 108 mph
    Stall: 54 mph
Transitions
    Takeoff over 50-foot obstacle: 1500 feet
    Ground run: 950 feet
    Landing over 50-foot obstacle: 1250 feet
    Ground roll: 500 feet
Weights
    Gross: 1650 pounds

Fig. 7-30. *Piper PA-22 Colt.*

Empty: 940 pounds
Dimensions
    Length: 20 feet
    Height: 6 feet, 3 inches
    Span: 30 feet
Other
    Fuel capacity: 36 gallons
    Rate of climb: 610 fpm
Seats: Two, side by side

Make: Piper
Model: PA-38 Tomahawk
Year: 1978–1982
Engine
    Make: Lycoming
    Model: O-235-L2C
    Horsepower: 112
    TBO: 2000 hours
Speeds
    Maximum: 125 mph
    Cruise: 122 mph
    Stall: 54 mph
Transitions
    Takeoff over 50-foot obstacle: 1340 feet
    Ground run: 810 feet
    Landing over 50-foot obstacle: 1520 feet
    Ground roll: 710 feet

Fig. 7-31. *Piper PA-38 Tomahawk.*

Weights
    Gross: 1670 pounds
    Empty: 1128 pounds
Dimensions
    Length: 23 feet, 1 inch
    Height: 9 feet, 1 inch
    Span: 34 feet
Other
    Fuel capacity: 30 gallons
    Rate of climb: 720 fpm
Seats: Two, side by side

PITTS

The Pitts is a specialized airplane designed for high-performance aerobatics. It is a biplane designed for fun in the air and serious aerobatic competition. It can handle load factors of + 6g and –5g, and has symmetrical wing airfoils for inverted flight and an airframe consisting of both a fabric-covered tubular fuselage (tube and fabric) and wings of a fabric-covered wood structure.

The Pitts was first conceived in 1945 by Curtis Pitts for aerobatic competition. The first flight was in 1946 and the airplane was powered by only an 85-hp engine. By the late 1950s, the little Pitts airplane had become very popular as a homebuilt, but it was not until 1973 that the Pitts became available as a factory-built airplane. By then it was powered by a 180-hp Lycoming engine.

Early models of the Pitts were homebuilt from factory-designed kits. These include models S1-C, D, and E. Models S1-S and T were built at the factory as certified airplanes.

The Pitts is currently produced as models S2-C (two seat) and S2-S (single seat) by Aviat Aircraft, Inc., of Afton, Wyoming under license from White International Ltd., of the United Kingdom.

A Pitts is not for everyone, as it is surely not a cross-country family airplane. In 2004 a new Pitts S-2C sold for $215,000 (Fig. 7-32).

Courtesy of Aviat

Fig. 7-32. *Pitts.*

Make: Pitts
Model: S2-S
Year: 1981–
Engine
    Make: Lycoming
    Model: IO-540-D4A5
    Horsepower: 260
    TBO: 1200 hours
Speeds
    Maximum: 187 mph
    Cruise: 154 mph
    Stall: 58 mph
Transitions
    Takeoff over 50-foot obstacle: NA
    Ground run: 480 feet
    Landing over 50-foot obstacle: NA
    Ground roll: 960 feet

Weights
    Gross: 1575 pounds
    Empty: 1100 pounds
Dimensions
    Length: 18 feet, 9 inches
    Height: 6 feet, 7 inches
    Span: 20 feet
Other
    Fuel capacity: 35 gallons
    Rate of climb: 2800 fpm
Seats: One

Make: Pitts
Model: S2-B
Year: 1983–1996
Engine
    Make: Lycoming
    Model: IO-540-D4A5
    Horsepower: 260
    TBO: 1200 hours
Speeds
    Maximum: 186 mph
    Cruise: 154 mph
    Stall: 63 mph
Transitions
    Takeoff over 50-foot obstacle: NA
    Ground run: 557 feet
    Landing over 50-foot obstacle: NA
    Ground roll: 1054 feet
Weights
    Gross: 1625 pounds
    Empty: 1150 pounds
Dimensions
    Length: 18 feet, 9 inches
    Height: 6 feet, 7 inches
    Span: 20 feet
Other
    Fuel capacity: 29 gallons
    Rate of climb: 2700 fpm
Seats: Two, tandem

Make: Pitts
Model: S-2C
Year: 1997–

Engine
    Make: Lycoming
    Model: AEIO-540
    Horsepower: 260
    TBO: 1200 hours
Speeds
    Maximum: 194 mph
    Cruise: 172 mph
    Stall: 64 mph
Transitions
    Takeoff over 50-foot obstacle: 860 feet
    Ground run: 540 feet
    Landing over 50-foot obstacle: 1200 feet
    Ground roll: 750 feet
Weights
    Gross: 1700 pounds
    Empty: 1155 pounds
Dimensions
    Length: 17 feet, 9 inches
    Height: 6 feet, 5 inches
    Span: 20 feet
Other
    Fuel capacity: 29 gallons
    Rate of climb: 2900 fpm
Seats: Two, tandem

SWIFT

The Swift airplane was far ahead of the other two-seat airplanes of its era. Built of all-metal construction, it had retractable conventional landing gear, a variable-pitch prop, and was meant to fly fast.

Developed by the Globe Aircraft Corporation, Swifts were introduced in the immediate post–World War II era. The factory was at one time cranking out Swifts on a 24-hour basis, and in one six-month period pushed 833 planes out the door. This number represents the bulk of those manufactured; the total number of Swifts built is around 1500.

The first Swifts were powered with 85-hp engines, but nearly all of those have since been repowered with larger engines. The B models were equipped with a 125-hp engine and many of these have also been modified with engines up to 225 hp.

Swifts were built strong; they were stressed to 7.2 Gs positive and 4.4 Gs negative. They are real hot-rods and take plenty of pilot skill to safely handle. A 1970 NTSB safety study showed the Swift to have the highest fatal accident rate of the 33 small airplane makes and models studied.

Swift airplanes never returned to production, but parts have been available through the Swift Museum Foundation of Athens, Tennessee. In December of 1996, Aviat Aircraft, builders of the Husky and Pitts, leased Swift production rights. Current plans call for reintroduction of the airplane in 2007 (Fig. 7-33).

Make: Swift
Model: GC-1B
Year: 1946–1950
Engine
    Make: Continental
    Model: C-125
    Horsepower: 125
    TBO: 1800 hours
Speeds
    Maximum: 150 mph
    Cruise: 140 mph
    Stall: 50 mph
Transitions
    Takeoff over 50-foot obstacle: 1185 feet
    Ground run: 830 feet
    Landing over 50-foot obstacle: 880 feet
    Ground roll: 650 feet
Weights
    Gross: 1710 pounds
    Empty: 1125 pounds
Dimensions
    Length: 20 feet, 9 inches
    Height: 6 feet, 1 inch
    Span: 29 feet, 3 inches

Fig. 7-33. *Swift GC-1B.*

Other
    Fuel capacity: 30 gallons
    Rate of climb: 1000 fpm
Seats: Two, side by side

TAYLORCRAFT

Taylorcraft airplanes have their historical roots with Piper Aircraft. Originally, C. Gilbert Taylor was in business with William Piper; however, there was a parting of the ways and Mr. Taylor set about on his own. Although built at the same time as the Piper Cub series, Taylor airplanes never enjoyed Cub-like popularity.

Taylorcraft's BC-12 was introduced in 1941, but it saw a short production run because of the start of World War II. After the war, Taylorcraft, like all the other aircraft companies, restarted production. The new plane was the BC-12D, an updated version of the prewar BC-12.

Despite the wider side-by-side seating fuselage and its inherent extra wind resistance, the BC-12D can outpace a J3 Cub by better than 20 mph. The cockpit has dual control wheels, but brakes only on the pilot's side. About 3000 BC-12Ds were built before production stopped in 1950.

By late 1950, the Taylorcraft Aviation Corporation of Alliance, Ohio was a thing of the past and a new company, Taylorcraft Inc., began in Conway, Pennsylvania. This new company introduced the model 19, called the Sportsman, which was powered with an 85-hp Continental engine. About 200 of these planes were built before Taylorcraft Inc. folded in 1957. Between then and 1965, the Univair Aircraft Corporation of Aurora, Colorado built and sold parts for Taylorcrafts. Univair did not build complete airplanes.

Production was restarted in 1965 at Alliance, Ohio with the Taylorcraft model F-19, which was powered by a Continental O-200 engine. In 1980, the Continental engine was dropped in favor of the 118-hp Lycoming, which called for a new model number, F-21. Other than slight updating, little had changed from the BC-12D airplanes, showing that it is difficult to improve on something as time-proven as the postwar Taylorcraft airplanes. In 1989, Taylorcraft's production rights were purchased by a new owner and manufacturing facilities were set up in Piper's old Lock Haven, Pennsylvania facility. Between 1989 and 1996, a tricycle-gear version with a 180-hp engine was approved for manufacture, but only the prototype ever flew.

A used Taylorcraft in good condition is about the most efficient (cheapest) airplane you can fly because they burn only a little over four gallons of fuel per hour (Figs. 7-34 through 7-36).

Make: Taylorcraft
Model: BC-12D
Year: 1946–1947

Fig. 7-34. *Taylorcraft BC-12D.*

Engine
    Make: Continental
    Model: A-65-8A
    Horsepower: 65
    TBO: 1800 hours
Speeds
    Maximum: 100 mph
    Cruise: 95 mph
    Stall: 38 mph
Transitions
    Takeoff over 50-foot obstacle: 700 feet
    Ground run: 350 feet
    Landing over 50-foot obstacle: 450 feet
    Ground roll: 300 feet
Weights
    Gross: 1200 pounds
    Empty: 730 pounds
Dimensions
    Length: 21 feet, 9 inches
    Height: 6 feet, 10 inches
    Span: 36 feet
Other
    Fuel capacity: 18 gallons
    Rate of climb: 500 fpm
Seats: Two, side by side

Make: Taylorcraft
Model: F-19
Year: 1974–1979
Engine
    Make: Continental

Model: O-200
Horsepower: 100
TBO: 1800 hours
Speeds
 Maximum: 127 mph
 Cruise: 115 mph
 Stall: 43 mph
Transitions
 Takeoff over 50-foot obstacle: 350 feet
 Ground run: 200 feet
 Landing over 50-foot obstacle: 350 feet
 Ground roll: 275 feet
Weights
 Gross: 1500 pounds
 Empty: 900 pounds
Dimensions
 Length: 22 feet, 1 inch
 Height: 6 feet, 10 inches
 Span: 36 feet
Other
 Fuel capacity: 24 gallons
 Rate of climb: 775 fpm
Seats: Two, side by side

Make: Taylorcraft
Model: F-21 & F-21A/B
Year: 1980–1990
Engine
 Make: Lycoming
 Model: O-235-L2C

Fig. 7-35. *Taylorcraft F-21B.*

Fig. 7-36. *This prototype of the never-produced 180-hp Taylorcraft F-22 tri-gear airplane shows that good designs never get old.*

    Horsepower: 118
    TBO: 2000 hours
Speeds
    Maximum: 125 mph
    Cruise: 120 mph
    Stall: 43 mph
Transitions
    Takeoff over 50-foot obstacle: 350 feet
    Ground run: 275 feet
    Landing over 50-foot obstacle: 350 feet
    Ground roll: 275 feet
Weights
    Gross: 1500 pounds
    Empty: 1040 pounds
Dimensions
    Length: 22 feet, 3 inches
    Height: 6 feet, 6 inches
    Span: 36 feet
Other
    Fuel capacity: 40 gallons
    Rate of climb: 875 fpm
Seats: Two, side by side

VARGA

The Varga is an all-metal, military-style airplane that somewhat resembles the Beech T-34 Mentor in outward appearance, but is much smaller. This airplane first appeared in 1958 as the Morrisey 2150. Only 12 were built before the Shinn Company started building them under license as the model 2150A. Of note, all 35 Shinn-built 2150As carry Morrisey Aviation, Inc. nameplates.

The 2150 has tricycle landing gear and tandem seating under a greenhouse-style canopy. A limited number were produced as taildraggers, and some of the tri-gear planes were converted to taildraggers. The original 2150 and 2150As were powered by a 150-hp engine. A 180-hp model, the 2180, was also produced.

In 1967, the production rights were sold to Varga Aviation. From 1977 until 1982, Varga built 140 of these airplanes. The Varga is currently an orphan airplane, however, and parts can be difficult to locate (Fig. 7-37).

Make: Varga
Model: 2150 Kachina
Year: 1977–1980
Engine
    Make: Lycoming
    Model: O-320-A2C
    Horsepower: 150
    TBO: 2000 hours
Speeds
    Maximum: 135 mph
    Cruise: 120 mph
    Stall: 52 mph
Transitions
    Takeoff over 50-foot obstacle: NA
    Ground run: 440 feet

Courtesy of Don Downie

Fig. 7-37. *Varga 2150.*

Landing over 50-foot obstacle: NA
Ground roll: 450 feet
Weights
Gross: 1817 pounds
Empty: 1125 pounds
Dimensions
Length: 21 feet, 3 inches
Height: 7 feet
Span: 30 feet
Other
Fuel capacity: 33 gallons
Rate of climb: 910 fpm
Seats: Two, tandem

Make: Varga
Model: 2180
Year: 1981–1982
Engine
Make: Lycoming
Model: O-360-A4AD
Horsepower: 180
TBO: 2000 hours
Speeds
Maximum: 150 mph
Cruise: 133 mph
Stall: 52 mph
Transitions
Takeoff over 50-foot obstacle: NA
Ground run: 400 feet
Landing over 50-foot obstacle: NA
Ground roll: 450 feet
Weights
Gross: 1817 pounds
Empty: 1175 pounds
Dimensions
Length: 21 feet, 3 inches
Height: 7 feet
Span: 30 feet
Other
Fuel capacity: 33 gallons
Rate of climb: 1310 fpm
Seats: Two, tandem

8

Four-Place Easy Fliers

THE FOUR-PLACE "EASY FLIER" AIRPLANE is by far the most airplane for the money. Over the years many have been produced by different companies, but there are only two basic designs: high-wing and low-wing.

This type of airplane can provide adequate transportation for most personal and business needs, although there are few complexities about the type. All have fixed landing gear, all have engines of 200 hp or less, and most come with fixed-pitch propellers (although a few have optional constant-speed props). They are, as a group, easily found, inexpensively maintained, quickly sold, and tolerant of most new equipment necessary to fulfill many specialized flying requirements. The majority of these planes are of all-metal construction, have tricycle landing gear, and are easy to fly.

A few of the easy-flier airplanes are considered classics because of their age, including the Luscombe and Aeronca Sedans, Stinsons, Tri-Pacers, and Pacers.

AERO COMMANDER

The Aero Commander Darter started life as the Volaire, and was made in Aliquippa, Pennsylvania. Only a few were made before Aero Commander, a division of Rockwell International, bought the design.

The airplane appears very similar to the Cessna 172, is all metal, and has similar performance data. The Darter was aimed at the Cessna 172 market, but sales failed and they were produced for only a few years.

The original Volaires were underpowered, having only a 135-hp engine. The model 100 Darters have a 150-hp engine.

The biggest drawback to these planes is the scarcity of parts. Additionally, the braking system uses a single-handle control, which is far less maneuverable than the more familiar toe brakes found in similar airplanes.

Aero Commander built a larger version of the 100, called the Lark, which was powered with a 180-hp engine. It was to compete with the Cessna 182, but like the Darter did not prove to be popular. Although they share a model number, the Lark does not visually resemble the Darter. All production halted in 1971.

Aero Commanders of this group may be listed under Aero Commander, Rockwell, or Volaire in classified advertisements (Figs. 8-1 and 8-2).

Fig. 8-1. *Aero Commander 100 Darter.*

Make: Aero Commander
Model: 100 Darter
Year: 1965–1969
Engine
    Make: Lycoming
    Model: O-320-A2B (O-290 if Volaire)
    Horsepower: 150 (135 if Volaire)
    TBO: 2000 hours
Speeds
    Maximum: 133 mph
    Cruise: 128 mph
    Stall: 55 mph
Transitions
    Takeoff over 50-foot obstacle: 1550 feet
    Ground run: 870 feet
    Landing over 50-foot obstacle: 1215 feet
    Ground roll: 655 feet
Weights
    Gross: 2250 pounds
    Empty: 1280 pounds

Dimensions
    Length: 22 feet, 6 inches
    Height: 9 feet, 4 inches
    Span: 35 feet
Other
    Fuel capacity: 44 gallons
    Rate of climb: 785 fpm
Seats: Four

Make: Aero Commander
Model: 100 Lark
Year: 1968–1971
Engine
    Make: Lycoming
    Model: O-360-A2F
    Horsepower: 180
    TBO: 2000 hours
Speeds
    Maximum: 138 mph
    Cruise: 132 mph
    Stall: 60 mph
Transitions
    Takeoff over 50-foot obstacle: 1575 feet
    Ground run: 875 feet
    Landing over 50-foot obstacle: 1280 feet
    Ground roll: 675 feet
Weights
    Gross: 2450 pounds
    Empty: 1450 pounds
Dimensions
    Length: 24 feet, 9 inches

Fig. 8-2. *Aero Commander 100 Lark.*

Height: 10 feet, 1 inch
Span: 35 feet
Other
Fuel capacity: 44 gallons
Rate of climb: 750 fpm
Seats: Four

AERONCA

In the postwar period from 1948 to 1950, Aeronca produced a four-place airplane called the 15AC Aeronca Sedan. Like other Aeronca airplanes of the period, the Sedan was a tube-and-fabric design. It did, however, have all-metal wings.

Although only 550 Sedans were built, nearly 250 remain in service today. Many are registered in Alaska where they see service for bush and float flying, which is a real testimonial to their worth and strength. A small number of Sedans were built specifically for float plane use and are officially called S15ACs; floats, however, can be installed on both versions.

In 1950, an Aeronca Sedan once set an in-the-air endurance record of 42 days. This feat required inflight refueling for both airplane and pilots (Fig. 8-3).

Make: Aeronca
Model: 15AC Sedan
Engine
Make: Continental
Model: C-145
Horsepower: 85
TBO: 1800 hours

Fig. 8-3. *Aeronca 15AC Sedan.*

Speeds
    Maximum: 120 mph
    Cruise: 105 mph
    Stall: 53 mph
Transitions
    Takeoff over 50-foot obstacle: 1509 feet
    Ground run: 900 feet
    Landing over 50-foot obstacle: 1826 feet
    Ground roll: 1300 feet
Weights
    Gross: 2050 pounds
    Empty: 1180 pounds
Dimensions
    Length: 25 feet, 3 inches
    Height: 7 feet, 4 inches
    Span: 37 feet, 5 inches
Other
    Fuel capacity: 36 gallons
    Rate of climb: 570 fpm
Seats: Four

BEECHCRAFT/RAYTHEON

All Beech four-place easy-flier airplanes have a low-wing design and all-metal construction. All have large, roomy cabins and give the appearance of more plane than they really are. Although these airplanes are stoutly constructed and usually well equipped with avionics, resale values remain low because of a general lack of popularity.

As with most manufacturers, Beech used an assortment of engines during various production runs:

- 160-hp Lycoming (1963)
- 165-hp fuel-injected Continental (1964)
- 180-hp Lycoming (1968)

To further confuse identification, Model 23s were built under different model names:

- Model 23 Musketeer (1963)
- Model A23 II Musketeer (1964–1965)
- Model A23 IIIA Custom (1966–1967)
- Model B23 Custom (1968–1969)
- Model C23 Custom (1970–1971)
- Model C23 Sundowner (1972–1983)

In 1964, Beech introduced the A23-19 as a four-place airplane. It really should be considered a two- or four-place airplane, meaning that if you carry four adult passengers, you can carry only partial fuel (because of the gross weight limitations). An optional aerobatic version was also available.

The 23-24 was introduced in 1966 as the Super III, was equipped with a 200-hp Lycoming fuel-injected engine, and easily carried four adults and full fuel.

Performance-wise, these planes cruise slow, are underpowered for their size, and have low safety ratings from NTSB (see App. A). They can prove expensive to maintain and appear poorly supported by Beechcraft (Figs. 8-4 through 8-6).

Make: Beechcraft
Model: 23 Musketeer
Year: 1963
Engine
    Make: Lycoming
    Model: O-320-D2B
    Horsepower: 160
    TBO: 2000 hours
Speeds
    Maximum: 144 mph
    Cruise: 135 mph
    Stall: 62 mph
Transitions
    Takeoff over 50-foot obstacle: 1320 feet
    Ground run: 925 feet

Courtesy of Beechcraft

Fig. 8-4. *Beechcraft C23 Musketeer.*

Landing over 50-foot obstacle: 1260 feet
Ground roll: 640 feet
Weights
Gross: 2300 pounds
Empty: 1300 pounds
Dimensions
Length: 25 feet
Height: 8 feet, 3 inches
Span: 32 feet, 9 inches
Other
Fuel capacity: 60 gallons
Rate of climb: 710 fpm
Seats: Four

Make: Beechcraft
Model: A23 II and IIIA Musketeer
Year: 1964–1968
Engine
Make: Continental
Model: IO-346-A
Horsepower: 165
TBO: 1500 hours
Speeds
Maximum: 146 mph
Cruise: 138 mph
Stall: 58 mph
Transitions
Takeoff over 50-foot obstacle: 1460 feet
Ground run: 990 feet
Landing over 50-foot obstacle: 1260 feet
Ground roll: 640 feet
Weights
Gross: 2350 pounds
Empty: 1375 pounds
Dimensions
Length: 25 feet
Height: 8 feet, 3 inches
Span: 32 feet, 9 inches
Other
Fuel capacity: 60 gallons
Rate of climb: 728 fpm
Seats: Four

Make: Beechcraft
Model: B23 and C23 Custom
Year: 1968–1971

Engine
    Make: Lycoming
    Model: O-360-A4K
    Horsepower: 180
    TBO: 2000 hours
Speeds
    Maximum: 151 mph
    Cruise: 143 mph
    Stall: 60 mph
Transitions
    Takeoff over 50-foot obstacle: 1380 feet
    Ground run: 950 feet
    Landing over 50-foot obstacle: 1275 feet
    Ground roll: 640 feet
Weights
    Gross: 2450 pounds
    Empty: 1416 pounds
Dimensions
    Length: 25 feet
    Height: 8 feet, 3 inches
    Span: 32 feet, 9 inches
Other
    Fuel capacity: 60 gallons
    Rate of climb: 820 fpm
Seats: Four

Make: Beechcraft
Model: C23 Sundowner
Year: 1972–1983
Engine
    Make: Lycoming
    Model: O-360-A4G
    Horsepower: 180
    TBO: 2000 hours
Speeds
    Maximum: 147 mph
    Cruise: 133 mph
    Stall: 59 mph
Transitions
    Takeoff over 50-foot obstacle: 1955 feet
    Ground run: 1130 feet
    Landing over 50-foot obstacle: 1484 feet
    Ground roll: 700 feet
Weights
    Gross: 2450 pounds

Fig. 8-5. *Beechcraft C23 Sundowner.*

Empty: 1494 pounds
Dimensions
    Length: 25 feet, 9 inches
    Height: 8 feet, 3 inches
    Span: 32 feet, 9 inches
Other
    Fuel capacity: 57 gallons
    Rate of climb: 792 fpm
Seats: Four

Make: Beechcraft
Model: A23-24 Super III
Year: 1966–1969
Engine
    Make: Lycoming
    Model: O-360-A2B
    Horsepower: 200
    TBO: 1800 hours
Speeds
    Maximum: 158 mph
    Cruise: 150 mph
    Stall: 61 mph
Transitions
    Takeoff over 50-foot obstacle: 1380 feet
    Ground run: 950 feet
    Landing over 50-foot obstacle: 1300 feet
    Ground roll: 660 feet
Weights
    Gross: 2550 pounds
    Empty: 1410 pounds

Dimensions
    Length: 25 feet
    Height: 8 feet, 3 inches
    Span: 32 feet, 9 inches
Other
    Fuel capacity: 60 gallons
    Rate of climb: 880 fpm
Seats: Four

Make: Beechcraft
Model: 23-19/19A Sport and Sport III
Year: 1966–1967
Engine
    Make: Lycoming
    Model: O-320-E2C
    Horsepower: 150
    TBO: 2000 hours
Speeds
    Maximum: 140 mph
    Cruise: 131 mph
    Stall: 56 mph
Transitions
    Takeoff over 50-foot obstacle: 1320 feet
    Ground run: 885 feet
    Landing over 50-foot obstacle: 1220 feet
    Ground roll: 590 feet
Weights
    Gross: 2250 pounds

Fig. 8-6. *Beechcraft 23-19 Sport.*

Empty: 1374 pounds
Dimensions
    Length: 25 feet, 1 inch
    Height: 8 feet, 2 inches
    Span: 32 feet, 7 inches
Other
    Fuel capacity: 60 gallons
    Rate of climb: 700 fpm
Seats: Four (not with full fuel)

Make: Beechcraft
Model: B-19 Sport
Year: 1968–1978
Engine
    Make: Lycoming
    Model: O-320-E2C
    Horsepower: 150
    TBO: 2000 hours
Speeds
    Maximum: 127 mph
    Cruise: 123 mph
    Stall: 57 mph
Transitions
    Takeoff over 50-foot obstacle: 1635 feet
    Ground run: 1030 feet
    Landing over 50-foot obstacle: 1690 feet
    Ground roll: 825 feet
Weights
    Gross: 2150 pounds
    Empty: 1414 pounds
Dimensions
    Length: 25 feet, 9 inches
    Height: 8 feet, 3 inches
    Span: 32 feet, 9 inches
Other
    Fuel capacity: 57 gallons
    Rate of climb: 680 fpm
Seats: Four

CESSNA

The modern Cessna line of four-place airplanes started in 1948 with the Model 170's introduction. The 170 had a metal fuselage, fabric-covered wings, and conventional landing gear. The later 170-B models are all metal.

A good 170 will take you just about anywhere and do so economically. Although considered a classic, based on the dates of production and age, the 170 is as modern as today.

In 1956, the model 172 was brought to the market. It would later prove to be the most popular four-seat airplane ever manufactured. No doubt you can see the resemblance between the Cessna 170 and 172. The advent of the 172's nosewheel rang the death knell for the 170, and in 1956 production of the model 170 ceased. Production of the 172s, however, continued into the 1980s when all of general aviation faltered.

The 172 has seen many refinements since entering production in 1956:

- Swept-back tail (1960)
- Omni-vision (1963)
- Electric flaps (1964)
- Lycoming O-320-E2D engine (1968)
- Conical wing tips (1970)
- Tubular landing gear (1971)
- The 160-hp O-320-H2AD low-lead engine (1977)
- Lycoming O-320-D2J engine (1981)

The 172 is also known as the Skyhawk or Skyhawk 100, depending on factory-installed equipment. In 1977, the 172 Hawk XP was introduced with a 195-hp IO-360-KB fuel-injected Continental engine. The last 172 built before the financial bubble burst was a 1986 model. In 1996, Cessna announced that production of the 172 would restart in 1997.

On schedule, Cessna reintroduced the 172 Skyhawk as the model 172R. The 1997 sticker price was nearly $150,000. Although many "experts" said the price was too high and they would not sell—sell they did. By early 1998 the model number had been slipped to 172S and by 1999 more than 1000 new 172 Skyhawks had been sold. The 2004 list price for a 172SP was $195,000.

Also in 1998, Cessna rolled out an improved version of the Skyhawk. Called the Skyhawk SP, the biggest single improvement was a 180-hp engine.

Cessna brought out the 175 Skylark in 1958 with a GO-300 Continental engine. It never achieved the popularity of the 172 because of chronic engine problems, however. The 175's GO-300 engine is geared and develops higher horsepower by operating at a higher rpm, which leads to early wear problems. Some 175s have had their GO-300 engine replaced with an O-320 or O-360, and these planes can be good buys because they are stigmatized by the 175 model number.

The Cessna model 177 Cardinals appeared in 1968. The first 177s were underpowered with only a 150-hp Lycoming engine; 1969 and later models have 180-hp engines. A new airfoil in 1970 attempted to improve low-speed handling, and some models have constant-speed propellers. Cardinals remained in production until 1978.

Cessna airplanes in the easy-flier category are proven safe airplanes that will not drag you to the bank for maintenance and parts. If the plane is properly cared for, your initial investment will easily be returned at resale. In fact, you can often make a good profit (Figs. 8-7 through 8-14).

Make: Cessna
Model: 170 (all)
Year: 1948–1956
Engine
    Make: Continental
    Model: C-145-2
    Horsepower: 145
    TBO: 1800 hours
Speeds
    Maximum: 135 mph
    Cruise: 120 mph
    Stall: 52 mph
Transitions
    Takeoff over 50-foot obstacle: 1820 feet
    Ground run: 700 feet
    Landing over 50-foot obstacle: 1145 feet
    Ground roll: 500 feet
Weights
    Gross: 2200 pounds
    Empty: 1260 pounds
Dimensions
    Length: 25 feet
    Height: 6 feet, 5 inches
    Span: 36 feet
Other
    Fuel capacity: 42 gallons
    Rate of climb: 660 fpm
Seats: Four

Courtesy of Cessna

Fig. 8-7. *1956 Cessna 170B.*

Fig. 8-8. *1956 Cessna 172.*

Make: Cessna
Model: 172 Skyhawk
Year: 1956–1967
Engine
    Make: Continental
    Model: O-300 A (D after 1960)
    Horsepower: 145
    TBO: 1800 hours
Speeds
    Maximum: 138 mph
    Cruise: 130 mph
    Stall: 49 mph
Transitions
    Takeoff over 50-foot obstacle: 1525 feet
    Ground run: 865 feet
    Landing over 50-foot obstacle: 1250 feet
    Ground roll: 520 feet
Weights
    Gross: 2300 pounds
    Empty: 1260 pounds
Dimensions
    Length: 26 feet, 6 inches
    Height: 8 feet, 11 inches
    Span: 36 feet, 2 inches
Other
    Fuel capacity: 42 gallons
    Rate of climb: 645 fpm
Seats: Four

Fig. 8-9. *1968 Cessna Skyhawk.*

Make: Cessna
Model: 172 Skyhawk
Year: 1968–1976
Engine
    Make: Lycoming
    Model: O-320-E2D
    Horsepower: 150
    TBO: 2000 hours
Speeds
    Maximum: 139 mph
    Cruise: 131 mph
    Stall: 49 mph
Transitions
    Takeoff over 50-foot obstacle: 1525 feet
    Ground run: 865 feet
    Landing over 50-foot obstacle: 1250 feet
    Ground roll: 520 feet
Weights
    Gross: 2300 pounds
    Empty: 1265 pounds
Dimensions
    Length: 26 feet, 6 inches
    Height: 8 feet, 11 inches
    Span: 36 feet, 2 inches
Other
    Fuel capacity: 42 gallons
    Rate of climb: 645 fpm
Seats: Four

Fig. 8-10. *1978 Cessna Hawk XP II.*

Make: Cessna
Model: 172 Skyhawk
Year: 1977–1980
Engine
    Make: Lycoming
    Model: O-320-H2AD
    Horsepower: 160
    TBO: 2000 hours
Speeds
    Maximum: 141 mph
    Cruise: 138 mph
    Stall: 51 mph
Transitions
    Takeoff over 50-foot obstacle: 1825 feet
    Ground run: 890 feet
    Landing over 50-foot obstacle: 1280 feet
    Ground roll: 540 feet
Weights
    Gross: 2400 pounds
    Empty: 1414 pounds
Dimensions
    Length: 26 feet, 6 inches
    Height: 8 feet, 11 inches
    Span: 36 feet, 2 inches
Other
    Fuel capacity: 43 gallons
    Rate of climb: 700 fpm
Seats: Four

Fig. 8-11. *The utility of the 172 is shown with this Cessna T-41, a military trainer version of the model.*

Make: Cessna
Model: 172P II
Year: 1981–1986
Engine
 Make: Lycoming
 Model: O-320-D2J
 Horsepower: 160
 TBO: 2000 hours
Speeds
 Maximum: 141 mph
 Cruise: 138 mph
 Stall: 53 mph
Transitions
 Takeoff over 50-foot obstacle: 1625 feet
 Ground run: 890 feet
 Landing over 50-foot obstacle: 1280 feet
 Ground roll: 540 feet
Weights
 Gross: 2400 pounds
 Empty: 1454 pounds
Dimensions
 Length: 26 feet, 11 inches
 Height: 8 feet, 10 inches
 Span: 35 feet, 10 inches
Other
 Fuel capacity: 43 gallons
 Rate of climb: 700 fpm
Seats: Four

Make: Cessna
Model: 172 Cutlass Q
Year: 1983–1984
Engine
    Make: Continental
    Model: IO-360-A4N
    Horsepower: 180
    TBO: 1500 hours
Speeds
    Maximum: 141 mph
    Cruise: 140 mph
    Stall: 55 mph
Transitions
    Takeoff over 50-foot obstacle: 1690 feet
    Ground run: 960 feet
    Landing over 50-foot obstacle: 1335 feet
    Ground roll: 575 feet
Weights
    Gross: 2550 pounds
    Empty: 1480 pounds
Dimensions
    Length: 26 feet, 11 inches
    Height: 8 feet, 10 inches
    Span: 36 feet, 1 inch
Other
    Fuel capacity: 54 gallons
    Rate of climb: 680 fpm
Seats: Four

Make: Cessna
Model: 172 Skyhawk XP
Year: 1977–1981
Engine
    Make: Continental
    Model: IO-360-K (KB after 1977)
    Horsepower: 195
    TBO: 1500 hours (2000 on KB)
Speeds
    Maximum: 153 mph
    Cruise: 150 mph
    Stall: 54 mph
Transitions
    Takeoff over 50-foot obstacle: 1360 feet
    Ground run: 800 feet
    Landing over 50-foot obstacle: 1345 feet
    Ground roll: 635 feet

Weights
    Gross: 2550 pounds
    Empty: 1546 pounds
Dimensions
    Length: 26 feet, 6 inches
    Height: 8 feet, 11 inches
    Span: 36 feet, 2 inches
Other
    Fuel capacity: 52 gallons
    Rate of climb: 870 fpm
Seats: Four

Make: Cessna
Model: 172 Skyhawk
Year: 1997–
Engine
    Make: Lycoming
    Model: IO-360-L2A
    Horsepower: 160
    TBO: 2000 hours
Speeds
    Maximum: 141 mph
    Cruise: 140 mph
    Stall: 54 mph

Courtesy of Cessna

Fig. 8-12. *1999 Cessna 172 Skyhawk.*

Transitions
    Takeoff over 50-foot obstacle: 1685 feet
    Ground run: 945 feet
    Landing over 50-foot obstacle: 1295 feet
    Ground roll: 550 feet
Weights
    Gross: 2457 pounds
    Empty: 1640 pounds
Dimensions
    Length: 27 feet, 2 inches
    Height: 8 feet, 11 inches
    Span: 36 feet, 1 inch
Other
    Fuel capacity: 56 gallons
    Rate of climb: 720 fpm
Seats: Four

Make: Cessna
Model: 172 Skyhawk SP
Year: 1998–
Engine
    Make: Lycoming
    Model: IO-360-L2A
    Horsepower: 180
    TBO: 2000 hours
Speeds
    Maximum: 145 mph
    Cruise: 143 mph
    Stall: 55 mph
Transitions
    Takeoff over 50-foot obstacle: 1630 feet
    Ground run: 960 feet
    Landing over 50-foot obstacle: 1335 feet
    Ground roll: 575 feet
Weights
    Gross: 2558 pounds
    Empty: 1665 pounds
Dimensions
    Length: 27 feet, 2 inches
    Height: 8 feet, 11 inches
    Span: 36 feet, 1 inch
Other
    Fuel capacity: 56 gallons
    Rate of climb: 730 fpm
Seats: Four

Fig. 8-13. *1959 Cessna Skylark 175.*

Make: Cessna
Model: 175 Skylark and Powermatic
Year: 1958–1962
Engine
    Make: Continental
    Model: GO-300-E
    Horsepower: 175
    TBO: 1200 hours
Speeds
    Maximum: 139 mph
    Cruise: 131 mph
    Stall: 50 mph
Transitions
    Takeoff over 50-foot obstacle: 1340 feet
    Ground run: 735 feet
    Landing over 50-foot obstacle: 1155 feet
    Ground roll: 590 feet
Weights
    Gross: 2300 pounds
    Empty: 1330 pounds
Dimensions
    Length: 25 feet
    Height: 8 feet, 5 inches
    Span: 36 feet
Other
    Fuel capacity: 52 gallons
    Rate of climb: 850 fpm
Seats: Four

Make: Cessna
Model: 177 Cardinal (150)
Year: 1968
Engine
    Make: Lycoming
    Model: O-320-E2D
    Horsepower: 150
    TBO: 2000 hours
Speeds
    Maximum: 144 mph
    Cruise: 134 mph
    Stall: 53 mph
Transitions
    Takeoff over 50-foot obstacle: 1575 feet
    Ground run: 845 feet
    Landing over 50-foot obstacle: 1135 feet
    Ground roll: 400 feet
Weights
    Gross: 2350 pounds
    Empty: 1415 pounds
Dimensions
    Length: 27 feet, 3 inches
    Height: 8 feet, 7 inches
    Span: 35 feet, 7 inches
Other
    Fuel capacity: 49 gallons
    Rate of climb: 670 fpm
Seats: Four

Make: Cessna
Model: 177 Cardinal (180)
Year: 1969–1978
Engine
    Make: Lycoming
    Model: O-360-A1F6 (A1F6D after 1975)
    Horsepower: 180
    TBO: 2000 hours (1800 on A1F6D w/o mod)
Speeds
    Maximum: 150 mph
    Cruise: 139 mph
    Stall: 53 mph
Transitions
    Takeoff over 50-foot obstacle: 1400 feet
    Ground run: 750 feet

Fig. 8-14. *Cessna Cardinal 177.*

Landing over 50-foot obstacle: 1220 feet
Ground roll: 600 feet
Weights
Gross: 2500 pounds
Empty: 1430 pounds
Dimensions
Length: 27 feet, 3 inches
Height: 8 feet, 7 inches
Span: 35 feet, 7 inches
Other
Fuel capacity: 50 gallons
Rate of climb: 840 fpm
Seats: Four

CIRRUS

Cirrus airplanes were first introduced in 1984 as a kit airplane manufactured in Baraboo, Wisconsin. In 1994, Cirrus moved to larger facilities in Duluth, Minnesota and began research and development of the SR20.

The SR20, made of composite construction, was awarded an FAA Type Certification in 1998. It incorporates a flat-panel, multifunction display; state-of-the-art safety innovations; and the Cirrus Airframe Parachute System (CAPS).

By 2003 Cirrus was producing airplanes at the rate of 467 for the year. In 2004, Cirrus laid claim to being the world's second largest manufacturer of single-engine, piston-powered aircraft.

The SR22 model is included in this chapter, although powered with a 310-hp engine and a slightly larger airframe, due to the relative ease of flight. For example, the single lever for engine and propeller control dubbed "Simple"—for single movement power lever—is found in the SR22 as well as the lower-powered models. Further, this model remains a fixed landing gear airplane. The performance is, however, greatly enhanced with the larger engine.

Many Cirrus airplanes have digital instrument displays, commonly called "all glass."

Prices in 2004 started at $189,900 for the SRV-G2, $236,700 for the SR20-G2, and $334,700 for the SR22-G2 (Figs. 8-15 through 8-17).

Make: Cirrus
Model: SRV-G2
Year: 1999–
Engine
    Make: Continental
    Model: IO-360-ES
    Horsepower: 200
    TBO: 2000 hours
Speeds
    Maximum: NA
    Cruise: 173 mph
    Stall: 62 mph
Transitions
    Takeoff over 50 foot obstacle: 1958 feet
    Ground run: 1341 feet
    Landing over 50 foot obstacle: 2040 feet
    Ground roll: 1014 feet
Weights
    Gross: 3000 pounds
    Empty: 2050 pounds
Dimensions
    Length: 26 feet
    Height: 8 feet, 6 inches
    Span: 35 feet, 7 inches
Other
    Fuel capacity: 56 gallons
    Rate of climb: 900 fpm
Seats: Four

Make: Cirrus
Model: SR20-G2
Year: 1999–
Engine
    Make: Continental
    Model: IO-360-ES
    Horsepower: 200
    TBO: 2000 hours
Speeds
    Maximum: NA
    Cruise: 180 mph
    Stall: 62 mph

Fig. 8-15. *Cirrus SR20.*

Transitions
    Takeoff over 50 foot obstacle: 1958 feet
    Ground run: 1341 feet
    Landing over 50 foot obstacle: 2040 feet
    Ground roll: 1014 feet
Weights
    Gross: 3000 pounds
    Empty: 2070 pounds
Dimensions
    Length: 26 feet
    Height: 8 feet, 6 inches
    Span: 35 feet, 7 inches
Other
    Fuel capacity: 56 gallons
    Rate of climb: 900 fpm
Seats: Four

Make: Cirrus
Model: SR22-G2
Year: 2001–
Engine
    Make: Continental
    Model: IO-550-N

Fig. 8-16. *Cirrus SR22.*

    Horsepower: 310
    TBO: 2000 hours
Speeds
    Maximum: NA
    Cruise: 213 mph
    Stall: 68 mph
Transitions
    Takeoff over 50 foot obstacle: 1575 feet
    Ground run: 1020 feet
    Landing over 50 foot obstacle: 2325 feet
    Ground roll: 1140 feet
Weights
    Gross: 3400 pounds
    Empty: 2250 pounds
Dimensions
    Length: 26 feet
    Height: 8 feet, 7 inches
    Span: 38 feet, 6 inches
Other
    Fuel capacity: 81 gallons
    Rate of climb: 1400 fpm
Seats: Four

Courtesy of Cirrus

Fig. 8-17. *Glass panel installation.*

DIAMOND

Diamond Aircraft Industries Inc. is the largest general aviation manufacturer of single-engine aircraft in Canada and the third largest in North America.

In early 2001 Diamond introduced the all-composite DA40 Diamond Star, powered by the Lycoming 180-hp IO-360 engine. Diamond claimed all-around performance, a 147-kt cruise speed, and a 9-gph fuel burn at a price of $189,900, equipped for IFR flight. The Diamond aircraft use a flip-up canopy system for entry into the aircraft. A separate flip-up side door is used for the rear seat. The cargo compartment is generous size at nearly 25 cu. ft.

In 2004 the DA40 with an all-glass instrument panel sold for $229,500, and an airplane equipped with a standard panel sold for $196,600 (Figs. 8-18 and 8-19).

Make: Diamond
Model: DA40-180
Year: 2001–
Engine
    Make: Lycoming
    Model: IO-360-M1A
    Horsepower: 180
    TBO: 2000 hours

Fig. 8-18. *Diamond DA40.*

Fig. 8-19. *Unique canopy design of the Diamond DA40.*

Speeds
    Maximum: NA
    Cruise: 167 mph
    Stall: 56 mph
Transitions
    Takeoff over 50 foot obstacle: 1150 feet
    Ground run: NA

Landing over 50 foot obstacle: NA
Ground roll: NA
Weights
    Gross: 2535 pounds
    Empty: NA
Dimensions
    Length: 26 feet, 4 inches
    Height: 6 feet, 7 inches
    Span: 39 feet, 5 inches
Other
    Fuel capacity: 41 gallons
    Rate of climb: 1070 fpm
Seats: Four

GULFSTREAM

Gulfstream built two airplanes in the four-place easy-flier category under three names. The AA-5 and AA-5A Traveler and Cheetah models are powered by a 150-hp engine, and the AA-5B Tiger has a 180-hp engine.

Gulfstream airplanes are unusual because they have space-age wing construction, with no rivets to hold the skin in place. Adhesive holds the skin to the honeycombed wing ribs. As with the two-place airplanes of similar construction, some rivets may have been installed to halt delamination.

Gulfstreams have a sliding canopy; you enter by stepping over the sidewall of the cabin and onto the seat. Unfortunately, this means a complete cabin interior washdown if it's opened during a rainstorm.

Directional control on the ground is via differential braking and a swiveling nosewheel. The AA-5 airplanes are good performers, although somewhat hot on landings. They have a short propeller clearance and are not recommended for soft field work. The laminated fiberglass landing gear is noted for its strength and ability to make botched landings look good. Gulfstream production of both models stopped in 1979.

In 1991, American Aircraft Corporation restarted Tiger production and continued until 1993. The base price for a new Tiger in 1991 was $94,250. A typically equipped version sold for about $103,000.

Tiger Aircraft LLC is the current producer of these airplanes, and the model being built is the AG-5B, which now sells at a base price of $235,800.

Production numbers indicate that 834 AA-5s, 900 AA-5As, and 1473 AA-5Bs were built (Figs. 8-20 through 8-22).

Make: Gulfstream
Model: AA-5/AA-5A Traveler/Cheetah
Year: 1972–1979

Fig. 8-20. *Gulfstream AA-5 Cheetah.*

Engine
    Make: Lycoming
    Model: O-320-E2G
    Horsepower: 150
    TBO: 2000 hours
Speeds
    Maximum: 150 mph
    Cruise: 140 mph
    Stall: 58 mph
Transitions
    Takeoff over 50-foot obstacle: 1600 feet
    Ground run: 880 feet
    Landing over 50-foot obstacle: 1100 feet
    Ground roll: 380 feet
Weights
    Gross: 2200 pounds
    Empty: 1200 pounds
Dimensions
    Length: 22 feet
    Height: 8 feet
    Span: 32 feet, 6 inches
Other
    Fuel capacity: 38 gallons
    Rate of climb: 660 fpm
Seats: Four

Make: Gulfstream/American Aircraft
Model: AA-5B Tiger
Year: 1975–1993
Engine
    Make: Lycoming
    Model: O-360-A4K

Fig. 8-21. *Gulfstream AA-5 Tiger.*

    Horsepower: 180
    TBO: 2000 hours
Speeds
    Maximum: 170 mph
    Cruise: 160 mph
    Stall: 61 mph
Transitions
    Takeoff over 50-foot obstacle: 1550 feet
    Ground run: 865 feet
    Landing over 50-foot obstacle: 1120 feet
    Ground roll: 410 feet
Weights
    Gross: 2400 pounds
    Empty: 1285 pounds
Dimensions
    Length: 22 feet
    Height: 8 feet
    Span: 31 feet, 6 inches
Other
    Fuel capacity: 51 gallons
    Rate of climb: 850 fpm
Seats: Four

Make: Tiger Aircraft LLC
Model: AG-5B
Year: 2000–
Engine
    Make: Lycoming
    Model: O-360-A4K
    Horsepower: 180
    TBO: 2000 hours

Fig. 8-22. *Tiger AG-5B.*

Speeds
    Maximum: 170 mph
    Cruise: 165 mph
    Stall: 61 mph
Transitions
    Takeoff over 50 foot obstacle: 1550 feet
    Ground run: NA
    Landing over 50 foot obstacle: 1120 feet
    Ground roll: NA
Weights
    Gross: 2400 pounds
    Empty: 1500 pounds
Dimensions
    Length: 22 feet
    Height: 8 feet
    Span: 31 feet 6 inches
Other
    Fuel capacity: 52.6 gallons
    Rate of climb: 850 fpm
Seats: Four

LUSCOMBE

The Luscombe 11A Sedan is an all-metal, high-wing airplane with conventional landing gear. Ahead of their time, the Sedans had a rear window, something Cessna didn't discover until 20 years later. The wide landing gear stance makes the Sedan unusually easy to handle.

Although the number of Sedans built was not high, they frequently appear on the used market. They are a classic, but they are also an orphan. Parts replacement could be a serious problem.

The airplane's modern lines have attracted renewed interest; a prototype of a tricycle gear version of the Sedan was being flown in 1996, and production was planned for 1999. Production did not start, however. The company is still planning, though, to build the new version at a yet-to-be-announced time (Fig. 8-23).

Fig. 8-23. *Luscombe 11A Sedan.*

Make: Luscombe
Model: 11A Sedan
Year: 1948–1950
Engine
    Make: Continental
    Model: C-165
    Horsepower: 165
    TBO: 1800 hours
Speeds
    Maximum: 140 mph
    Cruise: 130 mph
    Stall: 55 mph
Transitions
    Takeoff over 50-foot obstacle: 1540 feet
    Ground run: 800 feet
    Landing over 50-foot obstacle: 1310 feet
    Ground roll: 500 feet

Weights
    Gross: 2280 pounds
    Empty: 1280 pounds
Dimensions
    Length: 23 feet, 6 inches
    Height: 6 feet, 10 inches
    Span: 38 feet
Other
    Fuel capacity: 42 gallons
    Rate of climb: 900 fpm
Seats: Four

MAULE

The Maule M-4 was conceived as a homebuilt called the Bee Dee M4. Rather than market it as a homebuilt, designer B. D. Maule decided to commercially produce the airplane himself.

The M-4s are built of a fiberglass-covered tubular fuselage and have all-metal wings. Maule airplanes exhibit excellent short-field capabilities, yet have relatively good cruise speeds. They are fine examples in the art of matching engine power to wing design.

An interesting innovation from Maule is the addition of tricycle landing gear on their latest model M-7 series 180-hp planes. First introduced in 1989, the standard tri-gear model currently sells for $126,888.

A stripped-down version of the M-7 was introduced in 1993 with a 160-hp engine and only the bare minimum VFR instrumentation. It was advertised for $44,995 at that time. In 1996 it sold for about $90,000 and in 1999 for $108,000.

In 1996, the 2000th Maule airplane was rolled out of the factory. Several additional models of Maule airplanes are available on the market, generally for use as utility airplanes, all having more engine power. They are reviewed in Chapter 10 (Figs. 8-24 and 8-25).

Make: Maule
Model: M-4 Jeteson
Year: 1962–1967
Engine
    Make: Continental
    Model: O-300-A
    Horsepower: 145
    TBO: 1800 hours
Speeds
    Maximum: 180 mph
    Cruise: 150 mph
    Stall: 40 mph

Fig. 8-24. *Maule M-4-145.*

Transitions
    Takeoff over 50-foot obstacle: 900 feet
    Ground run: 700 feet
    Landing over 50-foot obstacle: 600 feet
    Ground roll: 450 feet
Weights
    Gross: 2100 pounds
    Empty: 1100 pounds
Dimensions
    Length: 22 feet
    Height: 6 feet, 2 inches
    Span: 29 feet, 8 inches
Other
    Fuel capacity: 42 gallons
    Rate of climb: 700 fpm
Seats: Four

Make: Maule
Model: MX-7/160
Year: 1993–
Engine
    Make: Lycoming
    Model: O-320-B2D
    Horsepower: 160
    TBO: 2000 hours
Speeds
    Maximum: 140 mph
    Cruise: 135 mph
    Stall: 40 mph
Transitions
    Takeoff over 50-foot obstacle: 1180 feet
    Ground run: NA

Landing over 50-foot obstacle: 500 feet
Ground roll: NA
Weights
Gross: 2200 pounds
Empty: 1330 pounds
Dimensions
Length: 23 feet, 6 inches
Height: 6 feet, 4 inches
Span: 32 feet, 10 inches
Other
Fuel capacity: 40 gallons
Rate of climb: 825 fpm
Seats: Four

Make: Maule
Model: MX-7/180 Conventional gear
Year: 1985–
Engine
Make: Lycoming
Model: O-320-C1F
Horsepower: 180
TBO: 2000 hours
Speeds
Maximum: 180 mph
Cruise: 145 mph
Stall: 40 mph
Transitions
Takeoff over 50-foot obstacle: 600 feet
Ground run: 125 feet
Landing over 50-foot obstacle: 500 feet
Ground roll: 275 feet
Weights
Gross: 2500 pounds
Empty: 1365 pounds

Courtesy of Maule

Fig. 8-25. *Maule M7-180.*

Dimensions
    Length: 23 feet, 6 inches
    Height: 6 feet, 4 inches
    Span: 30 feet, 10 inches
Other
    Fuel capacity: 70 gallons
    Rate of climb: 1200 fpm
Seats: Four

Make: Maule
Model: MX-7/180 and MXT-7/180 Tri-gear
Year: 1985–
Engine
    Make: Lycoming
    Model: O-360-C1F
    Horsepower: 180
    TBO: 2000 hours
Speeds
    Maximum: NA
    Cruise: 140 mph
    Stall: 40 mph
Transitions
    Takeoff over 50-foot obstacle: 600 feet
    Ground run: NA
    Landing over 50-foot obstacle: 500 feet
    Ground roll: NA
Weights
    Gross: 2500 pounds
    Empty: 1410 pounds
Dimensions
    Length: 23 feet, 6 inches
    Height: 8 feet, 4 inches
    Span: 33 feet
Other
    Fuel capacity: 70 gallons
    Rate of climb: 1200 fpm
Seats: Four

PIPER

Piper started production of four-place airplanes with the Family Cruiser PA-14. It was small and underpowered. Only 236 Cruisers were manufactured before being replaced by the PA-16 Clipper, a slightly larger plane.

In 1950, the PA-20 Pacer series appeared. Like the PA-14s and 16s, the original PA-20s were of tube-and-fabric construction and had conventional landing gear. They had small engines and did not display eye-dazzling performance numbers.

Evolution of the Pacer resulted in the Tri-Pacer, billed as an "anyone can fly it" airplane because of tricycle landing gear. It became a success and soon the Tri-Pacer's sales far outstripped those of the Pacer. General aviation customers were demanding easier-to-handle airplanes and they found what they wanted in the Tri-Pacer. PA-22s, as Tri-Pacers are officially known, were built with several different engines. Power varied from 125 to 160 hp.

Piper introduced the PA-28 series in 1961. Unlike previous Piper products, this new airplane was of all-metal construction and had low wings. The PA-28 series is the backbone for the entire line of Piper's all-metal, single-engine products. All of the early PA-28 series have the "Hershey bar" wing, making them very docile to handle.

The PA-28 was initially produced as the model 140, powered with a 150-hp Lycoming engine. An optional version, a PA-28 160 with a 160-hp engine, was also available.

In 1964, the Cherokee 140 was introduced as a two-/four-place trainer. Initially delivered with a 140-hp engine, a 150-hp Lycoming was available for increased load carrying. The model 140 in slightly different versions is called the Cruiser or Fliteliner and was produced from 1964 to 1977.

The PA-28-180 was built from 1963 until 1975 and is a true four-place airplane, able to carry four adults and full fuel. Later called the Challenger or Archer, it had a 180-hp engine.

In 1974, Piper changed the wing design of the entire PA-28 series. The Warrior wing, as the newly designed wing was called, provides increased load-carrying abilities and very gentle stall characteristics.

The wing was added to the PA-28 Challenger/Archer fuselage and designated the model PA-28-151 Warrior. It had a 150-hp engine and was built from 1974 through 1977. It was replaced by the model 161 Warrior, with a 160-hp low-lead engine, in late 1977. The PA-28-181 was introduced in 1976 as the Archer II with a 180-hp Lycoming engine.

In the early 1990s, Piper fell victim to hard financial times and reorganized into The New Piper Aircraft, Inc. The company is currently producing the PA-28 series airplanes. In 1997 the suggested price for a new PA-28-161 was $134,900 and in 2004 was $180,600. These airplanes may be listed under Piper or New Piper.

Piper airplanes have always been reliable and affordable. They make good purchases that are easy to fly and economical to maintain (Figs. 8-26 through 8-32).

Make: Piper
Model: PA-14 Cruiser
Year: 1948–1949

Engine
  Make: Lycoming
  Model: O-235-C1
  Horsepower: 115
  TBO: 2000 hours
Speeds
  Maximum: 123 mph
  Cruise: 110 mph
  Stall: 46 mph
Transitions
  Takeoff over 50-foot obstacle: 1770 feet
  Ground run: 720 feet
  Landing over 50-foot obstacle: 1410 feet
  Ground roll: 470 feet
Weights
  Gross: 1850 pounds
  Empty: 1020 pounds
Dimensions
  Length: 23 feet, 2 inches
  Height: 6 feet, 7 inches
  Span: 35 feet, 6 inches
Other
  Fuel capacity: 38 gallons
  Rate of climb: 540 fpm
Seats: Four

Make: Piper
Model: PA-16 Clipper
Year: 1949

Courtesy of Piper

Fig. 8-26. *Piper PA-16.*

Engine
    Make: Lycoming
    Model: O-235-C1
    Horsepower: 115
    TBO: 2000 hours
Speeds
    Maximum: 125 mph
    Cruise: 112 mph
    Stall: 50 mph
Transitions
    Takeoff over 50-foot obstacle: 1910 feet
    Ground run: 720 feet
    Landing over 50-foot obstacle: 1440 feet
    Ground roll: 600 feet
Weights
    Gross: 1650 pounds
    Empty: 850 pounds
Dimensions
    Length: 20 feet, 1 inch
    Height: 6 feet, 2 inches
    Span: 29 feet, 3 inches
Other
    Fuel capacity: 36 gallons
    Rate of climb: 580 fpm
Seats: Four

Make: Piper
Model: PA-20 Pacer
Year: 1951–1952
Engine
    Make: Lycoming
    Model: O-290-D

Fig. 8-27. *Piper PA-20 Pacer.*

Horsepower: 125
TBO: 2000 hours
Speeds
  Maximum: 134 mph
  Cruise: 119 mph
  Stall: 47 mph
Transitions
  Takeoff over 50-foot obstacle: 1725 feet
  Ground run: 1210 feet
  Landing over 50-foot obstacle: 1280 feet
  Ground roll: 780 feet
Weights
  Gross: 1800 pounds
  Empty: 970 pounds
Dimensions
  Length: 20 feet, 4 inches
  Height: 6 feet, 1 inch
  Span: 29 feet, 3 inches
Other
  Fuel capacity: 36 gallons
  Rate of climb: 550 fpm
Seats: Four

Make: Piper
Model: PA-20 Pacer
Year: 1952–1954
Engine
  Make: Lycoming
  Model: O-290-D2 (optional C/S propeller)
  Horsepower: 135
  TBO: 1500 hours
Speeds
  Maximum: 139 mph
  Cruise: 125 mph
  Stall: 48 mph
Transitions
  Takeoff over 50-foot obstacle: 1600 feet
  Ground run: 1120 feet
  Landing over 50-foot obstacle: 1280 feet
  Ground roll: 780 feet
Weights
  Gross: 1920 pounds
  Empty: 1020 pounds
Dimensions
  Length: 20 feet, 4 inches

Height: 6 feet, 1 inch
Span: 29 feet, 3 inches
Other
Fuel capacity: 36 gallons
Rate of climb: 620 fpm
Seats: Four

Fig. 8-28. *Piper PA-22 Tri-Pacer.*

Make: Piper
Model: PA-22 Tri-Pacer
Year: 1951–1952
Engine
Make: Lycoming
Model: O-290-D
Horsepower: 125
TBO: 2000 hours
Speeds
Maximum: 134 mph
Cruise: 130 mph
Stall: 48 mph
Transitions
Takeoff over 50-foot obstacle: 1600 feet
Ground run: 1120 feet
Landing over 50-foot obstacle: 1280 feet
Ground roll: 650 feet
Weights
Gross: 1850 pounds
Empty: 1060 pounds

Dimensions
    Length: 20 feet, 4 inches
    Height: 8 feet, 3 inches
    Span: 29 feet, 3 inches
Other
    Fuel capacity: 36 gallons
    Rate of climb: 550 fpm
Seats: Four

Make: Piper
Model: PA-22 Tri-Pacer
Year: 1952–1954
Engine
    Make: Lycoming
    Model: O-290-D2
    Horsepower: 135
    TBO: 1500 hours
Speeds
    Maximum: 137 mph
    Cruise: 132 mph
    Stall: 48 mph
Transitions
    Takeoff over 50-foot obstacle: 1550 feet
    Ground run: 1080 feet
    Landing over 50-foot obstacle: 1280 feet
    Ground roll: 650 feet
Weights
    Gross: 1850 pounds
    Empty: 1060 pounds
Dimensions
    Length: 20 feet, 4 inches
    Height: 8 feet, 3 inches
    Span: 29 feet, 3 inches
Other
    Fuel capacity: 36 gallons
    Rate of climb: 620 fpm
Seats: Four

Make: Piper
Model: PA-22 Tri-Pacer
Year: 1955–1960
Engine
    Make: Lycoming
    Model: O-320-A1A
    Horsepower: 150
    TBO: 2000 hours

Speeds
  Maximum: 139 mph
  Cruise: 132 mph
  Stall: 49 mph
Transitions
  Takeoff over 50-foot obstacle: 1500 feet
  Ground run: 1050 feet
  Landing over 50-foot obstacle: 1280 feet
  Ground roll: 650 feet
Weights
  Gross: 2000 pounds
  Empty: 1100 pounds
Dimensions
  Length: 20 feet, 4 inches
  Height: 8 feet, 3 inches
  Span: 29 feet, 3 inches
Other
  Fuel capacity: 36 gallons
  Rate of climb: 725 fpm
Seats: Four

Make: Piper
Model: PA-22 Tri-Pacer
Year: 1958–1960
Engine
  Make: Lycoming
  Model: O-320-B2A
  Horsepower: 160
  TBO: 2000 hours
Speeds
  Maximum: 141 mph
  Cruise: 133 mph
  Stall: 48 mph
Transitions
  Takeoff over 50-foot obstacle: 1480 feet
  Ground run: 1035 feet
  Landing over 50-foot obstacle: 1280 feet
  Ground roll: 650 feet
Weights
  Gross: 2000 pounds
  Empty: 1110 pounds
Dimensions
  Length: 20 feet, 5 inches
  Height: 8 feet, 3 inches
  Span: 29 feet, 3 inches

Other
    Fuel capacity: 36 gallons
    Rate of climb: 800 fpm
Seats: Four

Make: Piper
Model: PA-28-140
Year: 1964–1977
Engine
    Make: Lycoming
    Model: O-320-E2A
    Horsepower: 150
    TBO: 2000 hours
Speeds
    Maximum: 139 mph
    Cruise: 130 mph
    Stall: 53 mph
Transitions
    Takeoff over 50-foot obstacle: 1750 feet
    Ground run: 800 feet
    Landing over 50-foot obstacle: 1890 feet
    Ground roll: 535 feet
Weights
    Gross: 2150 pounds
    Empty: 1205 pounds
Dimensions
    Length: 23 feet, 3 inches
    Height: 7 feet, 3 inches
    Span: 30 feet
Other
    Fuel capacity: 36 gallons
    Rate of climb: 660 fpm
Seats: Four

Courtesy of Piper Aviation Museum Foundation

Fig. 8-29. *Piper PA-28-140 Cruiser.*

Make: Piper
Model: PA-28 Cherokee
Year: 1962–1967
Engine
    Make: Lycoming
    Model: O-320-B2B
    Horsepower: 160
    TBO: 2000 hours
Speeds
    Maximum: 141 mph
    Cruise: 132 mph
    Stall: 55 mph
Transitions
    Takeoff over 50-foot obstacle: 1700 feet
    Ground run: 775 feet
    Landing over 50-foot obstacle: 1890 feet
    Ground roll: 550 feet
Weights
    Gross: 2200 pounds
    Empty: 1210 pounds
Dimensions
    Length: 23 feet, 3 inches
    Height: 7 feet, 3 inches
    Span: 30 feet
Other
    Fuel capacity: 36 gallons
    Rate of climb: 700 fpm
Seats: Four

Make: Piper
Model: PA-28-180 Cherokee/Challenger/Archer
Year: 1963–1975

Fig. 8-30. *Piper PA-28-180 Challenger.*

Engine
    Make: Lycoming
    Model: O-360-A3A
    Horsepower: 180
    TBO: 2000 hours
Speeds
    Maximum: 150 mph
    Cruise: 141 mph
    Stall: 57 mph
Transitions
    Takeoff over 50-foot obstacle: 1620 feet
    Ground run: 725 feet
    Landing over 50-foot obstacle: 1150 feet
    Ground roll: 600 feet
Weights
    Gross: 2400 pounds
    Empty: 1225 pounds
Dimensions
    Length: 23 feet, 3 inches
    Height: 7 feet, 3 inches
    Span: 30 feet
Other
    Fuel capacity: 50 gallons
    Rate of climb: 720 fpm
Seats: Four

Make: Piper
Model: PA-28-151 Warrior
Year: 1974–1977
Engine
    Make: Lycoming
    Model: O-320-E3D
    Horsepower: 150
    TBO: 2000 hours
Speeds
    Maximum: 134 mph
    Cruise: 126 mph
    Stall: 58 mph
Transitions
    Takeoff over 50-foot obstacle: 1760 feet
    Ground run: 1065 feet
    Landing over 50-foot obstacle: 1115 feet
    Ground roll: 595 feet
Weights
    Gross: 2325 pounds

Empty: 1301 pounds
Dimensions
    Length: 23 feet, 8 inches
    Height: 7 feet, 3 inches
    Span: 35 feet
Other
    Fuel capacity: 48 gallons
    Rate of climb: 649 fpm
Seats: Four

Fig. 8-31. *Piper PA-28-161 Warrior.*

Make: Piper
Model: PA-28-161 Warrior
Year: 1977–1986
Engine
    Make: Lycoming
    Model: O-320-D3G
    Horsepower: 160
    TBO: 2000 hours
Speeds
    Maximum: 145 mph
    Cruise: 140 mph
    Stall: 57 mph
Transitions
    Takeoff over 50-foot obstacle: 1490 feet
    Ground run: 975 feet

Landing over 50-foot obstacle: 1115 feet
Ground roll: 595 feet
Weights
Gross: 2325 pounds
Empty: 1353 pounds
Dimensions
Length: 23 feet, 8 inches
Height: 7 feet, 3 inches
Span: 35 feet
Other
Fuel capacity: 48 gallons
Rate of climb: 710 fpm
Seats: Four

Make: Piper
Model: PA-28-161 Warrior II
Year: 1987–1995
Engine
Make: Lycoming
Model: O-320-D3G
Horsepower: 160
TBO: 2000 hours
Speeds
Maximum: 145 mph
Cruise: 135 mph
Stall: 51 mph
Transitions
Takeoff over 50-foot obstacle: 1650 feet
Ground run: 975 feet
Landing over 50-foot obstacle: 1160 feet
Ground roll: 595 feet
Weights
Gross: 2325 pounds
Empty: 1344 pounds
Dimensions
Length: 23 feet, 8 inches
Height: 7 feet, 3 inches
Span: 35 feet
Other
Fuel capacity: 48 gallons
Rate of climb: 644 fpm
Seats: Four

Make: New Piper
Model: PA-28-161 Warrior III
Year: 1994–

Engine
  Make: Lycoming
  Model: O-320-D3G
  Horsepower: 160
  TBO: 2000 hours
Speeds
  Maximum: 135 mph
  Cruise: 132 mph
  Stall: 51 mph
Transitions
  Takeoff over 50-foot obstacle: 1620 feet
  Ground run: NA
  Landing over 50-foot obstacle: 1160 feet
  Ground roll: NA
Weights
  Gross: 2440 pounds
  Empty: 1527 pounds
Dimensions
  Length: 23 feet, 8 inches
  Height: 7 feet, 3 inches
  Span: 35 feet
Other
  Fuel capacity: 48 gallons
  Rate of climb: NA
Seats: Four

Make: Piper
Model: PA-28-181 Archer
Year: 1976–1995
Engine
  Make: Lycoming
  Model: O-360-A4M
  Horsepower: 180
  TBO: 2000 hours
Speeds
  Maximum: 154 mph
  Cruise: 148 mph
  Stall: 61 mph
Transitions
  Takeoff over 50-foot obstacle: 1625 feet
  Ground run: 870 feet
  Landing over 50-foot obstacle: 1390 feet
  Ground roll: 925 feet
Weights
  Gross: 2550 pounds

Fig. 8-32. *Piper PA-28-181 Archer.*

Empty: 1413 pounds
Dimensions
    Length: 23 feet, 8 inches
    Height: 7 feet, 3 inches
    Span: 35 feet
Other
    Fuel capacity: 48 gallons
    Rate of climb: 735 fpm
Seats: Four

Make: New Piper
Model: PA-28-181 Archer III
Year: 1996–
Engine
    Make: Lycoming
    Model: O-360-A4M
    Horsepower: 180
    TBO: 2000 hours
Speeds
    Maximum: 153 mph
    Cruise: 147 mph
    Stall: 52 mph
Transitions
    Takeoff over 50-foot obstacle: 1608 feet
    Ground run: 1135 feet

Landing over 50-foot obstacle: 1400 feet
Ground roll: 920 feet
Weights
Gross: 2550 pounds
Empty: 1703 pounds
Dimensions
Length: 24 feet
Height: 7 feet, 3 inches
Span: 35 feet, 6 inches
Other
Fuel capacity: 48 gallons
Rate of climb: NA
Seats: Four

SOCATA

Socata is the light aviation subsidiary of the French government-owned Aerospatiale Group.

These airplanes sold in the United States all carry the Socata name and are often called "Caribbean planes," because of their model names (Trinidad, Tobago, Tampico). Only the TB-9 Tampico and TB-10 Tobago are reviewed in this chapter. The remaining models are complex airplanes and thus are included in the next chapter.

All the airplanes in this series share a common fuselage, including the retractable landing gear versions seen in the next chapter. These airplanes also share wings and empennages. Socata claims that 80 percent of the systems and components are manufactured in the United States. The airframe is not.

The cabins are very roomy and the rear seat of the TB-10 has three seatbelts. It is not a five-place airplane, however, because of weight limitations. The TB-9 is a popular training airplane.

Although the line of airplanes were first introduced in Europe in the 1970s, they were not certified by the FAA until the 1980s (Figs. 8-33 and 8-34).

Make: Socata
Model: TB-9 Tampico Club
Year: 1988–
Engine
Make: Lycoming
Model: O-320-D2A
Horsepower: 160
TBO: 2000 hours
Speeds
Maximum: 130 mph
Cruise: 122 mph
Stall: 58 mph

Fig. 8-33. *Socata TB-9.*

Transitions
>Takeoff over 50-foot obstacle: 1706 feet
>Ground run: NA
>Landing over 50-foot obstacle: 1378 feet
>Ground roll: NA

Weights
>Gross: 2337 pounds
>Empty: 1416 pounds

Dimensions
>Length: 25 feet, 3 inches
>Height: 9 feet
>Span: 32 feet

Other
>Fuel capacity: 41 gallons
>Rate of climb: 738 fpm

Seats: Four

Make: Socata
Model: TB-10 Tobago
Year: 1985–
Engine
>Make: Lycoming
>Model: O-360-A1AD
>Horsepower: 180
>TBO: 2000 hours

Speeds
>Maximum: 152 mph
>Cruise: 135 mph
>Stall: 59 mph

Transitions
>Takeoff over 50-foot obstacle: 1657 feet
>Ground run: 1066 feet
>Landing over 50-foot obstacle: 1394 feet
>Ground roll: 623 feet

Fig. 8-34. *Socata TB-10.*

Weights
    Gross: 2535 pounds
    Empty: 1477 pounds
Dimensions
    Length: 25 feet
    Height: 10 feet, 6 inches
    Span: 32 feet
Other
    Fuel capacity: 54 gallons
    Rate of climb: 790 fpm
Seats: Four

STINSON

All Stinson airplanes were originally built of tube and fabric, but many have since been metallized, that is, covered with a metal skin in place of the fabric. All have conventional landing gear.

Franklin engines, both heavy case and light case, were installed on these airplanes. Only the heavy-case engine is acceptable because the light case did not stand up well. It is very unlikely you would ever encounter a light-case engine at this late date.

The 108-1 models have 150-hp engines, and models 108-2 and 108-3 use 165-hp engines. Many of these airplanes have been modified with Lycoming or Continental engines ranging from 200 to 250 hp. Stinsons make good seaplanes and are often seen in western Canada and Alaska in float configuration.

Piper bought out Stinson in 1948 and continued to produce the model 108s, but the numbers built by Piper were few and production was soon halted. Any Stinson with a serial number above 4231 is a Piper-built airplane. A total of 5260 Stinsons were built.

Although long out of production, new parts are available from Univair Aircraft Corporation. Stinsons are roomy and strong airplanes (Fig. 8-35).

Make: Stinson
Model: 108-1
Year: 1946–1947
Engine
    Make: Franklin
    Model: 6A4-150-B23
    Horsepower: 150
    TBO: 1200 hours
Speeds
    Maximum: 130 mph
    Cruise: 117 mph
    Stall: 57 mph
Transitions
    Takeoff over 50-foot obstacle: 1750 feet
    Ground run: 945 feet
    Landing over 50-foot obstacle: 1400 feet
    Ground roll: 940 feet
Weights
    Gross: 2230 pounds
    Empty: 1206 pounds
Dimensions
    Length: 24 feet
    Height: 7 feet
    Span: 33 feet, 11 inches
Other
    Fuel capacity: 50 gallons
    Rate of climb: 700 fpm
Seats: Four

Fig. 8-35. *Stinson 108.*

Make: Stinson
Model: 108-2/3 Voyager and Station Wagon
Year: 1947–1948
Engine
    Make: Franklin
    Model: 6A4-165-B3
    Horsepower: 165
    TBO: 1200 hours
Speeds
    Maximum: 133 mph
    Cruise: 125 mph
    Stall: 61 mph
Transitions
    Takeoff over 50-foot obstacle: 1400 feet
    Ground run: 980 feet
    Landing over 50-foot obstacle: 1680 feet
    Ground roll: 940 feet
Weights
    Gross: 2400 pounds
    Empty: 1300 pounds
Dimensions
    Length: 24 feet
    Height: 7 feet
    Span: 33 feet, 11 inches
Other
    Fuel capacity: 50 gallons
    Rate of climb: 750 fpm
Seats: Four

9

Complex Airplanes

COMPLEX AIRPLANES REPRESENT THE PINNACLE of single-engine aircraft design and capabilities. They are real people-movers and are used extensively by businesspeople and families requiring fast and reliable transportation.

Complex airplane cruise speeds are higher, ranges are longer, and load capacities are greater than simpler four-place airplanes. They have more powerful engines and more features, such as retractable landing gear and constant-speed propellers. Many complex airplanes offer up to six-place seating.

Purchase and maintenance expenses of these airplanes are considerably higher than for simpler planes. If you can justify a need for this class of airplane, however, then the expenses will not be out of line.

Most of these airplanes are IFR-equipped. Perhaps this is an indication of business usage, where reliable transportation is a requirement rather than a pleasure, or an indication of owner-required completeness.

Airplane ownership can offer tax advantages for a business. It's recommended, however, that you first consult with an accountant if you plan to use an airplane for business purposes.

BEECHCRAFT/RAYTHEON

The name Beechcraft goes back further, historically, than most other manufacturers found in this book. Although many early Beech airplanes are not relevant to this book, a particular large, single-engine Beech plane is not only a capable people-mover, it is a historic airplane.

Staggerwing Beech airplanes were built from the 1930s through the 1940s and are classics in the truest sense of the word. Of tube-and-fabric construction, the cabin is larger than any airplane in the same class today. Its engine was a radial design and, although expensive to maintain, makes the right sound (you'll know it when you hear it, and never forget it).

Staggerwings are very expensive to purchase, and also are expensive to operate and maintain.

The all-metal, V-tailed model 35 Bonanza has probably captured the imagination of more pilots and nonpilots over the years than any other light plane. The basic style has been used for over 45 years. It is the plane most people visualize when the name Beechcraft is mentioned.

The first model 35s were powered with a 185-hp engine, had a wooden prop, and seated four people. The model 35 Bonanza remained in production through the early 1980s and many improvements, refinements, and changes were made to the model down through the years.

The final Bonanza model 35s were powered with a 285-hp engine and seated six. Today, it's difficult to find an early model 35 in stock configuration. Most have been updated with regard to appearance, avionics, and of course power.

The model 33 Debonair was introduced in 1960 as a four-seat conventional tail airplane powered with a 225-hp engine. An optional 285-hp version became available in 1966. The Debonair model name was dropped in 1968, but the model 33 line continued as a Bonanza.

The model 36 Bonanza was introduced in 1968. It was basically a conventional-tailed version of the model 35 V-tailed airplane. The model 36 was followed by the model A36, which continues in current production. In 1984 an upgrade to 300-hp was made to the model. All told, more than 3000 model 36 airplanes have been built.

The A36 series airplanes are still manufactured, and in 2004 a new A36 sold for a base price of over $650,000.

Beechcraft introduced the Sierra in 1970 as the model 24R. It had a 200-hp engine, constant-speed prop, and retractable landing gear. The Sierra was the top of a special line of airplanes built for Beech Aero Centers, which were set up to sell and rent lower-end Beech products to the flying public.

Beech airplanes are generally thought of as the "cream of the crop." They are tough and hold their values well, but, as with anything complex, they are expensive to maintain (Figs. 9-1 through 9-8).

Make: Beechcraft
Model: D17S Staggerwing (unofficial name)
Year: 1937–1948
Engine
    Make: Pratt & Whitney
    Model: R-985
    Horsepower: 450
    TBO: NA
Speeds
    Maximum: 212 mph
    Cruise: 202 mph
    Stall: 60 mph

Fig. 9-1. *Beechcraft 17 Staggerwing.*

Transitions
    Takeoff over 50-foot obstacle: 1130 feet
    Ground run: 610 feet
    Landing over 50-foot obstacle: 980 feet
    Ground roll: 750 feet
Weights
    Gross: 4250 pounds
    Empty: 2540 pounds
Dimensions
    Length: 26 feet, 10 inches
    Height: 8 feet
    Span: 32 feet
Other
    Fuel capacity: 124 gallons
    Rate of climb: 1500 fpm
Seats: Four

Make: Beechcraft
Model: 24R Sierra
Year: 1970–1983
Engine
    Make: Lycoming
    Model: IO-360-A1B6
    Horsepower: 200
    TBO: 1800 hours

Fig. 9-2. *Beechcraft 24R Sierra.*

Speeds
    Maximum: 170 mph
    Cruise: 162 mph
    Stall: 66 mph
Transitions
    Takeoff over 50-foot obstacle: 1980 feet
    Ground run: 1260 feet
    Landing over 50-foot obstacle: 1670 feet
    Ground roll: 752 feet
Weights
    Gross: 2750 pounds
    Empty: 1610 pounds
Dimensions
    Length: 25 feet, 9 inches
    Height: 8 feet, 3 inches
    Span: 32 feet, 9 inches
Other
    Fuel capacity: 59 gallons
    Rate of climb: 862 fpm
Seats: Four

Make: Beechcraft
Model: 33 Debonair/Bonanza
Year: 1960–1970
Engine
    Make: Continental
    Model: IO-470-J
    Horsepower: 225
    TBO: 1500 hours

Fig. 9-3. *Beechcraft F33A Bonanza.*

Speeds
    Maximum: 195 mph
    Cruise: 185 mph
    Stall: 60 mph
Transitions
    Takeoff over 50-foot obstacle: 1235 feet
    Ground run: 940 feet
    Landing over 50-foot obstacle: 1282 feet
    Ground roll: 635 feet
Weights
    Gross: 3000 pounds
    Empty: 1745 pounds
Dimensions
    Length: 25 feet, 6 inches
    Height: 8 feet, 3 inches
    Span: 32 feet, 10 inches
Other
    Fuel capacity: 50 gallons
    Rate of climb: 960 fpm
Seats: Four to five

Make: Beechcraft
Model: 33 Debonair/Bonanza
Year: 1966–1990
Engine
    Make: Continental
    Model: IO-520-BA
    Horsepower: 285
    TBO: 1700 hours

Speeds
    Maximum: 208 mph
    Cruise: 200 mph
    Stall: 63 mph
Transitions
    Takeoff over 50-foot obstacle: 1873 feet
    Ground run: 1091 feet
    Landing over 50-foot obstacle: 1500 feet
    Ground roll: 795 feet
Weights
    Gross: 3400 pounds
    Empty: 1965 pounds
Dimensions
    Length: 25 feet, 6 inches
    Height: 8 feet, 3 inches
    Span: 33 feet, 5 inches
Other
    Fuel capacity: 50 gallons
    Rate of climb: 1136 fpm
Seats: Five to six

Make: Beechcraft
Model: 35-A35 Bonanza
Year: 1947–1949
Engine
    Make: Continental
    Model: E-185-1
    Horsepower: 185 (205 and 225 optional)
    TBO: 1500 hours

Fig. 9-4. *Beechcraft 35 Bonanza.*

Speeds
    Maximum: 184 mph
    Cruise: 172 mph
    Stall: 55 mph
Transitions
    Takeoff over 50-foot obstacle: 1440 feet
    Ground run: 1200 feet
    Landing over 50-foot obstacle: 925 feet
    Ground roll: 580 feet
Weights
    Gross: 2550 pounds
    Empty: 1458 pounds
Dimensions
    Length: 25 feet, 1 inch
    Height: 6 feet, 6 inches
    Span: 32 feet, 9 inches
Other
    Fuel capacity: 39 gallons
    Rate of climb: 950 fpm
Seats: Four

Make: Beechcraft
Model: B35 Bonanza
Year: 1950
Engine
    Make: Continental
    Model: E-185-8
    Horsepower: 196
    TBO: 1500 hours
Speeds
    Maximum: 184 mph
    Cruise: 170 mph
    Stall: 56 mph
Transitions
    Takeoff over 50-foot obstacle: 1515 feet
    Ground run: 1275 feet
    Landing over 50-foot obstacle: 950 feet
    Ground roll: 625 feet
Weights
    Gross: 2650 pounds
    Empty: 1575 pounds
Dimensions
    Length: 25 feet, 1 inch
    Height: 6 feet, 6 inches
    Span: 32 feet, 9 inches

Other
    Fuel capacity: 39 gallons
    Rate of climb: 890 fpm
Seats: Four

Make: Beechcraft
Model: C35/D35 Bonanza
Year: 1951–1953
Engine
    Make: Continental
    Model: E-185-11
    Horsepower: 205
    TBO: 1500 hours
Speeds
    Maximum: 190 mph
    Cruise: 175 mph
    Stall: 55 mph
Transitions
    Takeoff over 50-foot obstacle: 1500 feet
    Ground run: 1250 feet
    Landing over 50-foot obstacle: 975 feet
    Ground roll: 625 feet
Weights
    Gross: 2700 pounds
    Empty: 1650 pounds
Dimensions
    Length: 25 feet, 1 inch
    Height: 6 feet, 6 inches
    Span: 32 feet, 9 inches
Other
    Fuel capacity: 39 gallons
    Rate of climb: 1100 fpm
Seats: Four

Make: Beechcraft
Model: E35/G35 Bonanza
Year: 1954–1956
Engine
    Make: Continental
    Model: E-225-8
    Horsepower: 225
    TBO: 1500 hours
Speeds
    Maximum: 194 mph
    Cruise: 184 mph
    Stall: 55 mph

Transitions
    Takeoff over 50-foot obstacle: 1270 feet
    Ground run: 1060 feet
    Landing over 50-foot obstacle: 1025 feet
    Ground roll: 680 feet
Weights
    Gross: 2775 pounds
    Empty: 1722 pounds
Dimensions
    Length: 25 feet, 1 inch
    Height: 6 feet, 6 inches
    Span: 32 feet, 9 inches
Other
    Fuel capacity: 39 gallons
    Rate of climb: 1300 fpm
Seats: Four

Make: Beechcraft
Model: H35 Bonanza
Year: 1957
Engine
    Make: Continental
    Model: O-470-G
    Horsepower: 240
    TBO: 1500 hours
Speeds
    Maximum: 206 mph
    Cruise: 196 mph
    Stall: 57 mph
Transitions
    Takeoff over 50-foot obstacle: 1260 feet
    Ground run: 1050 feet
    Landing over 50-foot obstacle: 1050 feet
    Ground roll: 710 feet
Weights
    Gross: 2900 pounds
    Empty: 1833 pounds
Dimensions
    Length: 25 feet, 1 inch
    Height: 6 feet, 6 inches
    Span: 32 feet, 9 inches
Other
    Fuel capacity: 39 gallons
    Rate of climb: 1250 fpm
Seats: Four

Make: Beechcraft
Model: J35/M35 Bonanza
Year: 1958–1960
Engine
Make: Continental
Model: O-470-C
Horsepower: 250
TBO: 1500 hours
Speeds
Maximum: 210 mph
Cruise: 195 mph
Stall: 57 mph
Transitions
Takeoff over 50-foot obstacle: 1185 feet
Ground run: 950 feet
Landing over 50-foot obstacle: 1050 feet
Ground roll: 710 feet
Weights
Gross: 2900 pounds
Empty: 1820 pounds
Dimensions
Length: 25 feet, 1 inch
Height: 6 feet, 6 inches
Span: 32 feet, 9 inches
Other
Fuel capacity: 39 gallons
Rate of climb: 1250 fpm
Seats: Four

Make: Beechcraft
Model: N35/P35 Bonanza
Year: 1961–1963
Engine
Make: Continental
Model: IO-470-N
Horsepower: 260
TBO: 1500 hours
Speeds
Maximum: 205 mph
Cruise: 190 mph
Stall: 60 mph
Transitions
Takeoff over 50-foot obstacle: 1260 feet
Ground run: 1050 feet

Fig. 9-5. *Beechcraft N35 Bonanza.*

    Landing over 50-foot obstacle: 1100 feet
    Ground roll: 650 feet
Weights
    Gross: 3125 pounds
    Empty: 1855 pounds
Dimensions
    Length: 25 feet, 1 inch
    Height: 6 feet, 6 inches
    Span: 32 feet, 9 inches
Other
    Fuel capacity: 49 gallons
    Rate of climb: 1150 fpm
Seats: Five

Make: Beechcraft
Model: S35/V35 Bonanza
Year: 1965–1984
Engine
    Make: Continental
    Model: IO-520-B
    Horsepower: 285
    TBO: 1700 hours
Speeds
    Maximum: 210 mph
    Cruise: 203 mph
    Stall: 63 mph
Transitions
    Takeoff over 50-foot obstacle: 1320 feet
    Ground run: 965 feet
    Landing over 50-foot obstacle: 1177 feet
    Ground roll: 647 feet

Fig. 9-6. *Beechcraft V35B Bonanza.*

Weights
    Gross: 3400 pounds
    Empty: 1970 pounds
Dimensions
    Length: 25 feet, 1 inch
    Height: 6 feet, 6 inches
    Span: 32 feet, 9 inches
Other
    Fuel capacity: 50 gallons
    Rate of climb: 1136 fpm
Seats: Five to six

Make: Beechcraft
Model: V35TC Turbo Bonanza
Year: 1966–1970
Engine
    Make: Continental
    Model: TSIO-520-D
    Horsepower: 285
    TBO: 1400 hours
Speeds
    Maximum: 240 mph
    Cruise: 224 mph
    Stall: 63 mph
Transitions
    Takeoff over 50-foot obstacle: 1320 feet
    Ground run: 950 feet
    Landing over 50-foot obstacle: 1177 feet
    Ground roll: 647 feet

Weights
    Gross: 3400 pounds
    Empty: 2027 pounds
Dimensions
    Length: 25 feet, 1 inch
    Height: 6 feet, 6 inches
    Span: 32 feet, 9 inches
Other
    Fuel capacity: 50 gallons
    Rate of climb: 1225 fpm
Seats: Six

Make: Beechcraft
Model: 36 Bonanza
Year: 1968–1969
Engine
    Make: Continental
    Model: IO-520-B
    Horsepower: 285
    TBO: 1700 hours
Speeds
    Maximum: 204 mph
    Cruise: 196 mph
    Stall: 64 mph
Transitions
    Takeoff over 50-foot obstacle: 1525 feet
    Ground run: 1112 feet
    Landing over 50-foot obstacle: 1240 feet
    Ground roll: 683 feet
Weights
    Gross: 3600 pounds
    Empty: 1980 pounds
Dimensions
    Length: 26 feet, 4 inches
    Height: 8 feet, 5 inches
    Span: 32 feet, 10 inches
Other
    Fuel capacity: 50 gallons
    Rate of climb: 1015 fpm
Seats: Four

Make: Beechcraft
Model: A36 Bonanza
Year: 1971–1983

Fig. 9-7. *Beechcraft A36 Bonanza.*

Engine
> Make: Continental
> Model: IO-520-BA
> Horsepower: 285
> TBO: 1700 hours

Speeds
> Maximum: 206 mph
> Cruise: 193 mph
> Stall: 60 mph

Transitions
> Takeoff over 50-foot obstacle: 2040 feet
> Ground run: 1140 feet
> Landing over 50-foot obstacle: 1450 feet
> Ground roll: 840 feet

Weights
> Gross: 3600 pounds
> Empty: 2195 pounds

Dimensions
> Length: 27 feet, 6 inches
> Height: 8 feet, 5 inches
> Span: 33 feet, 6 inches

Other
> Fuel capacity: 74 gallons
> Rate of climb: 1030 fpm

Seats: Four

Make: Beechcraft
Model: A36 Bonanza
Year: 1984–
Engine
> Make: Continental
> Model: IO-550-B

Horsepower: 300
TBO: 1700 hours
Speeds
Maximum: 212 mph
Cruise: 194 mph
Stall: 68 mph
Transitions
Takeoff over 50-foot obstacle: 1913 feet
Ground run: 971 feet
Landing over 50-foot obstacle: 1473 feet
Ground roll: 913 feet
Weights
Gross: 3650 pounds
Empty: 2247 pounds
Dimensions
Length: 27 feet, 6 inches
Height: 8 feet, 5 inches
Span: 33 feet, 6 inches
Other
Fuel capacity: 74 gallons
Rate of climb: 1210 fpm
Seats: Four

Make: Beechcraft
Model: A36 TC Bonanza
Year: 1979–
Engine
Make: Continental
Model: TSIO-520-UB
Horsepower: 300
TBO: 1600 hours
Speeds
Maximum: 246 mph
Cruise: 219 mph
Stall: 66 mph
Transitions
Takeoff over 50-foot obstacle: 2012 feet
Ground run: 1176 feet
Landing over 50-foot obstacle: 1449 feet
Ground roll: 721 feet
Weights
Gross: 3650 pounds
Empty: 2278 pounds
Dimensions
Length: 27 feet, 6 inches

Height: 8 feet, 5 inches
Span: 33 feet, 6 inches
Other
Fuel capacity: 74 gallons
Rate of climb: 1165 fpm
Seats: Four

Make: Beechcraft
Model: B36 TC Bonanza
Year: 1982–
Engine
Make: Continental
Model: TSIO-520-UB
Horsepower: 300
TBO: 1600 hours
Speeds
Maximum: 245 mph
Cruise: 219 mph
Stall: 66 mph
Transitions
Takeoff over 50-foot obstacle: 2364 feet
Ground run: 1156 feet
Landing over 50-foot obstacle: 1692 feet
Ground roll: 976 feet
Weights
Gross: 3850 pounds
Empty: 2338 pounds

Fig. 9-8. *Raytheon Beech B36 TC.*

Dimensions
    Length: 27 feet, 6 inches
    Height: 8 feet, 5 inches
    Span: 37 feet, 10 inches
Other
    Fuel capacity: 102 gallons
    Rate of climb: 1049 fpm
Seats: Four

BELLANCA

Bellanca airplanes have been good performers through the years, but there aren't as many in numbers as the more popular makes. Bellanca has been in and out of business several times, but this is no mark against the airplanes themselves. Northern Aircraft Company, Downer Aircraft, Inter-Aire, Bellanca Sales, and Miller Flying Service have all been associated with manufacturing Bellanca airplanes at one time or another.

The Bellanca fuselage is fabric-covered and the wings are wooden. This type of construction might be part of the reason for the general lack of popularity of these planes and is, in fact, a point of concern. Wood rot has been encountered in the wing structures and is the subject of AD required inspections.

The tail configurations have changed considerably over the years. Prior to 1964, Bellancas had three surfaces (similar to the Lockheed Constellation). Later models have standard tails with a single vertical surface. Early pre-Viking Bellancas are considered classics.

Bellanca continues to produce limited numbers of the Super Viking and supports the older models. A used Bellanca in good condition can be a lot of airplane for the dollar. They are fast and roomy, but beware of a plane that requires fabric re-covering, as that is a sure way to spend several thousand dollars. A new Super Viking cost $235,000 in 1999 (Figs. 9-9, 9-10, and 9-11).

Make: Bellanca
Model: 14-13 Cruisemaster
Year: 1950–1951
Engine
    Make: Franklin
    Model: 6A4-150-B3
    Horsepower: 150
    TBO: 1200 hours
Speeds
    Maximum: 169 mph
    Cruise: 150 mph
    Stall: 44 mph

Fig. 9-9. *Bellanca 14-13 Cruisemaster.*

Courtesy of Carl Schuppel, EAA

Transitions
    Takeoff over 50-foot obstacle: 1350 feet
    Ground run: 606 feet
    Landing over 50-foot obstacle: 975 feet
    Ground roll: 437 feet
Weights
    Gross: 2100 pounds
    Empty: 1520 pounds
Dimensions
    Length: 21 feet, 2 inches
    Height: 6 feet, 2 inches
    Span: 34 feet, 2 inches
Other
    Fuel capacity: 40 gallons
    Rate of climb: 1100 fpm
Seats: Four

Make: Bellanca
Model: 14-19 Cruisemaster
Year: 1950–1951
Engine
    Make: Lycoming
    Model: O-435-A
    Horsepower: 190
    TBO: 1200 hours
Speeds
    Maximum: 200 mph
    Cruise: 180 mph
    Stall: 44 mph
Transitions
    Takeoff over 50-foot obstacle: 1270 feet
    Ground run: 850 feet
    Landing over 50-foot obstacle: 1025 feet
    Ground roll: 450 feet

Fig. 9-10. *Bellanca 14-19-3A.*

Weights
    Gross: 2600 pounds
    Empty: 1575 pounds
Dimensions
    Length: 23 feet
    Height: 6 feet, 2 inches
    Span: 34 feet, 2 inches
Other
    Fuel capacity: 40 gallons
    Rate of climb: 1250 fpm
Seats: Four

Make: Bellanca
Model: 14-19-2 Cruisemaster
Year: 1957–1959
Engine
    Make: Continental
    Model: O-470-K
    Horsepower: 230
    TBO: 1500 hours
Speeds
    Maximum: 206 mph
    Cruise: 196 mph
    Stall: 46 mph
Transitions
    Takeoff over 50-foot obstacle: 1025 feet
    Ground run: 760 feet
    Landing over 50-foot obstacle: 1150 feet
    Ground roll: 470 feet
Weights
    Gross: 2700 pounds
    Empty: 1640 pounds

Dimensions
    Length: 23 feet
    Height: 6 feet, 2 inches
    Span: 34 feet, 2 inches
Other
    Fuel capacity: 40 gallons
    Rate of climb: 1500 fpm
Seats: Four

Make: Bellanca
Model: 14-19-3 (A-C)
Year: 1959–1968
Engine
    Make: Continental
    Model: IO-470-F
    Horsepower: 260
    TBO: 1500 hours
Speeds
    Maximum: 208 mph
    Cruise: 203 mph
    Stall: 62 mph
Transitions
    Takeoff over 50-foot obstacle: 1000 feet
    Ground run: 340 feet
    Landing over 50-foot obstacle: 800 feet
    Ground roll: 400 feet
Weights
    Gross: 3000 pounds
    Empty: 1850 pounds
Dimensions
    Length: 23 feet, 6 inches
    Height: 6 feet, 5 inches
    Span: 34 feet, 2 inches
Other
    Fuel capacity: 40 gallons
    Rate of climb: 1500 fpm
Seats: Four

Make: Bellanca
Model: 17-30 Viking
Year: 1967–1970
Engine
    Make: Continental
    Model: IO-520-K1A
    Horsepower: 300
    TBO: 1700

Fig. 9-11.. *Bellanca Viking.*

Speeds
    Maximum: 192 mph
    Cruise: 188 mph
    Stall: 62 mph
Transitions
    Takeoff over 50-foot obstacle: 908 feet
    Ground run: 450 feet
    Landing over 50-foot obstacle: 1050 feet
    Ground roll: 575 feet
Weights
    Gross: 3200 pounds
    Empty: 1900 pounds
Dimensions
    Length: 23 feet, 7 inches
    Height: 7 feet, 4 inches
    Span: 34 feet, 2 inches
Other
    Fuel capacity: 58 gallons
    Rate of climb: 1840 fpm
Seats: Four

Make: Bellanca
Model: 17-30 A/B Super Viking
Year: 1970 –
Engine
    Make: Continental
    Model: IO-520-K1A

Horsepower: 300
TBO: 1700
Speeds
Maximum: 208 mph
Cruise: 202 mph
Stall: 70 mph
Transitions
Takeoff over 50-foot obstacle: 1420 feet
Ground run: NA
Landing over 50-foot obstacle: 1340 feet
Ground roll: NA
Weights
Gross: 3325 pounds
Empty: 2185 pounds
Dimensions
Length: 26 feet, 4 inches
Height: 7 feet, 4 inches
Span: 34 feet, 2 inches
Other
Fuel capacity: 68 gallons
Rate of climb: 1210 fpm
Seats: Four

Make: Bellanca
Model: 17-31/A
Year: 1969–1978
Engine
Make: Lycoming
Model: IO-540-K1E5
Horsepower: 300
TBO: 2000
Speeds
Maximum: 200 mph
Cruise: 190 mph
Stall: 70 mph
Transitions
Takeoff over 50-foot obstacle: 1420 feet
Ground run: 980
Landing over 50-foot obstacle: 1340 feet
Ground roll: 835
Weights
Gross: 3325 pounds
Empty: 2247 pounds
Dimensions
Length: 26 feet, 4 inches

Height: 7 feet, 4 inches
Span: 34 feet, 2 inches
Other
    Fuel capacity: 68 gallons
    Rate of climb: 1170 fpm
Seats: Four

Make: Bellanca
Model: 17-31 TC/ATC (turbo)
Year: 1969–1979
Engine
    Make: Continental
    Model: IO-520-K1A (Rayjay Turbo charging)
    Horsepower: 300
    TBO: 1700
Speeds
    Maximum: 222 mph
    Cruise: 215 mph
    Stall: 62 mph
Transitions
    Takeoff over 50-foot obstacle: 890 feet
    Ground run: 460 feet
    Landing over 50-foot obstacle: 1100 feet
    Ground roll: 575 feet
Weights
    Gross: 3200 pounds
    Empty: 2010 pounds
Dimensions
    Length: 23 feet, 7 inches
    Height: 7 feet, 4 inches
    Span: 34 feet, 2 inches
Other
    Fuel capacity: 72 gallons
    Rate of climb: 1800 fpm
Seats: Four

CESSNA

Along with the Beech Staggerwing, another historic airplane is suitable for inclusion into this chapter: the Cessna 190. The 190 and subsequent 195 airplanes, although produced later than the Beech Staggerwings, are also powered with radial engines. They are very roomy, can land nearly anywhere, and their all-metal construction and fixed landing gear make them less expensive to maintain than the Staggerwing.

Cessna's model 180, covered in the next chapter, was the ancestor of the popular Cessna 182. The 182 is a real workhorse, able to carry a full complement of passengers, fuel, and baggage. Although without retractable landing gear, it is considered a complex airplane because of gross weight and the 230-hp engine. Changes to the 182 help identify the various models, produced in the following years:

1960: Swept tail

1962: Rear window

1972: Tubular landing gear

1977: 100-octane engine

In 1960, the model 210 Centurion entered production as a four-seat plane with a 260-hp engine and retractable gear; the engine power was increased to 285-hp in 1964.

Cessna added retractable landing gear to several planes in the easy-flier category, resulting in the 172-RG and the 177-RG. Even the 182 got retractable landing gear and was renamed 182-RG. Cessna plans to revive the 172-RG at a price of $200,000 and the 182-RG at $250,000.

In 1996, Cessna announced that the model 182 would again be produced, possibly as early as 1997. The company was right on target when the new 182 appeared in April 1997. In 1999, a new 182 Skylane listed for about $225,000. The 2004 base price was $250,000.

Cessna airplanes are generally well thought of, can be maintained without excess expense, and fly easily. The high wings afford excellent downward visibility (Figs. 9-12 through 9-19).

Make: Cessna
Model: 190/195
Year: 1947–1954
Engine
    Make: Jacobs
    Model: R-755 (Continental R-670 240-hp in the 190)
    Horsepower: 245 to 300
    TBO: 1000 hours
Speeds
    Maximum: 176 mph
    Cruise: 170 mph
    Stall: 64 mph
Transitions
    Takeoff over 50-foot obstacle: 1670 feet
    Ground run: NA
    Landing over 50-foot obstacle: 1495 feet
    Ground roll: NA
Weights
    Gross: 3350 pounds

Fig. 9-12. *Cessna 1949 195.*

　　　Empty: 2030 pounds
Dimensions
　　　Length: 27 feet, 3 inches
　　　Height: 7 feet, 2 inches
　　　Span: 36 feet, 2 inches
Other
　　　Fuel capacity: 80 gallons
　　　Rate of climb: 1050 feet
Seats: Five

Make: Cessna
Model: 182 Skylane
Year: 1956–1974
Engine
　　　Make: Continental
　　　Model: O-470 (O-470-U after 1976)
　　　Horsepower: 230
　　　TBO: 1500 hours
Speeds
　　　Maximum: 165 mph
　　　Cruise: 157 mph
　　　Stall: 57 mph
Transitions
　　　Takeoff over 50-foot obstacle: 1350 feet
　　　Ground run: 705 feet
　　　Landing over 50-foot obstacle: 1350 feet
　　　Ground roll: 590 feet
Weights
　　　Gross: 2950 pounds
　　　Empty: 1595 pounds
Dimensions
　　　Length: 28 feet, 2 inches

Fig. 9-13. *Cessna 1956 182.*

Height: 9 feet, 2 inches
Span: 35 feet, 10 inches
Other
Fuel capacity: 56 gallons
Rate of climb: 890 fpm
Seats: Four

Make: Cessna
Model: 182 Skylane II
Year: 1975–1986
Engine
Make: Continental
Model: O-470-U
Horsepower: 230
TBO: 2000 hours
Speeds
Maximum: 168 mph
Cruise: 163 mph
Stall: 56 mph
Transitions
Takeoff over 50-foot obstacle: 1515 feet
Ground run: 805 feet
Landing over 50-foot obstacle: 1350 feet
Ground roll: 590 feet
Weights
Gross: 3100 pounds
Empty: 1775 pounds
Dimensions
Length: 28 feet

Fig. 9-14. *Cessna 1983 182.*

Fig. 9-15. *1999 Cessna 182 Skylane.*

    Height: 9 feet, 3 inches
    Span: 35 feet, 10 inches
Other
    Fuel capacity: 92 gallons
    Rate of climb: 865 fpm
Seats: Four

Make: Cessna
Model: T-182 II Turbo Skylane
Year: 1981–1985
Engine
    Make: Lycoming
    Model: O-540-L3C5D
    Horsepower: 235
    TBO: 2000 hours
Speeds
    Maximum: 193 mph
    Cruise: 181 mph
    Stall: 56 mph

Transitions
    Takeoff over 50-foot obstacle: 1475 feet
    Ground run: 790 feet
    Landing over 50-foot obstacle: 1350 feet
    Ground roll: 590 feet
Weights
    Gross: 3100 pounds
    Empty: 1781 pounds
Dimensions
    Length: 28 feet, 5 inches
    Height: 9 feet, 3 inches
    Span: 35 feet, 10 inches
Other
    Fuel capacity: 92 gallons
    Rate of climb: 965 fpm
Seats: Four

Make: Cessna
Model: 182 Turbo Skylane
Year: 1997–
Engine
    Make: Textron Lycoming
    Model: IO-540-AB1A5
    Horsepower: 230
    TBO: 2000 hours
Speeds
    Maximum: 167 mph
    Cruise: 161 mph
    Stall: 56 mph
Transitions
    Takeoff over 50-foot obstacle: 1514 feet
    Ground run: 795 feet
    Landing over 50-foot obstacle: 1350 feet
    Ground roll: 590 feet
Weights
    Gross: 3110 pounds
    Empty: 1882 pounds
Dimensions
    Length: 28 feet
    Height: 9 feet, 3 inches
    Span: 36 feet, 1 inch
Other
    Fuel capacity: 92 gallons
    Rate of climb: 924 fpm
Seats: Four

Make: Cessna
Model: 210
Year: 1960–1963
Engine
    Make: Continental
    Model: IO-470-E
    Horsepower: 260
    TBO: 1500 hours
Speeds
    Maximum: 198 mph
    Cruise: 189 mph
    Stall: 60 mph
Transitions
    Takeoff over 50-foot obstacle: 1210 feet
    Ground run: 695 feet
    Landing over 50-foot obstacle: 1110 feet
    Ground roll: 725 feet
Weights
    Gross: 3000 pounds
    Empty: 1780 pounds
Dimensions
    Length: 27 feet, 9 inches
    Height: 9 feet, 9 inches
    Span: 36 feet, 7 inches
Other
    Fuel capacity: 84 gallons
    Rate of climb: 1270 fpm
Seats: Four

Make: Cessna
Model: 210 Centurion
Year: 1964–1973
Engine
    Make: Continental

Courtesy of Cessna

Fig. 9-16. *Cessna 1972 210 Centurion.*

Model: IO-520-A
Horsepower: 285
TBO: 1700 hours
Speeds
    Maximum: 200 mph
    Cruise: 188 mph
    Stall: 65 mph
Transitions
    Takeoff over 50-foot obstacle: 1900 feet
    Ground run: 1100 feet
    Landing over 50-foot obstacle: 1500 feet
    Ground roll: 765 feet
Weights
    Gross: 3800 pounds
    Empty: 2134 pounds
Dimensions
    Length: 27 feet, 9 inches
    Height: 9 feet, 9 inches
    Span: 36 feet, 7 inches
Other
    Fuel capacity: 84 gallons
    Rate of climb: 860 fpm
Seats: Six

Make: Cessna
Model: 210 Turbo Centurion
Year: 1974–1976
Engine
    Make: Continental
    Model: TSIO-520
    Horsepower: 300
    TBO: 1400 hours
Speeds
    Maximum: 200 mph
    Cruise: 196 mph
    Stall: 65 mph
Transitions
    Takeoff over 50-foot obstacle: 2050 feet
    Ground run: 1215 feet
    Landing over 50-foot obstacle: 1585
    Ground roll: 815 feet
Weights
    Gross: 3850 pounds
    Empty: 2220 pounds

Dimensions
    Length: 28 feet, 2 inches
    Height: 9 feet, 8 inches
    Span: 36 feet, 9 inches
Other
    Fuel capacity: 84 gallons
    Rate of climb: 1060 fpm
Seats: Six

Fig. 9-17. *Cessna 1983 172 Cutlass RG.*

Make: Cessna
Model: 172-RG
Year: 1980–1985
Engine
    Make: Lycoming
    Model: O-360-F1A6
    Horsepower: 180
    TBO: 2000 hours
Speeds
    Maximum: 167 mph
    Cruise: 161 mph
    Stall: 58 mph
Transitions
    Takeoff over 50-foot obstacle: 1775 feet
    Ground run: 1060 feet
    Landing over 50-foot obstacle: 1340 feet
    Ground roll: 625 feet
Weights
    Gross: 2650 pounds
    Empty: 1555 pounds
Dimensions
    Length: 27 feet, 5 inches

Height: 8 feet, 10 inches
Span: 35 feet, 10 inches
Other
Fuel capacity: 66 gallons
Rate of climb: 800 fpm
Seats: Four

Fig. 9-18. *Cessna 177-RG Cardinal.*

Make: Cessna
Model: 177-RG Cardinal
Year: 1971–1978
Engine
Make: Lycoming
Model: IO-360-A1B6D
Horsepower: 200
TBO: 1800 hours
Speeds
Maximum: 180 mph
Cruise: 171 mph
Stall: 57 mph
Transitions
Takeoff over 50-foot obstacle: 1585 feet
Ground run: 890 feet
Landing over 50-foot obstacle: 1350 feet
Ground roll: 730 feet
Weights
Gross: 2800 pounds
Empty: 1645 pounds
Dimensions
Length: 27 feet, 3 inches
Height: 8 feet, 7 inches
Span: 35 feet, 6 inches

Other
    Fuel capacity: 60 gallons
    Rate of climb: 925 fpm
Seats: Four

Make: Cessna
Model: 182-RG
Year: 1978–1986
Engine
    Make: Lycoming
    Model: O-540-J3C5D
    Horsepower: 235
    TBO: 2000 hours
Speeds
    Maximum: 184 mph
    Cruise: 180 mph
    Stall: 58 mph
Transitions
    Takeoff over 50-foot obstacle: 1570 feet
    Ground run: 820 feet
    Landing over 50-foot obstacle: 1320 feet
    Ground roll: 600 feet
Weights
    Gross: 3100 pounds
    Empty: 1750 pounds
Dimensions
    Length: 28 feet, 7 inches
    Height: 8 feet, 11 inches
    Span: 35 feet, 10 inches
Other
    Fuel capacity: 56 gallons
    Rate of climb: 1140 fpm
Seats: Four

Fig. 9-19. *Cessna 1983 182 Skylane RG.*

COMMANDER

The Commander line of airplanes began life as Rockwell products, and were designed with the serious user in mind. The model 112s were first introduced in 1972 and were powered with a 200-hp fuel-injected engine. The manufacturer claimed that the cabin was the most spacious in its class. The 112TC Alpine version, with a turbo-charged 210-hp engine, came out in 1976. The same airplane, powered with a 260-hp fuel-injected engine, is called the model 114 Gran Turismo, and was produced from 1976 to 1979.

Airworthiness Directives plagued the early models of this series. Unusual as it might seem in the airplane world, the manufacturer eventually made good on much of the cost of repairing these airplanes, which came to over $12,000 per plane.

In 1988, the Commander Aircraft Company, the majority of which was owned by KuwAm (a Kuwaiti company), purchased the production rights to this series of airplanes and in 1992 introduced the 114B. A 1999 model 114B sells for about $400,000 new.

In 1995, Commander introduced a turbo-charged airplane, the 114TC. It offers improved performance over the normally aspirated engine model. The sticker price for a new 114TC in 1999 was about $450,000.

As of this writing, Commander airplanes are no longer being produced.

Search for these planes under Commander, Aero Commander, Rockwell, and Gulfstream in classified ads (Figs. 9-20 through 9-23).

Make: Rockwell
Model: 112
Year: 1972–1977
Engine
    Make: Lycoming
    Model: IO-360-C1B6
    Horsepower: 200
    TBO: 1800 hours

Fig. 9-20. *Commander 112.*

Speeds
  Maximum: 175 mph
  Cruise: 165 mph
  Stall: 61 mph
Transitions
  Takeoff over 50-foot obstacle: 1460 feet
  Ground run: 880 feet
  Landing over 50-foot obstacle: 1310 feet
  Ground roll: 680 feet
Weights
  Gross: 2550 pounds
  Empty: 1530 pounds
Dimensions
  Length: 24 feet, 11 inches
  Height: 8 feet, 5 inches
  Span: 32 feet, 9 inches
Other
  Fuel capacity: 60 gallons
  Rate of climb: 1000 fpm
Seats: Four

Make: Rockwell
Model: 112TC Alpine
Year: 1976–1979
Engine
  Make: Lycoming
  Model: TO-360-C1A6D
  Horsepower: 210
  TBO: 1600 hours
Speeds
  Maximum: 196 mph
  Cruise: 187 mph
  Stall: 61 mph
Transitions
  Takeoff over 50-foot obstacle: 1750 feet
  Ground run: 930 feet
  Landing over 50-foot obstacle: 1250 feet
  Ground roll: 680 feet
Weights
  Gross: 2950 pounds
  Empty: 2035 pounds
Dimensions
  Length: 24 feet, 11 inches
  Height: 8 feet, 5 inches
  Span: 32 feet, 9 inches

Other
    Fuel capacity: 60 gallons
    Rate of climb: 900 fpm
Seats: Four

Make: Rockwell
Model: 114 series Gran Turismo
Year: 1976–1979
Engine
    Make: Lycoming
    Model: IO-540-T3B5D
    Horsepower: 260
    TBO: 2000 hours
Speeds
    Maximum: 191 mph
    Cruise: 181 mph
    Stall: 63 mph
Transitions
    Takeoff over 50-foot obstacle: 2150 feet
    Ground run: NA
    Landing over 50-foot obstacle: 1200 feet
    Ground roll: NA
Weights
    Gross: 3260 pounds
    Empty: 2070 pounds
Dimensions
    Length: 24 feet, 11 inches
    Height: 8 feet, 5 inches
    Span: 32 feet, 9 inches
Other
    Fuel capacity: 68 gallons
    Rate of climb: 1030 fpm
Seats: Four

Make: Commander
Model: 114B
Year: 1992–1999
Engine
    Make: Lycoming
    Model: IO-540-T4B5
    Horsepower: 260
    TBO: 2000 hours
Speeds
    Maximum: 188 mph
    Cruise: 184 mph
    Stall: 64 mph

Fig. 9-21. *Commander 114B.*

Transitions
    Takeoff over 50-foot obstacle: 2000 feet
    Ground run: 1040 feet
    Landing over 50-foot obstacle: 1200 feet
    Ground roll: 720 feet
Weights
    Gross: 3250 pounds
    Empty: 2044 pounds
Dimensions
    Length: 24 feet, 11 inches
    Height: 8 feet, 5 inches
    Span: 32 feet, 9 inches
Other
    Fuel capacity: 70 gallons
    Rate of climb: 1070 fpm
Seats: Four

Make: Commander
Model: 114TC
Year: 1995–1999
Engine
    Make: Textron Lycoming
    Model: IO-540-T4B5
    Horsepower: 270
    TBO: 2000 hours
Speeds
    Maximum: 227 mph
    Cruise: 204 mph
    Stall: 68 mph
Transitions
    Takeoff over 50-foot obstacle: 2223 feet
    Ground run: 1408 feet

Fig. 9-22. *Commander's plush interior.*

Fig. 9-23. *Commander 114TC.*

Landing over 50-foot obstacle: 1312 feet
Ground roll: 734 feet
Weights
Gross: 3305 pounds
Empty: 2151 pounds
Dimensions
Length: 24 feet, 11 inches
Height: 8 feet, 5 inches
Span: 32 feet, 9 inches

Other
    Fuel capacity: 90 gallons
    Rate of climb: 1050 fpm
Seats: Four

LAKE

If you like the idea of having lunch in a quiet sheltered cove of a lake or a river, then perhaps the Lake Amphibian is for you. (Amphibian means that the plane can take off from and land on both land and water.) You can take off from a paved runway, fly hundreds of miles, and land on a secluded lake.

Some pilots consider amphibian aircraft to be lackluster in performance. They should consider the size of the engine and the fact this is a four-place airplane, then look at the performance data. The Lake is no slouch and performs as well or better than plenty of land-bound airplanes.

First produced with a 150-hp engine in 1957, the LA-4 was quickly upgraded to 180 hp in 1958. In 1970, a 200-hp engine was added and the model was designated the LA-4-200 Buccaneer (Fig. 9-24). The LA-250 Renegade, powered with a 250-hp engine, was introduced in 1984.

As of this writing, Lake airplanes are no longer being produced.

Make: Lake
Model: LA-4
Year: 1958–1971
Engine
    Make: Lycoming
    Model: O-360-A1A
    Horsepower: 180
    TBO: 2000 hours
Speeds
    Maximum: 135 mph
    Cruise: 131 mph
    Stall: 51 mph
Transitions (add 40 percent on water)
    Takeoff over 50-foot obstacle: 1275 feet
    Ground run: 650 feet
    Landing over 50-foot obstacle: 900 feet
    Ground roll: 475 feet
Weights
    Gross: 2400 pounds
    Empty: 1550 pounds
Dimensions
    Length: 24 feet, 11 inches
    Height: 9 feet, 4 inches
    Span: 38 feet

Other
    Fuel capacity: 40 gallons
    Rate of climb: 800 fpm
Seats: Four

Make: Lake
Model: LA-4 200/200EP Buccaneer
Year: 1970–1990
Engine
    Make: Lycoming
    Model: IO-360-A1B
    Horsepower: 200
    TBO: 1800 hours
Speeds
    Maximum: 154 mph
    Cruise: 150 mph
    Stall: 54 mph
Transitions (add 40 percent on water)
    Takeoff over 50-foot obstacle: 1100 feet
    Ground run: 600 feet
    Landing over 50-foot obstacle: 900 feet
    Ground roll: 475 feet
Weights
    Gross: 2690 pounds
    Empty: 1555 pounds

Fig. 9-24. *Lake Amphibian LA-4 200.*

Dimensions
    Length: 24 feet, 11 inches
    Height: 9 feet, 4 inches
    Span: 38 feet
Other
    Fuel capacity: 40 gallons
    Rate of climb: 1200 fpm
Seats: Four

Make: Lake
Model: LA-250
Year: 1984–1998
Engine
    Lycoming
    Model: IO-540-C4B5
    Horsepower: 250
    TBO: 2000 hours
Speeds
    Maximum: NA
    Cruise: 140 mph
    Stall: 61 mph
Transitions (add 40 percent on water)
    Takeoff over 50-foot obstacle: NA
    Ground run: 980 feet
    Landing over 50-foot obstacle: NA
    Ground roll: NA
Weights
    Gross: 3050 pounds
    Empty: 1850 pounds
Dimensions
    Length: 28 feet, 1 inch
    Height: 10 feet
    Span: 38 feet
Other
    Fuel capacity: 85 gallons
    Rate of climb: 900 fpm
Seats: Six

MEYERS

The Meyers 200 is a sleek, low-wing, very fast airplane. The cabin is smaller than the older Beech Bonanzas but not too cramped for the average family. It was first produced with a 260-hp engine, and models built after 1964 have a 285-hp engine. Production ceased in 1967.

Speed is the 200's forte, but load is not. Typically, a 200 can pass nearly anything in its class, but it cannot fly with full fuel, baggage, and four passengers.

It is interesting to note that no ADs have been issued against the Meyers 200 airframe, although a few have been issued against the engines used in them.

The manufacture of Meyers airplanes may be resurrected in the future by the New Meyers Aircraft Corporation of Fort Pierce, FL. Look for these airplanes to be listed under Meyers, Aero Commander, or Rockwell (Fig. 9-25).

Fig. 9-25. *Meyers 200.*

Make: Meyers
Model: 200 A/B/C
Year: 1959–1964
Engine
    Make: Continental
    Model: IO-470-D
    Horsepower: 260
    TBO: 1500 hours
Speeds
    Maximum: 216 mph
    Cruise: 195 mph
    Stall: 62 mph
Transitions
    Takeoff over 50-foot obstacle: 1260 feet
    Ground run: 1010 feet
    Landing over 50-foot obstacle: 1150 feet
    Ground roll: 850 feet
Weights
    Gross: 3000 pounds
    Empty: 1975 pounds

Dimensions
    Length: 24 feet, 4 inches
    Height: 8 feet, 6 inches
    Span: 30 feet, 5 inches
Other
    Fuel capacity: 40 gallons
    Rate of climb: 1245 fpm
Seats: Four

Make: Meyers
Model: 200 D
Year: 1965–1967
Engine
    Make: Continental
    Model: IO-520 A
    Horsepower: 285
    TBO: 1700 hours
Speeds
    Maximum: 215 mph
    Cruise: 210 mph
    Stall: 64 mph
Transitions
    Takeoff over 50-foot obstacle: 1150 feet
    Ground run: 900 feet
    Landing over 50-foot obstacle: 1150 feet
    Ground roll: 850 feet
Weights
    Gross: 3000 pounds
    Empty: 1990 pounds
Dimensions
    Length: 24 feet, 4 inches
    Height: 8 feet, 6 inches
    Span: 30 feet, 5 inches
Other
    Fuel capacity: 40 gallons
    Rate of climb: 1450 fpm
Seats: Four

MOONEY

Mooney airplanes are best known for their ability to extract maximum performance from the available horsepower with a minimum of fuel consumption. All, except a few very early versions, have retractable landing gear. The first Mooney retractables used a Johnson bar to manually retract and extend the gear; later versions were equipped with electrically operated landing gear.

Mooney airplanes have undergone many power and name changes over the years, often leading to considerable confusion:

M-20: 1955–1957, 150 hp (laminated wood wing and tail surfaces)

M-20A: 1958–1960, 180 hp

M-20B/C Mark 21 (Ranger): 1958–1978, 180 hp, all metal

M-20D Master: 1963–1965, 180 hp

M-20E Chaparral (Super 21): 1964–1975, 200 hp

M-20F Executive 21: 1967–1977, 200-hp (10 inches longer than the M-20E)

M-20G Mark 21 (Statesman): 1968–1970, 180 hp, all metal

M-20J Mooney 201: 1977–1991, 200 hp (sloped windshield)

M-20J Mooney MSE: 1990– , 200 hp (sloped windshield)

M-20K Mooney 231: 1979–1985, 210 hp (turbocharged)

M-20K Mooney 252 TSE: 1986–1990

M-20L Mooney PFM: 1988–1989, 217 hp (powered by Porsche engine)

M-20M Mooney TLS: 1989–

M-20R Mooney Ovation: 1994–

M-22 Mooney Mustang: 1967–1970 (pressurized five-place)

A real plus for Mooney airplanes is factory support. The company is still in the business of making airplanes, an enviable position in general aviation today.

Today's glass panel top-of-the-line Mooney M-20M airplane sells for $460,000 (Figs. 9-26 through 9-32).

Make: Mooney
Model: M-20
Year: 1955–1957
Engine
    Make: Lycoming
    Model: O-320
    Horsepower: 150
    TBO: 2000 hours
Speeds
    Maximum: 171 mph
    Cruise: 165 mph
    Stall: 57 mph
Transitions
    Takeoff over 50-foot obstacle: 1150 feet
    Ground run: 850 feet
    Landing over 50-foot obstacle: 1100 feet
    Ground roll: 600 feet

Fig. 9-26. *Mooney M-20.*

Weights
    Gross: 2450 pounds
    Empty: 1415 pounds
Dimensions
    Length: 23 feet, 1 inch
    Height: 8 feet, 3 inches
    Span: 35 feet
Other
    Fuel capacity: 35 gallons
    Rate of climb: 900 fpm
Seats: Four

Make: Mooney
Model: M-20 A/B/C/D/G
Year: 1958–1978
Engine
    Make: Lycoming
    Model: O-360
    Horsepower: 180
    TBO: 2000 hours
Speeds
    Maximum: 185 mph
    Cruise: 180 mph
    Stall: 57 mph
Transitions
    Takeoff over 50-foot obstacle: 1525 feet
    Ground run: 890 feet
    Landing over 50-foot obstacle: 1365 feet
    Ground roll: 550 feet

Fig. 9-27. *Mooney M-20B.*

Weights
    Gross: 2575 pounds
    Empty: 1525 pounds
Dimensions
    Length: 23 feet, 2 inches
    Height: 8 feet, 4 inches
    Span: 35 feet
Other
    Fuel capacity: 52 gallons
    Rate of climb: 1010 fpm
Seats: Four

Make: Mooney
Model: M-20 E/F
Year: 1965–1977
Engine
    Make: Lycoming
    Model: IO-360-A1A
    Horsepower: 200
    TBO: 1800 hours
Speeds
    Maximum: 190 mph
    Cruise: 184 mph
    Stall: 57 mph
Transitions
    Takeoff over 50-foot obstacle: 1550 feet
    Ground run: 760 feet
    Landing over 50-foot obstacle: 1550 feet
    Ground roll: 595 feet
Weights
    Gross: 2575 pounds
    Empty: 1600 pounds

Fig. 9-28. *Mooney M-20E.*

Fig. 9-29. *Mooney M-20E Chaparral.*

Dimensions
    Length: 23 feet, 2 inches
    Height: 8 feet, 4 inches
    Span: 35 feet
Other
    Fuel capacity: 52 gallons
    Rate of climb: 1125 fpm
Seats: Four

Make: Mooney
Model: M-20J 201
Year: 1977–1989

Courtesy of Mooney

Fig. 9-30. *Mooney 201/231.*

Engine
    Make: Lycoming
    Model: IO-360-A3B6D
    Horsepower: 200
    TBO: 1800 hours
Speeds
    Maximum: 202 mph
    Cruise: 195 mph
    Stall: 63 mph
Transitions
    Takeoff over 50-foot obstacle: 1517 feet
    Ground run: 850 feet
    Landing over 50-foot obstacle: 1610 feet
    Ground roll: 770 feet
Weights
    Gross: 2740 pounds
    Empty: 1640 pounds
Dimensions
    Length: 24 feet, 8 inches
    Height: 8 feet, 4 inches
    Span: 36 feet, 1 inch

Other
    Fuel capacity: 64 gallons
    Rate of climb: 1030 fpm
Seats: Four

Make: Mooney
Model: M-20J MSE
Year: 1990–1998
Engine
    Make: Lycoming
    Model: IO-360-A3B6D
    Horsepower: 200
    TBO: 2000 hours
Speeds
    Maximum: 205 mph
    Cruise: 197 mph
    Stall: 62 mph
Transitions
    Takeoff over 50-foot obstacle: 1700 feet
    Ground run: 900 feet
    Landing over 50-foot obstacle: 1600 feet
    Ground roll: 677 feet
Weights
    Gross: 2740 pounds
    Empty: 1710 pounds
Dimensions
    Length: 24 feet, 8 inches
    Height: 8 feet, 4 inches
    Span: 36 feet, 1 inch
Other
    Fuel capacity: 64 gallons
    Rate of climb: 1060 fpm
Seats: Four

Make: Mooney
Model: M-20K 231
Year: 1979–1985
Engine
    Make: Continental
    Model: TSIO-360-GBA
    Horsepower: 210
    TBO: 1800 hours
Speeds
    Maximum: 231 mph
    Cruise: 220 mph
    Stall: 66 mph

Transitions
    Takeoff over 50-foot obstacle: 2060 feet
    Ground run: 1220 feet
    Landing over 50-foot obstacle: 2280 feet
    Ground roll: 1147 feet
Weights
    Gross: 2900 pounds
    Empty: 1800 pounds
Dimensions
    Length: 25 feet, 5 inches
    Height: 8 feet, 4 inches
    Span: 36 feet, 1 inch
Other
    Fuel capacity: 75 gallons
    Rate of climb: 1080 fpm
Seats: Four

Make: Mooney
Model: M-20K 252
Year: 1986–1990
Engine
    Make: Continental
    Model: TSIO-360-MB1
    Horsepower: 210
    TBO: 1800 hours
Speeds
    Maximum: 251 mph
    Cruise: 231 mph
    Stall: 68 mph
Transitions
    Takeoff over 50-foot obstacle: 2200 feet
    Ground run: 1250 feet
    Landing over 50-foot obstacle: 2300 feet
    Ground roll: 1140 feet
Weights
    Gross: 2900 pounds
    Empty: 1800 pounds
Dimensions
    Length: 25 feet, 5 inches
    Height: 8 feet, 4 inches
    Span: 36 feet, 1 inch
Other
    Fuel capacity: 75 gallons
    Rate of climb: 1080 fpm
Seats: Four

Make: Mooney
Model: PFM
Year: 1988–1989
Engine
    Make: Porsche
    Model: PFM3200-NO3
    Horsepower: 217
    TBO: 1800 hours
Speeds
    Maximum: 185 mph
    Cruise: 178 mph
    Stall: 65 mph
Transitions
    Takeoff over 50-foot obstacle: 2160 feet
    Ground run: 1220 feet
    Landing over 50-foot obstacle: 2280 feet
    Ground roll: 1147 feet
Weights
    Gross: 2900 pounds
    Empty: 1800 pounds
Dimensions
    Length: 26 feet, 11 inches
    Height: 8 feet, 4 inches
    Span: 36 feet, 1 inch
Other
    Fuel capacity: 61 gallons
    Rate of climb: 1030 fpm
Seats: Four

Make: Mooney
Model: M-20M TLS
Year: 1989–1998
Engine
    Make: Lycoming
    Model: TIO-540-AF1A (B after 1997)
    Horsepower: 270
    TBO: 2000 hours
Speeds
    Maximum: 271 mph
    Cruise: 256 mph
    Stall: 68 mph
Transitions
    Takeoff over 50-foot obstacle: 2050 feet
    Ground run: 1080 feet
    Landing over 50-foot obstacle: 2600 feet

Ground roll: 1200 feet
Weights
    Gross: 3368 pounds
    Empty: 2353 pounds
Dimensions
    Length: 26 feet, 11 inches
    Height: 8 feet, 4 inches
    Span: 36 feet, 1 inch
Other
    Fuel capacity: 96 gallons
    Rate of climb: 1230 fpm
Seats: Four

Fig. 9-31. *Mooney M-20R Ovation.*

Make: Mooney
Model: M-20R Ovation
Year: 1994–
Engine
    Make: Continental
    Model: IO-550-G
    Horsepower: 280
    TBO: 2000 hours
Speeds
    Maximum: 227 mph
    Cruise: 220 mph
    Stall: 69 mph

Transitions
Takeoff over 50-foot obstacle: 1700 feet
Ground run: 900 feet
Landing over 50-foot obstacle: 1600 feet
Ground roll: 1000 feet
Weights
Gross: 3368 pounds
Empty: 2200 pounds
Dimensions
Length: 26 feet, 9 inches
Height: 8 feet, 4 inches
Span: 36 feet, 1 inch
Other
Fuel capacity: 95 gallons
Rate of climb: 1200 fpm
Seats: Four

Make: Mooney
Model: Mark 22 Mustang
Year: 1967–1970
Engine
Make: Lycoming
Model: TSIO-540-A1A
Horsepower: 310
TBO: 1300 hours
Speeds
Maximum: 250 mph
Cruise: 230 mph
Stall: 69 mph

Courtesy of Mooney

Fig. 9-32. *Mooney Mark 22 Mustang.*

Transitions
    Takeoff over 50-foot obstacle: 2079 feet
    Ground run: 1142 feet
    Landing over 50-foot obstacle: 1549 feet
    Ground roll: 958 feet
Weights
    Gross: 3680 pounds
    Empty: 2380 pounds
Dimensions
    Length: 26 feet, 1 inch
    Height: 9 feet, 1 inch
    Span: 35 feet
Other
    Fuel capacity: 92 gallons
    Rate of climb: 1120 fpm
Seats: Five

NAVION

Of all the complex airplanes, there is no other like the Navion. It was placed into production by North American Aviation in 1946, just before the Beechcraft Bonanza came out. Sadly, though, the Navion has bounced around from one manufacturer to another and has been in and out of production until 1976, never enjoying the popularity of the Bonanza.

The Navion is a strong and capable short-field plane; unimproved strips don't seem to bother it. Navions were built with 185-, 205-, and 225-hp engines. Rangemasters, a later model, were available with a 260- or 285-hp engine. The following versions were built:

A: 205-hp Continental E-185-3

B: 260-hp Lycoming GO-435-C2

D: 240-hp Continental O-470-P

E/F: 260-hp Continental IO-470-C

Servicing the smaller Continental engines is difficult, as they are no longer supported by their manufacturer. Few all-original Navions exist today, as most have been modified and updated. A wise purchaser will contact the American Navion Society (see App. D) and request their pamphlet about buying a used Navion before making a choice; they can also answer questions about modifications and maintenance difficulties. Around 2400 Navions were built (Figs. 9-33 and 9-34).

Make: Navion
Model: Navion/Navion A/B
Year: 1946–1951

Fig. 9-33. *Navion.*

Engine
    Make: Continental
    Model: E-185-3/E-185-9
    Horsepower: 185/295
    TBO: 1500 hours
Speeds
    Maximum: 163 mph
    Cruise: 148 mph
    Stall: 60 mph
Transitions
    Takeoff over 50-foot obstacle: 1500 feet
    Ground run: 670 feet
    Landing over 50-foot obstacle: 1300 feet
    Ground roll: 500 feet
Weights
    Gross: 2750 pounds
    Empty: 1700 pounds
Dimensions
    Length: 27 feet, 3 inches
    Height: 8 feet, 7 inches
    Span: 33 feet, 4 inches
Other
    Fuel capacity: 40 gallons
    Rate of climb: 750 fpm
Seats: Four

Make: Navion
Model: Navion D/E/F
Year: 1958–1960

Engine
    Make: Continental
    Model: O-470/IO-470
    Horsepower: 240/260
    TBO: 1500 hours
Speeds
    Maximum: 184 mph
    Cruise: 179 mph
    Stall: 58 mph
Transitions
    Takeoff over 50-foot obstacle: 980 feet
    Ground run: 785 feet
    Landing over 50-foot obstacle: 980 feet
    Ground roll: 425 feet
Weights
    Gross: 3315 pounds
    Empty: 1925 pounds
Dimensions
    Length: 27 feet, 3 inches
    Height: 8 feet, 7 inches
    Span: 34 feet, 5 inches
Other
    Fuel capacity: 40 gallons (108 optional)
    Rate of climb: 1150 fpm
Seats: Four

Make: Navion
Model: Rangemaster G/H
Year: 1961–1976
Engine
    Make: Continental
    Model: IO-520-B
    Horsepower: 285
    TBO: 1700 hours
Speeds
    Maximum: 203 mph
    Cruise: 191 mph
    Stall: 55 mph
Transitions
    Takeoff over 50-foot obstacle: 950 feet
    Ground run: 740 feet
    Landing over 50-foot obstacle: 980 feet
    Ground roll: 760 feet
Weights
    Gross: 3315 pounds

Fig. 9-34. *Navion Rangermaster.*

Empty: 2000 pounds
Dimensions
    Length: 27 feet, 3 inches
    Height: 8 feet, 7 inches
    Span: 33 feet, 4 inches
Other
    Fuel capacity: 40 gallons
    Rate of climb: 1375 fpm
Seats: Four

PIPER

Piper entered the complex airplane market in 1958 with the PA-24 Comanche. It was an all-metal, low-wing craft that became popular as a market trend-setter. Built with several different engines, production ceased in 1972 when Piper's Lock Haven, Pennsylvania manufacturing facility was flooded during a hurricane.

Piper's PA-28-235 is a rugged and honest airplane, yet easy to fly and quite inexpensive to maintain. The 235 will carry a load equivalent to its weight and, in later versions, do it in style. The 235 was replaced by the 236 in 1979, re-named the Dakota, and is a meld of the Warrior wing and the Archer fuselage.

The success of the PA-28 series no doubt moved Piper, as it had Cessna, to install retractable landing gear and aim toward a different market. Thus, in 1967 the PA-28R Arrow series arose as an extension to the Cherokee 180. The new Arrow had the same fuel burn as the Cherokee, yet cruised about 25 mph faster. They were shown to be efficient and, as simplicity was desired in these new retractables, even the responsibility of controlling the landing gear was removed from the pilot (automatic landing gear). Of course, the basic airplane saw different configurations and names:

1969: 200-hp Arrow 200

1973: Arrow II-200 (five-place)

1977: Arrow III-201 (turbo version is T201)

1979: Arrow IV with T-tail

1990: PA-28R-201 with conventional tail returns

In 1979, the PA-32R-300 Lance was introduced as a retractable version of the popular PA-32 Cherokee Six (see Chap. 10). The Lance first appeared with a conventional tail and, after the first half of 1978, with a T-tail. Many pilots and owners complained about the T-tail and its lack of authority at low speeds. In 1980, the conventional tail returned and the PA-32R-300 was renamed the Saratoga.

The latest complex entry in the piston-powered single-engine airplane class is the PA-46 Malibu and Malibu Mirage series. The early Malibu series was plagued with numerous ADs involving airframe failure, some of which were rescinded later. In 1989 the Mirage model, with a larger engine, was introduced and remains in current production. A cabin class single, the price for a new Malibu Mirage in 1999 exceeded $500,000.

The New Piper Aircraft Company, a reorganization of Piper Aircraft, continues to produce the PA-28R-201, PA-32-301, and PA-46 series of airplanes. It is common to see these airplanes listed under New Piper as well as Piper.

Although not as popular as other manufacturers' airplanes, Piper's complex airplanes offer fair values for the dollar. An excellent selection of them can be found on today's market (Figs. 9-35 through 9-46).

Make: Piper
Model: PA-24-180
Year: 1958–1964
Engine
    Make: Lycoming
    Model: O-360-A1A
    Horsepower: 180
    TBO: 2000 hours

Fig. 9-35. *Piper PA-24 Comanche.*

Speeds
  Maximum: 167 mph
  Cruise: 150 mph
  Stall: 61 mph
Transitions
  Takeoff over 50-foot obstacle: 2240 feet
  Ground run: 750 feet
  Landing over 50-foot obstacle: 1025 feet
  Ground roll: 600 feet
Weights
  Gross: 2550 pounds
  Empty: 1475 pounds
Dimensions
  Length: 24 feet, 9 inches
  Height: 7 feet, 5 inches
  Span: 36 feet
Other
  Fuel capacity: 60 gallons
  Rate of climb: 910 fpm
Seats: Four

Make: Piper
Model: PA-24-250
Year: 1958–1964
Engine
  Make: Lycoming
  Model: O-540-A1A5
  Horsepower: 250
  TBO: 2000 hours
Speeds
  Maximum: 190 mph
  Cruise: 181 mph
  Stall: 61 mph
Transitions
  Takeoff over 50-foot obstacle: 1650 feet
  Ground run: 750 feet
  Landing over 50-foot obstacle: 1025 feet
  Ground roll: 650 feet
Weights
  Gross: 2800 pounds
  Empty: 1690 pounds
Dimensions
  Length: 24 feet, 9 inches
  Height: 7 feet, 5 inches
  Span: 36 feet

Other
    Fuel capacity: 60 gallons
    Rate of climb: 1350 fpm
Seats: Four

Make: Piper
Model: PA-24-260
Year: 1965–1972
Engine
    Make: Lycoming
    Model: O-540 (IO-540 after 1969)
    Horsepower: 260
    TBO: 2000
Speeds
    Maximum: 195 mph
    Cruise: 185 mph
    Stall: 61 mph
Transitions
    Takeoff over 50-foot obstacle: 1400 feet
    Ground run: 820 feet
    Landing over 50-foot obstacle: 1200 feet
    Ground roll: 690 feet
Weights
    Gross: 3200 pounds
    Empty: 1773 pounds
Dimensions
    Length: 25 feet, 8 inches
    Height: 7 feet, 3 inches
    Span: 36 feet
Other
    Fuel capacity: 60 gallons
    Rate of climb: 1320 fpm
Seats: Four

Make: Piper
Model: PA-24-400
Year: 1964–1965
Engine
    Make: Lycoming
    Model: IO-720-A1A
    Horsepower: 400
    TBO: 1800
Speeds
    Maximum: 194 mph
    Cruise: 185 mph
    Stall: 59 mph

Transitions
    Takeoff over 50-foot obstacle: 1500 feet
    Ground run: 980 feet
    Landing over 50-foot obstacle: 1820 feet
    Ground roll: 1180 feet
Weights
    Gross: 3600 pounds
    Empty: 2110 pounds
Dimensions
    Length: 25 feet, 8 inches
    Height: 7 feet, 3 inches
    Span: 36 feet
Other
    Fuel capacity: 100 gallons
    Rate of climb: 1600 fpm
Seats: Four

Courtesy of Piper Aviation Museum Foundation

Fig. 9-36. *Piper PA-28-235 Cherokee.*

Make: Piper
Model: PA-28-235 Cherokee/Charger/Pathfinder
Year: 1964–1977
Engine
    Make: Lycoming
    Model: O-540-B4B5
    Horsepower: 235
    TBO: 2000 hours
Speeds
    Maximum: 166 mph
    Cruise: 156 mph
    Stall: 60 mph
Transitions
    Takeoff over 50-foot obstacle: 1360 feet
    Ground run: 800 feet

Landing over 50-foot obstacle: 1300 feet
Ground roll: 680 feet
Weights
　Gross: 2900 pounds
　Empty: 1410 pounds
Dimensions
　Length: 23 feet, 6 inches
　Height: 7 feet, 1 inch
　Span: 32 feet
Other
　Fuel capacity: 84 gallons
　Rate of climb: 825 fpm
Seats: Four

Make: Piper
Model: PA-28-236 Dakota
Year: 1979–1994
Engine
　Make: Lycoming
　Model: O-540-J3A5D
　Horsepower: 235
　TBO: 2000 hours
Speeds
　Maximum: 170 mph
　Cruise: 159 mph
　Stall: 64 mph
Transitions
　Takeoff over 50-foot obstacle: 1210 feet
　Ground run: 885 feet
　Landing over 50-foot obstacle: 1725 feet
　Ground roll: 825 feet

Courtesy of Piper

Fig. 9-37. *Piper PA-28-236 Dakota.*

Weights
    Gross: 3000 pounds
    Empty: 1634 pounds
Dimensions
    Length: 24 feet, 8 inches
    Height: 7 feet, 2 inches
    Span: 35 feet
Other
    Fuel capacity: 72 gallons
    Rate of climb: 1110 fpm
Seats: Four

Make: Piper
Model: PA-28R-180 Arrow
Year: 1967–1971
Engine
    Make: Lycoming
    Model: IO-360-B1E
    Horsepower: 180
    TBO: 2000 hours
Speeds
    Maximum: 170 mph
    Cruise: 162 mph
    Stall: 61 mph
Transitions
    Takeoff over 50-foot obstacle: 1240 feet
    Ground run: 820 feet
    Landing over 50-foot obstacle: 1340 feet
    Ground roll: 770 feet
Weights
    Gross: 2500 pounds
    Empty: 1380 pounds
Dimensions
    Length: 24 feet, 2 inches
    Height: 8 feet
    Span: 30 feet
Other
    Fuel capacity: 50 gallons
    Rate of climb: 875 fpm
Seats: Four

Make: Piper
Model: PA-28R-200
Year: 1969–1978
Engine
    Make: Lycoming

Fig. 9-38. *Piper PA-28R Arrow 200.*

Model: IO-360-C1C6
Horsepower: 200
TBO: 1800 hours
Speeds
Maximum: 176 mph
Cruise: 162 mph
Stall: 63 mph
Transitions
Takeoff over 50-foot obstacle: 1580 feet
Ground run: 780 feet
Landing over 50-foot obstacle: 1350 feet
Ground roll: 760 feet
Weights
Gross: 2750 pounds
Empty: 1601 pounds
Dimensions
Length: 24 feet, 2 inches
Height: 8 feet
Span: 30 feet
Other
Fuel capacity: 50 gallons
Rate of climb: 831 fpm
Seats: Four

Make: Piper
Model: PA-28R-200 Turbo Arrow
Year: 1977–1978
Engine
Make: Continental
Model: TSIO-360-F
Horsepower: 200
TBO: 1400 hours

Fig. 9-39. *Piper PA-28R Turbo Arrow IV.*

Speeds
    Maximum: 198 mph
    Cruise: 172 mph
    Stall: 63 mph
Transitions
    Takeoff over 50-foot obstacle: 1620 feet
    Ground run: 1120 feet
    Landing over 50-foot obstacle: 1555 feet
    Ground roll: 645 feet
Weights
    Gross: 2900 pounds
    Empty: 1638 pounds
Dimensions
    Length: 24 feet, 2 inches
    Height: 8 feet
    Span: 35 feet, 4 inches
Other
    Fuel capacity: 50 gallons
    Rate of climb: 940 fpm
Seats: Four

Make: Piper
Model: PA-28R-201 Arrow IV
Year: 1989–
Engine
    Make: Lycoming
    Model: IO-360-C1C6
    Horsepower: 200
    TBO: 2000 hours
Speeds
    Maximum: 167 mph
    Cruise: 158 mph
    Stall: 63 mph

Transitions
　　Takeoff over 50-foot obstacle: 1600 feet
　　Ground run: NA
　　Landing over 50-foot obstacle: 1520 feet
　　Ground roll: NA
Weights
　　Gross: 2750 pounds
　　Empty: 1783 pounds
Dimensions
　　Length: 24 feet, 8 inches
　　Height: 7 feet, 11 inches
　　Span: 35 feet, 5 inches
Other
　　Fuel capacity: 72 gallons
　　Rate of climb: NA
Seats: Four

Make: Piper
Model: PA-28RT-201T Turbo Arrow
Year: 1979–1990
Engine
　　Make: Continental
　　Model: TSIO-360-FB
　　Horsepower: 200
　　TBO: 1400 hours
Speeds
　　Maximum: 171 mph
　　Cruise: 158 mph
　　Stall: 63 mph

Courtesy of New Piper

Fig. 9-40. *1999 New Piper Arrow.*

Transitions
    Takeoff over 50-foot obstacle: 1600 feet
    Ground run: 1025 feet
    Landing over 50-foot obstacle: 1525 feet
    Ground roll: 615 feet
Weights
    Gross: 2750 pounds
    Empty: 1637 pounds
Dimensions
    Length: 24 feet, 7 inches
    Height: 8 feet
    Span: 35 feet, 4 inches
Other
    Fuel capacity: 72 gallons
    Rate of climb: 831 fpm
Seats: Four

Make: Piper
Model: PA-32R-300 Lance
Year: 1976–1979
Engine
    Make: Lycoming
    Model: IO-540 (opt. TC)
    Horsepower: 300
    TBO: 2000 hours
Speeds
    Maximum: 180 mph
    Cruise: 176 mph
    Stall: 60 mph
Transitions
    Takeoff over 50-foot obstacle: 1660 feet

Courtesy of Piper

Fig. 9-41. *Piper PA-32R-300 Lance.*

Ground run: 960 feet
Landing over 50-foot obstacle: 1708 feet
Ground roll: 880 feet
Weights
Gross: 3600 pounds
Empty: 1980 pounds
Dimensions
Length: 27 feet, 9 inches
Height: 9 feet
Span: 32 feet, 9 inches
Other
Fuel capacity: 98 gallons
Rate of climb: 1000 fpm
Seats: Seven

Make: Piper
Model: PA-32R-301 Saratoga SP
Year: 1980–1992
Engine
Make: Lycoming
Model: IO-540-K1G5
Horsepower: 300
TBO: 2000 hours
Speeds
Maximum: 188 mph
Cruise: 182 mph
Stall: 66 mph

Fig. 9-42. *Piper PA-32R-300 Lance II.*

Transitions
    Takeoff over 50-foot obstacle: 1573 feet
    Ground run: 1013 feet
    Landing over 50-foot obstacle: 1612 feet
    Ground roll: 732 feet
Weights
    Gross: 3600 pounds
    Empty: 1999 pounds
Dimensions
    Length: 27 feet, 8 inches
    Height: 8 feet, 2 inches
    Span: 36 feet, 2 inches
Other
    Fuel capacity: 102 gallons
    Rate of climb: 1010 fpm
Seats: Seven

Courtesy of Piper Aviation Museum Foundation

Fig. 9-43. *Piper PA-32R-301 Saratoga*

Make: Piper
Model: PA-32R-301 Saratoga II HP
Year: 1993–
Engine
    Make: Lycoming
    Model: IO-540-K1G5

Horsepower: 300
TBO: 2000 hours
Speeds
  Maximum: 201 mph
  Cruise: 191 mph
  Stall: 72 mph
Transitions
  Takeoff over 50-foot obstacle: 1770 feet
  Ground run: 1200 feet
  Landing over 50-foot obstacle: 1520 feet
  Ground roll: 640 feet
Weights
  Gross: 3600 pounds
  Empty: 2403 pounds
Dimensions
  Length: 27 feet, 11 inches
  Height: 8 feet, 6 inches
  Span: 36 feet, 3 inches
Other
  Fuel capacity: 102 gallons
  Rate of climb: 1116 fpm
Seats: Seven

Make: Piper
Model: PA-32R-301T Saratoga
Year: 1980–1987
Engine
  Make: Lycoming
  Model: TIO-540-K1G5
  Horsepower: 300
  TBO: 2000 hours
Speeds
  Maximum: 224 mph
  Cruise: 203 mph
  Stall: 65 mph
Transitions
  Takeoff over 50-foot obstacle: 1420 feet
  Ground run: 960 feet
  Landing over 50-foot obstacle: 1725 feet
  Ground roll: 732 feet
Weights
  Gross: 3600 pounds
  Empty: 2078 pounds
Dimensions
  Length: 28 feet, 2 inches

Height: 8 feet, 2 inches
Span: 36 feet, 2 inches
Other
Fuel capacity: 102 gallons
Rate of climb: 1120 fpm
Seats: Seven

Make: Piper
Model: PA-32R-301T Saratoga II TC
Year: 1988–
Engine
Make: Lycoming
Model: T IO-540-AH1A
Horsepower: 300
TBO: 1800 hours
Speeds
Maximum: 215 mph
Cruise: 213 mph
Stall: 72 mph
Transitions
Takeoff over 50-foot obstacle: 2480 feet
Ground run: 1550 feet
Landing over 50-foot obstacle: 1700 feet
Ground roll: 880 feet
Weights
Gross: 3600 pounds
Empty: 2478 pounds

Courtesy of New Piper

Fig. 9-44. *1999 New Piper Saratoga II TC.*

Dimensions
    Length: 27 feet, 11 inches
    Height: 8 feet, 6 inches
    Span: 36 feet, 3 inches
Other
    Fuel capacity: 102 gallons
    Rate of climb: 1119 fpm
Seats: Seven

Make: Piper
Model: PA-46-301P Malibu
Year: 1984–1988
Engine
    Make: Continental
    Model: TSIO-520-BE
    Horsepower: 310
    TBO: 2000 hours
Speeds
    Maximum: 269 mph
    Cruise: 235 mph
    Stall: 67 mph
Transitions
    Takeoff over 50-foot obstacle: 2025 feet
    Ground run: 1440 feet
    Landing over 50-foot obstacle: 1800 feet
    Ground roll: 1070 feet
Weights
    Gross: 4100 pounds
    Empty: 2460 pounds

Fig. 9-45. *Piper PA-46 Malibu.*

Dimensions
    Length: 28 feet, 10 inches
    Height: 11 feet, 4 inches
    Span: 43 feet
Other
    Fuel capacity: 120 gallons
    Rate of climb: 1143 fpm
Seats: Six

Make: Piper
Model: PA-46-350P Malibu Mirage
Year: 1989–
Engine
    Make: Lycoming
    Model: TIO-540-AE2A
    Horsepower: 350
    TBO: 2000 hours
Speeds
    Maximum: 267 mph
    Cruise: 231 mph
    Stall: 69 mph
Transitions
    Takeoff over 50-foot obstacle: 2375 feet
    Ground run: 1530 feet
    Landing over 50-foot obstacle: 1964 feet
    Ground roll: 1018 feet
Weights
    Gross: 4300 pounds
    Empty: 2790 pounds

Courtesy of New Piper

Fig. 9-46 . *1999 New Piper Malibu Mirage.*

Dimensions
    Length: 28 feet, 6 inches
    Height: 11 feet, 5 inches
    Span: 43 feet
Other
    Fuel capacity: 120 gallons
    Rate of climb: 1218 fpm
Seats: Six

REPUBLIC

Republic Aviation was famous for its World War II P-47 Thunderbolt fighters, and after the war entered the civilian aviation marketplace with the SeaBee, a four-place amphibian. Compared to some makes and models, few SeaBees were produced: about 500. In 1948, production was halted for all time.

SeaBees are stout aircraft, built to last. Most of those currently flying have been extensively modified—a few even to twin engine—hence performance data is sketchy. Today, a good SeaBee commands a high price, far more than the approximately $4800 they sold for when new.

SeaBee production rights are currently owned by W. E. Aerotech of Gig Harbor, Washington. Aerotech makes new parts and provides servicing for these airplanes (Fig. 9-47).

Make: Republic
Model: R-3 SeaBee
Year: 1948
Engine
    Make: Franklin
    Model: 6A-215-G8F

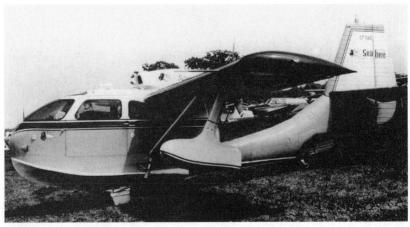

Fig. 9-47. *Republic SeaBee.*

Horsepower: 215
TBO: 1200
Speeds
    Maximum: 125 mph
    Cruise: 117 mph
    Stall: 58 mph
Transitions
    Takeoff over 50-foot obstacle: NA
    Ground run: NA
    Landing over 50-foot obstacle: NA
    Ground roll: NA
Weights
    Gross: 3150 pounds
    Empty: 1850 pounds
Dimensions
    Length: NA
    Height: NA
    Span: NA
Other
    Fuel capacity: 75 gallons
    Rate of climb: 700 fpm
Seats: Four

SOCATA

Socata is the light-aviation subsidiary of the French government-owned Aerospatiale Group. The Socata name is actually an abbreviation for "Société de Construction d'Avions de Tourisme et d'Affaires" (touring and business airplane manufacturing company).

The series of airplanes sold in the United States all carry the Socata name and are often called "Caribbean planes" because of their model names: Trinidad, Tobago, and Tampico. The TB-9 Tampico and TB-10 Tobago are easy fliers and are reviewed in Chap. 8.

The higher-power versions share a common fuselage, wings, and empennages with the lower-power members of the Caribbean series. The high-power entries to the series are designed to be comfortable and are meant for serious travel.

Although the line of airplanes were first introduced in Europe in the 1970s, they were not certified by the FAA until the 1980s (Fig. 9-48).

Make: Socata
Model: TB-20 Trinidad
Year: 1984–
Engine
    Make: Lycoming

Model: IO-540-C4D5D
Horsepower: 250
TBO: 2000 hours
Speeds
    Maximum: 192 mph
    Cruise: 184 mph
    Stall: 62 mph
Transitions
    Takeoff over 50-foot obstacle: 1953 feet
    Ground run: 1193 feet
    Landing over 50-foot obstacle: 1750 feet
    Ground roll: 755 feet
Weights
    Gross: 3083 pounds
    Empty: 1744 pounds
Dimensions
    Length: 25 feet, 4 inches
    Height: 9 feet, 4 inches
    Span: 32 feet, 6 inches
Other
    Fuel capacity: 86 gallons
    Rate of climb: 1260 fpm
Seats: Four

Make: Socata
Model: TB-21TC Trinidad
Year: 1986–
Engine
    Make: Lycoming
    Model: TIO-540-AB1AD
    Horsepower: 250
    TBO: 2000 hours

Courtesy of Socata

Fig. 9-48. *Socata TB-21.*

Speeds
    Maximum: 230 mph
    Cruise: 215 mph
    Stall: 62 mph
Transitions
    Takeoff over 50-foot obstacle: 1953 feet
    Ground run: 1193 feet
    Landing over 50-foot obstacle: 1750 feet
    Ground roll: 755 feet
Weights
    Gross: 3083 pounds
    Empty: 1795 pounds
Dimensions
    Length: 25 feet, 4 inches
    Height: 9 feet, 4 inches
    Span: 32 feet, 6 inches
Other
    Fuel capacity: 86 gallons
    Rate of climb: 1125 fpm
Seats: Four

10

Heavy Haulers

THE TERM *HEAVY HAULER* refers to the capabilities of the airplanes in this chapter, including carrying heavy loads—often exceeding that which they were designed to carry. The cargo ranges from people to livestock to all the assorted possible hardware and goods in between.

These airplanes will be found doing their everyday work in the outer reaches of the continental 48 states on ranches, reservations, or perhaps with the highway patrol. Alaska, Central America, Africa, and the Australian outback are also home to these airplanes.

Heavy haulers are built beefy, with ample power and large lifting capabilities. Most are high-wing planes, for clearance reasons as well as easier loading, and can be found on wheels, skis, or floats.

CESSNA

Cessna's 180 has become a legend; it's used everywhere that strength, reliability, and performance are required. From 1953 to 1981, when production ceased, the 180 remained essentially unchanged: equipped with a Continental 0-470 engine and conventional landing gear.

The 185 Skywagon, also with conventional gear, was introduced in 1961 equipped with a 260-hp engine and was upgraded in 1966 to a 300-hp engine. An optional under-fuselage cargo carrier is available that will hold up to 300 pounds.

The 1963 Cessna 205 was introduced as a fixed-gear version of the 210. The 205 is basically a tricycle-geared counterpart to the 185, and can also be fitted with the under-fuselage cargo carrier. The 205 was replaced in 1965 by the U206 Super Skywagon. The 206 has large double doors on the right side of the cabin to allow loading of awkward cargo items. It is powered by a 285-hp engine.

The P206 Super Skylane is a fancy version of the U206 and is designed to carry passengers. In 1969, the seven-place 207 Skywagon was introduced as an outgrowth of the 206 series.

A late Cessna entry to the heavy-hauler market is the Caravan, a turboprop, single-engine plane designed for short hauls of light cargo. It is the largest and most costly of all the heavy haulers, selling new for over $1 million.

Cessna has built an excellent reputation for work airplanes. They are seen all over the world, representing economy and dependability (Figs. 10-1 through 10-7).

Fig. 10-1. *Cessna 1953 180.*

Make: Cessna
Model: 180
Year: 1953–1981
Engine
    Make: Continental
    Model: O-470-K
    Horsepower: 230
    TBO: 1500 hours
Speeds
    Maximum: 170 mph
    Cruise: 162 mph
    Stall: 58 mph
Transitions
    Takeoff over 50-foot obstacle: 1225 feet
    Ground run: 625 feet
    Landing over 50-foot obstacle: 1365 feet
    Ground roll: 480 feet
Weights
    Gross: 2800 pounds
    Empty: 1545 pounds

Dimensions
  Length: 25 feet, 9 inches
  Height: 7 feet, 9 inches
  Span: 35 feet, 10 inches
Other
  Fuel capacity: 55 gallons
  Rate of climb: 1090 fpm
Seats: Four (six after 1963)

Make: Cessna
Model: 185 Skywagon (260)
Year: 1961–1966
Engine
  Make: Continental
  Model: IO-470-F
  Horsepower: 260
  TBO: 1500 hours
Speeds
  Maximum: 176 mph
  Cruise: 167 mph
  Stall: 62 mph
Transitions
  Takeoff over 50-foot obstacle: 1510 feet
  Ground run: 650 feet
  Landing over 50-foot obstacle: 1265 feet
  Ground roll: 610 feet
Weights
  Gross: 3200 pounds
  Empty: 1520 pounds
Dimensions
  Length: 25 feet, 9 inches
  Height: 7 feet, 9 inches
  Span: 35 feet, 10 inches
Other
  Fuel capacity: 65 gallons
  Rate of climb: 1000 fpm
Seats: Six

Make: Cessna
Model: 185 Skywagon (300)
Year: 1966–1985
Engine
  Make: Continental
  Model: IO-520-D
  Horsepower: 300
  TBO: 1700 hours

Fig. 10-2. *Cessna 1983 185 Skywagon.*

Speeds
 Maximum: 178 mph
 Cruise: 169 mph
 Stall: 59 mph
Transitions
 Takeoff over 50-foot obstacle: 1365 feet
 Ground run: 770 feet
 Landing over 50-foot obstacle: 1400 feet
 Ground roll: 480 feet
Weights
 Gross: 3350 pounds
 Empty: 1585 pounds
Dimensions
 Length: 25 feet, 9 inches
 Height: 7 feet, 9 inches
 Span: 35 feet, 10 inches
Other
 Fuel capacity: 65 gallons
 Rate of climb: 1010 fpm
Seats: Six

Make: Cessna
Model: 205
Year: 1963–1964
Engine
 Make: Continental
 Model: IO-470-S
 Horsepower: 260
 TBO: 1500 hours
Speeds
 Maximum: 167 mph
 Cruise: 159 mph
 Stall: 57 mph

Fig. 10-3. *Cessna 1963 205.*

Transitions
    Takeoff over 50-foot obstacle: 1465 feet
    Ground run: 685 feet
    Landing over 50-foot obstacle: 1510 feet
    Ground roll: 625 feet
Weights
    Gross: 3300 pounds
    Empty: 1750 pounds
Dimensions
    Length: 27 feet, 3 inches
    Height: 9 feet, 7 inches
    Span: 36 feet, 6 inches
Other
    Fuel capacity: 65 gallons
    Rate of climb: 965 fpm
Seats: Six

Make: Cessna
Model: U206/P206 Super Skylane/Skywagon
Year: 1965–1986
Engine
    Make: Continental
    Model: IO-520 (turbocharging optional)
    Horsepower: 285
    TBO: 1700 hours (1400 if turbocharged)
Speeds
    Maximum: 174 mph
    Cruise: 164 mph
    Stall: 61 mph

Fig. 10-4. *Cessna 1967 206 Super Skylane.*

Transitions
    Takeoff over 50-foot obstacle: 1265 feet
    Ground run: 675 feet
    Landing over 50-foot obstacle: 1340 feet
    Ground roll: 735 feet
Weights
    Gross: 3600 pounds
    Empty: 1750 pounds
Dimensions
    Length: 28 feet
    Height: 9 feet, 7 inches
    Span: 35 feet, 10 inches
Other
    Fuel capacity: 65 gallons
    Rate of climb: 920 fpm
Seats: Six

Make: Cessna
Model: U206/P206 Super Skylane/Skywagon
Year: 1967–1986
Engine
    Make: Continental
    Model: IO-520
    Horsepower: 300
    TBO: 1700 hours
Speeds
    Maximum: 179 mph
    Cruise: 169 mph
    Stall: 62 mph

Transitions
    Takeoff over 50-foot obstacle: 1780 feet
    Ground run: 900 feet
    Landing over 50-foot obstacle: 1395 feet
    Ground roll: 735 feet
Weights
    Gross: 3600 pounds
    Empty: 2000 pounds
Dimensions
    Length: 28 feet, 3 inches
    Height: 9 feet, 4 inches
    Span: 35 feet, 10 inches
Other
    Fuel capacity: 65 gallons
    Rate of climb: 920 fpm
Seats: Six

Make: Cessna
Model: 206 Stationair
Year: 1998–
Engine
    Make: Textron Lycoming
    Model: IO-540-AC1A
    Horsepower: 300
    TBO: 1800 hours
Speeds
    Maximum: 173 mph
    Cruise: 164 mph
    Stall: NA

Courtesy of Cessna

Fig. 10-5 . *1999 Cessna 206 Stationair.*

Transitions
    Takeoff over 50-foot obstacle: NA
    Ground run: 900 feet
    Landing over 50-foot obstacle: NA
    Ground roll: NA
Weights
    Gross: 3612 pounds
    Empty: 2146 pounds
Dimensions
    Length: 28 feet, 3 inches
    Height: 9 feet, 4 inches
    Span: 36 feet
Other
    Fuel capacity: 88 gallons
    Rate of climb: 920 fpm
Seats: Six

Make: Cessna
Model: 206 Turbo Stationair
Year: 1998–
Engine
    Make: Textron Lycoming
    Model: TIO-540-AJ1A
    Horsepower: 310
    TBO: 1800 hours
Speeds
    Maximum: 196 mph
    Cruise: 190 mph
    Stall: NA
Transitions
    Takeoff over 50-foot obstacle: NA
    Ground run: 835 feet
    Landing over 50-foot obstacle: NA
    Ground roll: NA
Weights
    Gross: 3616 pounds
    Empty: 2227 pounds
Dimensions
    Length: 28 feet, 3 inches
    Height: 9 feet, 4 inches
    Span: 36 feet
Other
    Fuel capacity: 88 gallons
    Rate of climb: 1010 fpm
Seats: Six

Fig. 10-6. *Cessna 1982 207 Stationair.*

Make: Cessna
Model: 207 Skywagon/Stationair
Year: 1969–1984
Engine
    Make: Continental
    Model: IO-520 (turbocharging optional)
    Horsepower: 300
    TBO: 1700 (1400 if turbocharged)
Speeds
    Maximum: 168 mph
    Cruise: 159 mph
    Stall: 67 mph
Transitions
    Takeoff over 50-foot obstacle: 1970 feet
    Ground run: 1100 feet
    Landing over 50-foot obstacle: 1500 feet
    Ground roll: 765 feet
Weights
    Gross: 3800 pounds
    Empty: 1890 pounds
Dimensions
    Length: 31 feet, 9 inches
    Height: 9 feet, 7 inches
    Span: 35 feet, 10 inches
Other
    Fuel capacity: 61 gallons
    Rate of climb: 810 fpm
Seats: Seven

Make: Cessna
Model: 208 Caravan
Year: 1985–1991

Fig. 10-7. *Cessna Caravan I.*

Engine
    Make: Pratt & Whitney
    Model: PT6A-114
    Horsepower: 600 (shaft)
    TBO: 3500
Speeds
    Maximum: 210 mph
    Cruise: 200 mph
    Stall: 69 mph
Transitions
    Takeoff over 50-foot obstacle: 1665 feet
    Ground run: 970 feet
    Landing over 50-foot obstacle: 1550 feet
    Ground roll: 645 feet
Weights
    Gross: 7300 pounds
    Empty: 3800 pounds
Dimensions
    Length: 37 feet, 7 inches
    Height: 14 feet, 2 inches
    Span: 51 feet, 8 inches
Other
    Fuel capacity: 332 gallons
    Rate of climb: 1215 fpm
Seats: Fourteen

HELIO

The Helio Courier is a STOL (short takeoff and landing) airplane designed to fly and remain fully controllable at speeds far slower than most other airplanes. To accomplish this feat, the wings have automatic Handley Page leading-edge slats and electrically operated slotted flaps. Frise ailerons work in conjunction with the arc type spoilers, which project upward from the wing.

Although a stall speed is listed in the specifications for Helios, the effect is that of mushing down in a parachute. The flaps extend nearly 75 percent of the length of the wings. A very few have tricycle landing gear, some from the factory and others via an STC.

Several different Helio models were produced, but most were built for special purposes and government operations, and cannot usually be found on the civilian market. Popular civilian model numbers and years of production are:

H-395: 1957–1965, 295-hp engine

H-295: 1965–1974, 295-hp engine (300-lb increase in useful load)

The Helio is a specialized airplane not often seen at the local airport. Parts and service can be a problem because there is no manufacture support. It is an orphan (Fig. 10-8).

Fig. 10-8. *Helio Courier.*

Make: Helio
Model: H-395/295 Super Courier
Year: 1957–1974
Engine
    Make: Lycoming
    Model: GO-480
    Horsepower: 295
    TBO: 1400 hours
Speeds
    Maximum: 167 mph
    Cruise: 162 mph
    Stall: 30 mph (min. fully controllable speed)

Transitions
Takeoff over 50-foot obstacle: 610 feet
Ground run: 335 feet
Landing over 50-foot obstacle: 520 feet
Ground roll: 270 feet
Weights
Gross: 3400 pounds
Empty: 2080 pounds
Dimensions
Length: 31 feet
Height: 8 feet, 10 inches
Span: 39 feet
Other
Fuel capacity: 60 gallons
Rate of climb: 1250 fpm
Seats: Six

MAULE

Maules are ruggedly constructed, simple airplanes specifically designed for short-field operation. With all-metal wings and a fiberglass-covered tubular fuselage, they have few moving or complex parts and can therefore be economically maintained.

Production started in 1962 with the M-4 (see Chap. 8), which was eventually built under several model names, each designating a different engine rating:

- 145-hp M-4
- 180-hp Astro Rocket
- 210-hp Rocket
- 220-hp Strata Rocket

In 1974, the M-5 Lunar Rocket series replaced the M-4 with a larger tail surface, a 30 percent increase in flap area, and a more powerful engine, making them capable of carrying larger loads almost anywhere.

The M-6, with longer wings and a 235-hp engine, was built from 1983 to 1985. Performance of the M-6 is remarkable; takeoff ground runs seem almost nonexistent and approaches are incredibly slow. Maule's most recent entry is the M-7, which is powered by a 180-hp engine, with an optional 235-hp or 260-hp engine available. Versions of the 160-hp and 180-hp M-7 are described in Chap.8.

Maule aircraft are known for having somewhat optimistic speed figures in their specifications, but they are hard-working airplanes, can be affordably maintained, and are usable nearly anywhere (Figs. 10-9 through 10-12).

Make: Maule
Model: M-4/180 Astro Rocket
Year: 1970–1971
Engine
    Make: Franklin
    Model: 6A-355-B1
    Horsepower: 180
    TBO: 2000 hours
Speeds
    Maximum: 170 mph
    Cruise: 155 mph
    Stall: 40 mph
Transitions
    Takeoff over 50-foot obstacle: 700 feet
    Ground run: 500 feet
    Landing over 50-foot obstacle: 600 feet
    Ground roll: 450 feet
Weights
    Gross: 2300 pounds
    Empty: 1250 pounds
Dimensions
    Length: 22 feet
    Height: 6 feet, 2 inches
    Span: 30 feet, 10 inches
Other
    Fuel capacity: 42 gallons
    Rate of climb: 1000 fpm
Seats: Four

Make: Maule
Model: M-4/210 Rocket
Year: 1965–1973
Engine
    Make: Continental
    Model: IO-360-A
    Horsepower: 210
    TBO: 1500 hours
Speeds
    Maximum: 180 mph
    Cruise: 165 mph
    Stall: 40 mph
Transitions
    Takeoff over 50-foot obstacle: 650 feet
    Ground run: 430 feet
    Landing over 50-foot obstacle: 600 feet

Ground roll: 390 feet
Weights
    Gross: 2300 pounds
    Empty: 1120 pounds
Dimensions
    Length: 22 feet
    Height: 6 feet, 2 inches
    Span: 30 feet, 10 inches
Other
    Fuel capacity: 42 gallons
    Rate of climb: 1250 fpm
Seats: Four

Fig. 10-9. *Maule M-4/220.*

Make: Maule
Model: M-4/220 Strata Rocket
Year: 1967–1973
Engine
    Make: Franklin
    Model: 6A-350-C1
    Horsepower: 220
    TBO: 1500 hours
Speeds
    Maximum: 180 mph
    Cruise: 165 mph
    Stall: 40 mph
Transitions
    Takeoff over 50-foot obstacle: 600 feet
    Ground run: 400 feet
    Landing over 50-foot obstacle: 600 feet
    Ground roll: 390 feet

Courtesy of Maule

Fig. 10-10. *Maule M-5/220C.*

Weights
    Gross: 2300 pounds
    Empty: 1220 pounds
Dimensions
    Length: 22 feet
    Height: 6 feet, 2 inches
    Span: 30 feet, 10 inches
Other
    Fuel capacity: 42 gallons
    Rate of climb: 1250 fpm
Seats: Four

Make: Maule
Model: M-5/180C
Year: 1979–1988
Engine
    Make: Continental
    Model: O-360-C1F
    Horsepower: 180
    TBO: 2000 hours
Speeds
    Maximum: NA
    Cruise: 156 mph
    Stall: 38 mph
Transitions
    Takeoff over 50-foot obstacle: 800 feet
    Ground run: 200 feet
    Landing over 50-foot obstacle: 600 feet
    Ground roll: NA
Weights
    Gross: 2300 pounds
    Empty: 1300 pounds

Dimensions
    Length: 22 feet, 9 inches
    Height: 6 feet, 4 inches
    Span: 30 feet, 10 inches
Other
    Fuel capacity: 40 gallons
    Rate of climb: 900 fpm
Seats: Four

Make: Maule
Model: M-5/210 Lunar Rocket
Year: 1974–1977
Engine
    Make: Continental
    Model: IO-360-D
    Horsepower: 210
    TBO: 1500 hours
Speeds
    Maximum: 180 mph
    Cruise: 158 mph
    Stall: 38 mph
Transitions
    Takeoff over 50-foot obstacle: 600 feet
    Ground run: 400 feet
    Landing over 50-foot obstacle: 600 feet
    Ground roll: 400 feet
Weights
    Gross: 2300 pounds
    Empty: 1350 pounds
Dimensions
    Length: 22 feet, 9 inches
    Height: 6 feet, 4 inches
    Span: 30 feet, 10 inches
Other
    Fuel capacity: 40 gallons
    Rate of climb: 1250 fpm
Seats: Four

Make: Maule
Model: M-5/235
Year: 1977–1988
Engine
    Make: Lycoming
    Model: O-540-J1A5D
    Horsepower: 235
    TBO: 2000 hours

Speeds
    Maximum: 185 mph
    Cruise: 172 mph
    Stall: 38 mph
Transitions
    Takeoff over 50-foot obstacle: 600 feet
    Ground run: 400 feet
    Landing over 50-foot obstacle: 600 feet
    Ground roll: 400 feet
Weights
    Gross: 2300 pounds
    Empty: 1400 pounds
Dimensions
    Length: 22 feet, 9 inches
    Height: 6 feet, 4 inches
    Span: 30 feet, 10 inches
Other
    Fuel capacity: 40 gallons
    Rate of climb: 1350 fpm
Seats: Four

Make: Maule
Model: M-6/235 Super Rocket
Year: 1981–1991
Engine
    Make: Lycoming
    Model: IO-540-J1A5D
    Horsepower: 235
    TBO: 2000 hours

Courtesy of Maule

Fig. 10-11. *Maule M-6/235C.*

Speeds
    Maximum: 180 mph
    Cruise: 148 mph
    Stall: 44 mph
Transitions
    Takeoff over 50-foot obstacle: 540 feet
    Ground run: 150 feet
    Landing over 50-foot obstacle: 440 feet
    Ground roll: 250 feet
Weights
    Gross: 2500 pounds
    Empty: 1450 pounds
Dimensions
    Length: 23 feet, 6 inches
    Height: 6 feet, 4 inches
    Span: 33 feet, 2 inches
Other
    Fuel capacity: 40 gallons
    Rate of climb: 1900 fpm
Seats: Four

Fig. 10-12. *Maule M-7/235.*

Make: Maule
Model: M-7/235 and MT-7/235
Year: 1985–
Engine
    Make: Lycoming
    Model: IO-540-W1A5D or O-540-J1A5D
    Horsepower: 235
    TBO: 2000 hours
Speeds
    Maximum: 180 mph
    Cruise: 160 mph
    Stall: 35 mph

Transitions
    Takeoff over 50-foot obstacle: 600 feet
    Ground run: 125 feet
    Landing over 50-foot obstacle: 500 feet
    Ground roll: 275 feet
Weights
    Gross: 2500 pounds
    Empty: 1500 pounds
Dimensions
    Length: 23 feet, 6 inches
    Height: 6 feet, 4 inches
    Span: 33 feet, 8 inches
Other
    Fuel capacity: 70 gallons
    Rate of climb: 2000 fpm
Seats: Four

Make: Maule
Model: MX-7/260
Year: 1998–
Engine
    Make: Lycoming
    Model: IO-540-V4A5
    Horsepower: 260
    TBO: 2000 hours
Speeds
    Maximum: 180 mph
    Cruise: 167 mph
    Stall: 35 mph
Transitions
    Takeoff over 50-foot obstacle: NA
    Ground run: NA
    Landing over 50-foot obstacle: NA
    Ground roll: NA
Weights
    Gross: 2500 pounds
    Empty: NA
Dimensions
    Length: 23 feet, 6 inches
    Height: 6 feet, 4 inches
    Span: 32 feet, 10 inches
Other
    Fuel capacity: 73 gallons
    Rate of climb: 1500 fpm (depending upon propeller installed)
Seats: Four

PIPER

Piper has only two airplanes fitting the heavy-hauler category and both defy the "high-wings only" rule for heavy haulers.

First is the low-winged Cherokee Six, which was manufactured from 1965 to 1979. Removable seats allow this craft to carry cargo, a stretcher, or livestock. A 260-hp engine was standard and a 300-hp engine was available as an option. The second Piper example in this class is the PA-32-301 Saratoga, which replaced the Cherokee Six in 1980. It was built with a normally aspirated engine, although a turbocharged engine was optionally available.

Both heavy-hauler Pipers are easy to fly, will operate reliably, and require little maintenance (Figs. 10-13 and 10-14).

In 1995, Piper reorganized and became The New Piper Aircraft, Inc. It is common to see these airplanes listed under New Piper as well as Piper.

Make: Piper
Model: PA-32-260 Cherokee Six
Year: 1965–1979
Engine
    Make: Lycoming
    Model: O-540-E4B5
    Horsepower: 260
    TBO: 2000 hours (1200 without modifications)
Speeds
    Maximum: 168 mph
    Cruise: 160 mph
    Stall: 63 mph
Transitions
    Takeoff over 50-foot obstacle: 1360 feet
    Ground run: 810 feet
    Landing over 50-foot obstacle: 1000 feet
    Ground roll: 630 feet
Weights
    Gross: 3400 pounds
    Empty: 1699 pounds
Dimensions
    Length: 27 feet, 9 inches
    Height: 7 feet, 11 inches
    Span: 32 feet, 9 inches
Other
    Fuel capacity: 50 gallons
    Rate of climb: 760 fpm
Seats: Seven

Fig. 10-13. *Piper PA-32-300 Cherokee Six.*

Make: Piper
Model: PA-32-300 Cherokee Six
Year: 1966–1979
Engine
    Make: Lycoming
    Model: O-540-K1A5B
    Horsepower: 300
    TBO: 2000 hours
Speeds
    Maximum: 174 mph
    Cruise: 168 mph
    Stall: 63 mph
Transitions
    Takeoff over 50-foot obstacle: 1140 feet
    Ground run: 700 feet
    Landing over 50-foot obstacle: 1000 feet
    Ground roll: 630 feet
Weights
    Gross: 3400 pounds
    Empty: 1846 pounds
Dimensions
    Length: 27 feet, 9 inches
    Height: 7 feet, 11 inches
    Span: 32 feet, 9 inches
Other
    Fuel capacity: 50 gallons
    Rate of climb: 1050 fpm
Seats: Seven

Make: Piper
Model: PA-32-301 Saratoga
Year: 1980–1990

Courtesy of Piper

Fig. 10-14. *Piper PA-32-301 Saratoga.*

Engine
    Make: Lycoming
    Model: IO-540-K1G5D (turbocharging is optional)
    Horsepower: 300
    TBO: 2000 hours
Speeds
    Maximum: 175 mph
    Cruise: 172 mph
    Stall: 67 mph
Transitions
    Takeoff over 50-foot obstacle: 1759 feet
    Ground run: 1183 feet
    Landing over 50-foot obstacle: 1612 feet
    Ground roll: 732 feet
Weights
    Gross: 3600 pounds
    Empty: 1940 pounds
Dimensions
    Length: 27 feet, 8 inches
    Height: 8 feet, 2 inches
    Span: 36 feet, 2 inches
Other
    Fuel capacity: 102 gallons
    Rate of climb: 990 fpm
Seats: Seven

11

Affordable Twins

SMALL TWIN-ENGINE AIRPLANES OFFER ADVANTAGES in safety, speed, and comfort over all the airplanes generally available for private ownership. Twins are well respected, are the ultimate in IFR operations, and make positive impressions in the flying community (at least that's what is inferred by manufacturers, sellers, and owners). They do make a statement and they offer no small amount of snob appeal.

Flying a twin-engine airplane requires pilot expertise and an FAA multi-engine rating to attest to that expertise. Sadly, many pilots obtain the rating and fly twins, yet don't maintain the high levels of proficiency required for emergency procedures. This lulls the pilot into a sense of false security until the eventual engine-out and the panic that follows.

A twin-engine airplane is always expensive to own and operate. When purchasing a used twin, keep in mind that they are very complex airplanes with retractable landing gear, constant-speed propellers, complicated fuel systems, and more. Of course there are two of just about everything that could ever require maintenance, which adds up to high maintenance costs.

AERO COMMANDER

Aero Commander twin-engine airplanes have a high wing, giving them a unique appearance and making them the easiest of all small twins to board. They make excellent air taxis and have seen service all over the world as the mainstay for many small airlines.

Attesting to the abilities of Aero Commander twins, a model 500 series once flew single-engine from Oklahoma to Washington, D.C., nonstop, and at maximum gross weight, taking off and landing on one engine. The inoperative engine's propeller was carried as baggage, inside the airplane.

Aero Commanders can be found in used airplane listings under Aero Commander, Gulfstream, or Rockwell (Fig. 11-1).

Fig. 11-1. *Aero Commander 500.*

Make: Aero Commander
Model: 500
Year: 1958–1959
Engines
    Make: Lycoming
    Model: O-540-A2B
    Horsepower: 250
    TBO: 2000 hours
Speeds
    Maximum: 218 mph
    Cruise: 205 mph
    Stall: 63 mph
Transitions
    Takeoff over 50-foot obstacle: 1250 feet
    Ground run: 1000 feet
    Landing over 50-foot obstacle: 1350 feet
    Ground roll: 950 feet
Weights
    Gross: 6000 pounds
    Empty: 3850 pounds
Dimensions
    Length: 35 feet, 1 inch
    Height: 14 feet, 5 inches
    Span: 49 feet
Other
    Fuel capacity: 156 gallons
    Rate of climb: 1400 fpm (290 single)
Seats: Seven

Make: Aero Commander
Model: 500A
Year: 1960–1963
Engines
    Make: Continental
    Model: IO-470-M
    Horsepower: 260
    TBO: 1500 hours
Speeds
    Maximum: 228 mph
    Cruise: 218 mph
    Stall: 62 mph
Transitions
    Takeoff over 50-foot obstacle: 1210 feet
    Ground run: 970 feet
    Landing over 50-foot obstacle: 1150 feet
    Ground roll: 865 feet
Weights
    Gross: 6250 pounds
    Empty: 4255 pounds
Dimensions
    Length: 35 feet, 1 inch
    Height: 14 feet, 5 inches
    Span: 49 feet, 5 inches
Other
    Fuel capacity: 156 gallons
    Rate of climb: 1400 fpm (320 single)
Seats: Seven

BEECHCRAFT/RAYTHEON

The oldest Beechcraft twin normally found on the used market is the Beech 18. It has radial engines, which make the "right sounds," and is a small airliner that for many years was flown by regional air carriers and corporations. Unfortunately, they now are old and very expensive to maintain.

The Twin Bonanza model 50 was built for nearly 10 years. The first production units were powered by 260-hp engines, and later versions used 295- and 340-hp engines. The T-bone, as the model 50 is often called, was not produced after 1962.

In 1958, the Travelair 95 was introduced. It was powered by two 180-hp engines with seats for four, and seating was increased to six on later models. Approximately 700 Travelairs were built before the end of production in 1968.

The first Beech Baron 55s were seen in 1961, seated six, and were powered by 260- to 340-hp engines. More than 5700 were built.

The last Beechcraft entry into the small-twin field was the Duchess model 76, a T-tailed craft with counter-rotating engines. Counter-rotating propellers take some danger out of engine-out operations by eliminating most of the critical-engine factor.

Beechcraft airplanes are tough and strong, and the roomiest of the small twins (Figs. 11-2 through 11-7). The product support is good but expensive.

Make: Beechcraft
Model: D-18S
Year: 1946–1969
Engines
    Make: Pratt & Whitney
    Model: R-985
    Horsepower: 450
    TBO: 1600 hours
Speeds
    Maximum: 230 mph
    Cruise: 214 mph
    Stall: 84 mph
Transitions
    Takeoff over 50-foot obstacle: 1980 feet
    Ground run: 1445 feet
    Landing over 50-foot obstacle: 1850 feet
    Ground roll: 1036 feet
Weights
    Gross: 9700 pounds
    Empty: 5910 pounds
Dimensions
    Length: 35 feet, 2 inches
    Height: 10 feet, 5 inches
    Span: 49 feet, 8 inches

Courtesy of Beechcraft

Fig. 11-2. *Beechcraft Model 18.*

Other
    Fuel capacity: 198 gallons
    Rate of climb: 1410 fpm (255 single)
Seats: Seven to nine (and crew)

Make: Beechcraft
Model: 50 and B/C-50
Year: 1952–1962
Engines
    Make: Lycoming
    Model: GO-435
    Horsepower: 260 (275 on C model)
    TBO: 1200 hours
Speeds
    Maximum: 203 mph
    Cruise: 183 mph
    Stall: 69 mph
Transitions
    Takeoff over 50-foot obstacle: 1344 feet
    Ground run: 1080 feet
    Landing over 50-foot obstacle: 1215 feet
    Ground roll: 975 feet
Weights
    Gross: 6000 pounds
    Empty: 3940 pounds
Dimensions
    Length: 31 feet, 5 inches
    Height: 11 feet, 5 inches
    Span: 45 feet, 2 inches
Other
    Fuel capacity: 134 gallons
    Rate of climb: 1450 fpm (300 single)
Seats: Six

Make: Beechcraft
Model: D-50
Year: 1956–1961
Engines
    Make: Lycoming
    Model: GO-480
    Horsepower: 295
    TBO: 1400 hours
Speeds
    Maximum: 214 mph
    Cruise: 203 mph
    Stall: 71 mph

Fig. 11-3. *Beechcraft D-50 Twin Bonanza.*

Transitions
    Takeoff over 50-foot obstacle: 1260 feet
    Ground run: 1000 feet
    Landing over 50-foot obstacle: 1455 feet
    Ground roll: 1010 feet
Weights
    Gross: 6300 pounds
    Empty: 4100 pounds
Dimensions
    Length: 31 feet, 5 inches
    Height: 11 feet, 5 inches
    Span: 45 feet, 9 inches
Other
    Fuel capacity: 134 gallons
    Rate of climb: 1450 fpm (300 single)
Seats: Seven

Make: Beechcraft
Model: E/F-50
Year: 1957–1958
Engines
    Make: Lycoming
    Model: GSO-480
    Horsepower: 340
    TBO: 1400 hours
Speeds
    Maximum: 240 mph
    Cruise: 212 mph
    Stall: 83 mph

Transitions
    Takeoff over 50-foot obstacle: 1250 feet
    Ground run: 975 feet
    Landing over 50-foot obstacle: 1840 feet
    Ground roll: 1250 feet
Weights
    Gross: 7000 pounds
    Empty: 4460 pounds
Dimensions
    Length: 31 feet, 5 inches
    Height: 11 feet, 5 inches
    Span: 45 feet, 9 inches
Other
    Fuel capacity: 180 gallons
    Rate of climb: 1320 fpm (325 single)
Seats: Seven

Fig. 11-4. *Beechcraft 95 Travelair.*

Make: Beechcraft
Model: 95 Travelair
Year: 1958–1968
Engines
    Make: Lycoming
    Model: O-360-A1A (fuel inj. after 1960)
    Horsepower: 180
    TBO: 2000 hours
Speeds
    Maximum: 210 mph
    Cruise: 200 mph
    Stall: 70 mph

Transitions
    Takeoff over 50-foot obstacle: 1280 feet
    Ground run: 1000 feet
    Landing over 50-foot obstacle: 1850 feet
    Ground roll: 1015 feet
Weights
    Gross: 4200 pounds
    Empty: 2635 pounds
Dimensions
    Length: 25 feet, 3 inches
    Height: 9 feet, 6 inches
    Span: 37 feet, 9 inches
Other
    Fuel capacity: 80 gallons
    Rate of climb: 1250 fpm (205 single)
Seats: Four (six after 1960)

Fig. 11-5. *Beechcraft B55 Baron.*

Make: Beechcraft
Model: A/B 55 Baron
Year: 1961–1982
Engines
    Make: Continental
    Model: IO-470-L
    Horsepower: 260
    TBO: 1500 hours
Speeds
    Maximum: 236 mph
    Cruise: 225 mph
    Stall: 78 mph

Transitions
    Takeoff over 50-foot obstacle: 1664 feet
    Ground run: 1339 feet
    Landing over 50-foot obstacle: 1853 feet
    Ground roll: 945 feet
Weights
    Gross: 5100 pounds
    Empty: 3070 pounds
Dimensions
    Length: 27 feet
    Height: 9 feet, 7 inches
    Span: 37 feet, 10 inches
Other
    Fuel capacity: 112 gallons
    Rate of climb: 1670 fpm (320 single)
Seats: Six

Make: Beechcraft
Model: C/D/E 55 Baron
Year: 1961–1982
Engines
    Make: Continental
    Model: IO-520-CB
    Horsepower: 285
    TBO: 1700 hours
Speeds
    Maximum: 239 mph
    Cruise: 224 mph
    Stall: 83 mph
Transitions
    Takeoff over 50-foot obstacle: 2050 feet
    Ground run: 1315 feet
    Landing over 50-foot obstacle: 2202 feet
    Ground roll: 1237 feet
Weights
    Gross: 5300 pounds
    Empty: 3291 pounds
Dimensions
    Length: 29 feet
    Height: 9 feet, 2 inches
    Span: 37 feet, 10 inches
Other
    Fuel capacity: 100 gallons
    Rate of climb: 1682 fpm (388 single)
Seats: Six

Make: Beechcraft
Model: 56 TC Baron
Year: 1967–1970
Engines
    Make: Lycoming
    Model: TIO-541-E1B4
    Horsepower: 380
    TBO: 1600 hours
Speeds
    Maximum: 290 mph
    Cruise: 284 mph
    Stall: 84 mph
Transitions
    Takeoff over 50-foot obstacle: 1420 feet
    Ground run: 1005 feet
    Landing over 50-foot obstacle: 2080 feet
    Ground roll: 1285 feet
Weights
    Gross: 5990 pounds
    Empty: 3650 pounds
Dimensions
    Length: 28 feet, 3 inches
    Height: 9 feet, 7 inches
    Span: 37 feet, 10 inches
Other
    Fuel capacity: 142 gallons
    Rate of climb: 2020 fpm
    Seats: Six

Make: Beechcraft
Model: 58 Baron
Year: 1970–1983
Engines
    Make: Continental
    Model: IO-520-C
    Horsepower: 285
    TBO: 1700 hours
Speeds
    Maximum: 240 mph
    Cruise: 224 mph
    Stall: 85 mph
Transitions
    Takeoff over 50-foot obstacle: 2101 feet
    Ground run: 1336 feet

Courtesy of Raytheon

Fig. 11-6. *Raytheon Beech 58 Baron.*

 Landing over 50-foot obstacle: 2498 feet
 Ground roll: 1439 feet
Weights
 Gross: 5400 pounds
 Empty: 3361 pounds
Dimensions
 Length: 29 feet, 10 inches
 Height: 9 feet, 6 inches
 Span: 37 feet, 10 inches
Other
 Fuel capacity: 136 gallons
 Rate of climb: 1660 fpm
 Seats: Six

Make: Beechcraft/Raytheon
Model: 58 Baron
Year: 1984–
Engines
 Make: Teledyne Continental
 Model: IO-550-C
 Horsepower: 300
 TBO: 1700 hours
Speeds
 Maximum: 240 mph
 Cruise: 224 mph
 Stall: 85 mph

Transitions
Takeoff over 50-foot obstacle: 2371 feet
Ground run: 1403 feet
Landing over 50-foot obstacle: 2498 feet
Ground roll: 1439 feet
Weights
Gross: 5500 pounds
Empty: 3443 pounds
Dimensions
Length: 29 feet, 10 inches
Height: 9 feet, 6 inches
Span: 37 feet, 10 inches
Other
Fuel capacity: 136 gallons
Rate of climb: 1750 fpm
Seats: Six

Make: Beechcraft
Model: 76 Duchess
Year: 1978–1982
Engines
Make: Lycoming
Model: O-360-A1G6D
Horsepower: 180
TBO: 2000 hp
Speeds
Maximum: 197 mph
Cruise: 182 mph
Stall: 69 mph

Fig. 11-7. *Beechcraft 76 Duchess.*

Transitions
    Takeoff over 50-foot obstacle: 2119 feet
    Ground run: 1017 feet
    Landing over 50-foot obstacle: 1880 feet
    Ground roll: 1000 feet
Weights
    Gross: 3900 pounds
    Empty: 2460 pounds
Dimensions
    Length: 29 feet
    Height: 9 feet, 6 inches
    Span: 38 feet
Other
    Fuel capacity: 100 gallons
    Rate of climb: 1248 fpm (235 single)
Seats: Four

CESSNA

Modern Cessna light twins have always been marked by their graceful lines, speed, and reliability. The two conventional Cessna twin models are the models 310 and sister ship 320. The 310 was introduced in 1954 and was powered by 240-hp engines. After that date, numerous changes were made as model years climbed:

1956: Additional window space

1959: 260-hp engines

1960: Swept tail

1963: Six seats

1969: 285-hp engines

The model 320, with turbocharged engines (260 hp each), entered production in 1962. In 1966, the engine size was increased to 285 hp and a sixth seat was added.

The second type of twin from Cessna is the model 336. It has centerline thrust, with one engine in front of the cabin and one in the rear. It was built to be simple to fly, with very easy engine-out procedures; it even had fixed landing gear. Cessna thought easy handling would make it a great seller, but it didn't. The next year, 1965, Cessna introduced the 337 (based on the 336) with improved engine cooling, quieter operation, and retractable landing gear.

The 337s are officially called the Skymaster, while in most hangars they are referred to as *mixmasters*. The 337s saw heavy use with military forces in Vietnam as the O-2.

Fig. 11-8. *Cessna 310, early.*

Cessna twins are plentiful and, as twins go, maintenance is reasonable. The high-wing 337s make excellent workhorses, while the 310 and 320 series make long trips quickly (Figs. 11-8 through 11-12).

Make: Cessna
Model: 310 and 310 A-B
Year: 1955–1958
Engines
    Make: Continental
    Model: O-470-B
    Horsepower: 240
    TBO: 1500 hours
Speeds
    Maximum: 232 mph
    Cruise: 213 mph
    Stall: 74 mph
Transitions
    Takeoff over 50-foot obstacle: 1410 feet
    Ground run: 810 feet
    Landing over 50-foot obstacle: 1720 feet
    Ground roll: 620 feet
Weights
    Gross: 4700 pounds
    Empty: 2900 pounds
Dimensions
    Length: 26 feet
    Height: 10 feet, 6 inches
    Span: 35 feet, 9 inches

Other
    Fuel capacity: 102 gallons
    Rate of climb: 1660 fpm (450 single)
Seats: Five

Make: Cessna
Model: 310 C-Q
Year: 1959–1974
Engines
    Make: Continental
    Model: IO-470
    Horsepower: 260
    TBO: 1500 hours
Speeds
    Maximum: 240 mph
    Cruise: 223 mph
    Stall: 76 mph
Transitions
    Takeoff over 50-foot obstacle: 1545 feet
    Ground run: 920 feet
    Landing over 50-foot obstacle: 1900 feet
    Ground roll: 690 feet
Weights
    Gross: 5100 pounds
    Empty: 3063 pounds
Dimensions
    Length: 29 feet, 5 inches
    Height: 9 feet, 11 inches
    Span: 36 feet, 11 inches
Other
    Fuel capacity: 102 gallons
    Rate of climb: 1690 fpm (380 single)
Seats: Six

Fig. 11-9. *Cessna 310, late.*

Make: Cessna
Model: 310 P-R
Year: 1969–1981
Engines
    Make: Continental
    Model: IO-520-M (turbocharger optional)
    Horsepower: 285
    TBO: 1700 hours
Speeds
    Maximum: 238 mph
    Cruise: 223 mph
    Stall: 81 mph
Transitions
    Takeoff over 50-foot obstacle: 1700 feet
    Ground run: 1335 feet
    Landing over 50-foot obstacle: 1790 feet
    Ground roll: 640
Weights
    Gross: 5500 pounds
    Empty: 3603 pounds
Dimensions
    Length: 31 feet, 11 inches
    Height: 10 feet, 8 inches
    Span: 36 feet, 11 inches
Other
    Fuel capacity: 102 gallons
    Rate of climb: 1662 fpm (370 single)
Seats: Six

Make: Cessna
Model: 320 (through C)
Year: 1962–1965
Engines
    Make: Continental
    Model: TSIO-470
    Horsepower: 260
    TBO: 1400 hours
Speeds
    Maximum: 265 mph
    Cruise: 235 mph
    Stall: 78 mph
Transitions
    Takeoff over 50-foot obstacle: 1890 feet
    Ground run: 870 feet

Landing over 50-foot obstacle: 2056 feet
Ground roll: 640 feet
Weights
Gross: 4990 pounds
Empty: 3190 pounds
Dimensions
Length: 29 feet, 5 inches
Height: 10 feet, 3 inches
Span: 36 feet, 9 inches
Other
Fuel capacity: 102 gallons
Rate of climb: 1820 fpm (400 single)
Seats: Six

Make: Cessna
Model: 320 D-F
Year: 1966–1968
Engines
Make: Continental
Model: TSIO-520
Horsepower: 285
TBO: 1400 hours
Speeds
Maximum: 275 mph
Cruise: 260 mph
Stall: 74 mph
Transitions
Takeoff over 50-foot obstacle: 1515 feet
Ground run: 1190 feet

Courtesy of Cessna

Fig. 11-10. *Cessna 320.*

Landing over 50-foot obstacle: 1736 feet
Ground roll: 614 feet
Weights
    Gross: 5300 pounds
    Empty: 3273 pounds
Dimensions
    Length: 29 feet, 5 inches
    Height: 10 feet, 3 inches
    Span: 36 feet, 9 inches
Other
    Fuel capacity: 102 gallons
    Rate of climb: 1924 fpm (475 single)
Seats: Six

Make: Cessna
Model: 336
Year: 1964
Engines
    Make: Continental
    Model: IO-360-C
    Horsepower: 210
    TBO: 1500 hours
Speeds
    Maximum: 183 mph
    Cruise: 173 mph
    Stall: 60 mph
Transitions
    Takeoff over 50-foot obstacle: 1145 feet
    Ground run: 625 feet
    Landing over 50-foot obstacle: 1395 feet
    Ground roll: 655 feet
Weights
    Gross: 3900 pounds
    Empty: 2340 pounds

Fig. 11-11. *Cessna 336.*

Dimensions
    Length: 29 feet, 7 inches
    Height: 9 feet, 4 inches
    Span: 38 feet
Other
    Fuel capacity: 93 gallons
    Rate of climb: 1340 fpm (355 single)
Seats: Four (six optional)

Fig. 11-12. *Cessna 337.*

Make: Cessna
Model: 337 (all)
Year: 1965–1980
Engines
    Make: Continental
    Model: IO-360-C
    Horsepower: 210
    TBO: 1500 hours
Speeds
    Maximum: 199 mph
    Cruise: 190 mph
    Stall: 70 mph
Transitions
    Takeoff over 50-foot obstacle: 1675 feet
    Ground run: 1000 feet
    Landing over 50-foot obstacle: 1650 feet
    Ground roll: 700 feet
Weights
    Gross: 4360 pounds
    Empty: 2695 pounds

Dimensions
    Length: 29 feet, 9 inches
    Height: 9 feet, 4 inches
    Span: 38 feet, 2 inches
Other
    Fuel capacity: 93 gallons
    Rate of climb: 1100 fpm (235 single)
Seats: Six

GULFSTREAM

The Gulfstream Cougar GA-7 was introduced in 1978 and went out of production the next year, as did its single-engine cousins (see Chapters 7 and 8). Although somewhat underpowered by most twin standards, with a pair of 160-hp Lycomings, the Cougar carried four with reasonable comfort and speed for a period in excess of five hours between fuel stops.

A total of only 115 Cougars were produced, so they are somewhat scarce on the market. The Cougar can be considered an orphan, meaning you can have a problem with parts and service. On the other hand, the Cougar represents a lot of airplane for the money and is generally priced well below older Piper Twin Comanches (Fig. 11-13).

Make: Gulfstream
Model: GA-7 Cougar
Year: 1978–1979
Engines
    Make: Lycoming
    Model: O-320-D1D
    Horsepower: 160
    TBO: 2000 hours
Speeds
    Maximum: 193 mph

Fig. 11-13. *Gulfstream GA-7.*

Cruise: 184 mph
Stall: 82 mph
Transitions
  Takeoff over 50-foot obstacle: 1850 feet
  Ground run: 1000 feet
  Landing over 50-foot obstacle: 1330 feet
  Ground roll: 710 feet
Weights
  Gross: 3800 pounds
  Empty: 2569 pounds
Dimensions
  Length: 29 feet, 7 inches
  Height: 10 feet, 4 inches
  Span: 36 feet, 9 inches
Other
  Fuel capacity: 118 gallons
  Rate of climb: 1160 fpm (200 single)
Seats: Four

PIPER

Piper entered the light twin market in 1954 with the PA-23 Apache. The original Apaches have reputations as unusually poor performers with ungainly looks. They are sometimes called *the flying potato*. Such comments might have some merit, but the small-engine versions are economical to operate. When flying with only one engine, the early Apache's pilot had to look for a place to land because the low power of the remaining engine did not allow the plane to climb well.

Several major changes, offered in the following years, improved the Apache's performance problems and increased the seating:

1958: 160-hp engines

1960: Five seats

1963: 235-hp engines

Apaches can be acquired cheap and, while the low-powered models have poor performance, they do represent a lot of plane for the money. The last year of production was 1963.

Piper introduced the Aztec in 1960 as a PA-23 (same model number as the Apache). It was a sleeker-looking airplane than the Apache and had a swept tail. Originally powered by 250-hp engines and seating five, it was soon updated to seat six. The Aztec is an excellent instrument plane and some are even equipped with de-icing equipment.

The Twin Comanche PA-30 was introduced in 1963. It was similar in appearance to the Cessna 310, was powered by 160-hp engines, and sat low to the ground. Unlike the 310, the Twin Comanche is quite docile in flight. It became the PA-39 in 1970 with the addition of counter-rotating propellers. Twin Comanche production ceased in 1972.

The Seneca was introduced in 1970 as the PA-34. Basically, the Seneca is a Cherokee Six with two engines.

The PA-44 Seminole was built from 1979 to 1982 as an entry-level twin, similar to the Beech Duchess.

In 1995, Piper reorganized and became The New Piper Aircraft, Inc. It is common to see these airplanes listed under New Piper as well as Piper.

Piper twins can appear to be a lot of airplane for the money, but the older models require continuing and expensive maintenance. The new models are somewhat more wallet-friendly (Figs. 11-14 through 11-19).

Make: Piper
Model: PA-23 Apache
Year: 1954–1957
Engines
    Make: Lycoming
    Model: O-320-A1A
    Horsepower: 150
    TBO: 2000 hours
Speeds
    Maximum: 180 mph
    Cruise: 170 mph
    Stall: 59 mph
Transitions
    Takeoff over 50-foot obstacle: 1600 feet
    Ground run: 900 feet
    Landing over 50-foot obstacle: 1360 feet
    Ground roll: 670 feet
Weights
    Gross: 3500 pounds
    Empty: 2180 pounds
Dimensions
    Length: 27 feet, 4 inches
    Height: 9 feet, 6 inches
    Span: 37 feet, 1 inch
Other
    Fuel capacity: 72 gallons
    Rate of climb: 1250 fpm (240 single)
Seats: Five

Make: Piper
Model: PA-23 D/E/F Apache
Year: 1957–1961

Fig. 11-14. *Piper PA-23 Apache.*

Engines
    Make: Lycoming
    Model: O-320-B3B
    Horsepower: 160
    TBO: 2000 hours
Speeds
    Maximum: 183 mph
    Cruise: 173 mph
    Stall: 61 mph
Transitions
    Takeoff over 50-foot obstacle: 1550 feet
    Ground run: 1190 feet
    Landing over 50-foot obstacle: 1360 feet
    Ground roll: 750 feet
Weights
    Gross: 3800 pounds
    Empty: 2230 pounds
Dimensions
    Length: 27 feet, 4 inches
    Height: 9 feet, 6 inches
    Span: 37 feet, 1 inch
Other
    Fuel capacity: 72 gallons
    Rate of climb: 1260 fpm (240 single)
Seats: Five

Make: Piper
Model: PA-23-235 Apache
Year: 1962–1965

Fig. 11-15. *Piper PA-23 Aztec, early.*

Engines
    Make: Lycoming
    Model: O-540-B1A5
    Horsepower: 235
    TBO: 2000 hours
Speeds
    Maximum: 202 mph
    Cruise: 191 mph
    Stall: 62 mph
Transitions
    Takeoff over 50-foot obstacle: 1280 feet
    Ground run: 830 feet
    Landing over 50-foot obstacle: 1360 feet
    Ground roll: 880 feet
Weights
    Gross: 4800 pounds
    Empty: 2735 pounds
Dimensions
    Length: 27 feet, 7 inches
    Height: 10 feet, 3 inches
    Span: 37 feet, 1 inch
Other
    Fuel capacity: 144 gallons
    Rate of climb: 1450 fpm (220 single)
Seats: Five

Courtesy of Piper Aviation Museum Foundation

Fig. 11-16. *Piper PA-23 Aztec, late.*

Make: Piper
Model: PA-23 Aztec
Year: 1960–1981
Engines
    Make: Lycoming
    Model: IO-540-C4B5 (turbocharger optional)
    Horsepower: 250
    TBO: 2000 hours (1800 if turbocharged)
Speeds
    Maximum: 215 mph
    Cruise: 205 mph
    Stall: 62 mph
Transitions
    Takeoff over 50-foot obstacle: 1100 feet
    Ground run: 750 feet
    Landing over 50-foot obstacle: 1260 feet
    Ground roll: 900 feet
Weights
    Gross: 4800 pounds
    Empty: 2900 pounds
Dimensions
    Length: 27 feet, 7 inches
    Height: 10 feet, 3 inches
    Span: 37 feet, 1 inch
Other
    Fuel capacity: 144 gallons
    Rate of climb: 1650 fpm (365 single)
Seats: Six

Make: Piper
Model: PA-30/39 Twin Comanche

Fig. 11-17. *Piper PA-39 Twin Comanche.*

Year: 1963–1972
Engines
    Make: Lycoming
    Model: IO-320-B1A (turbocharger optional)
    Horsepower: 160
    TBO: 2000 hours (1800 if turbocharged)
Speeds
    Maximum: 205 mph
    Cruise: 198 mph
    Stall: 70 mph
Transitions
    Takeoff over 50-foot obstacle: 1530 feet
    Ground run: 940 feet
    Landing over 50-foot obstacle: 1870 feet
    Ground roll: 700 feet
Weights
    Gross: 3600 pounds
    Empty: 2270 pounds
Dimensions
    Length: 25 feet, 2 inches
    Height: 8 feet, 3 inches
    Span: 36 feet
Other
    Fuel capacity: 90 gallons
    Rate of climb: 1460 fpm (260 single)
Seats: Four (six after 1965)

Make: Piper
Model: PA-34-200 Seneca
Year: 1972–1974
Engines
    Make: Lycoming
    Model: IO-360-C1E6

Horsepower: 200
TBO: 1800 hours
Speeds
    Maximum: 196 mph
    Cruise: 187 mph
    Stall: 67 mph
Transitions
    Takeoff over 50-foot obstacle: 1140 feet
    Ground run: 750 feet
    Landing over 50-foot obstacle: 1335 feet
    Ground roll: 705 feet
Weights
    Gross: 4000 pounds
    Empty: 2586 pounds
Dimensions
    Length: 28 feet, 6 inches
    Height: 9 feet, 10 inches
    Span: 38 feet, 10 inches
Other
    Fuel capacity: 100 gallons
    Rate of climb: 1460 fpm (190 single)
Seats: Six

Make: Piper
Model: PA-34-200T Seneca and Seneca II
Year: 1975–1981
Engines
    Make: Lycoming
    Model: TSIO-360-E
    Horsepower: 220
    TBO: 1800 hours
Speeds
    Maximum: 225 mph
    Cruise: 205 mph
    Stall: 74 mph
Transitions
    Takeoff over 50-foot obstacle: 1210 feet
    Ground run: 920 feet
    Landing over 50-foot obstacle: 2160 feet
    Ground roll: 1400 feet
Weights
    Gross: 4750 pounds
    Empty: 2852 pounds
Dimensions
    Length: 28 feet, 7 inches

Height: 9 feet, 11 inches
Span: 38 feet, 11 inches
Other
Fuel capacity: 93 gallons
Rate of climb: 1400 fpm (240 single)
Seats: Seven

Make: Piper
Model: PA-34-220T Seneca III, IV, V
Year: 1981–
Engines
Make: Teledyne Continental
Model: TSIO-360 (counter rotating)
Horsepower: 220
TBO: 1800 hours
Speeds
Maximum: 251 mph
Cruise: 219 mph
Stall: 70 mph
Transitions
Takeoff over 50-foot obstacle: 1707 feet
Ground run: 1143 feet
Landing over 50-foot obstacle: 2180 feet
Ground roll: 1400 feet
Weights
Gross: 4750 pounds
Empty: 3422 pounds
Dimensions
Length: 28 feet, 7 inches

Courtesy of Piper

Fig. 11-18. *Piper 1984 PA-34 Seneca.*

Height: 9 feet, 11 inches
Span: 38 feet, 11 inches
Other
Fuel capacity: 122 gallons
Rate of climb: 1400 fpm
Seats: Six

Make: Piper
Model: PA-44 Seminole
Year: 1979–1982
Engines
Make: Lycoming
Model: IO-360-E1A6D (turbocharger optional)
Horsepower: 180
TBO: 2000 hours
Speeds
Maximum: 193 mph
Cruise: 185 mph
Stall: 63 mph
Transitions
Takeoff over 50-foot obstacle: 1400 feet
Ground run: 880 feet
Landing over 50-foot obstacle: 1400 feet
Ground roll: 595 feet
Weights
Gross: 3800 pounds
Empty: 2406 pounds
Dimensions
Length: 27 feet, 7 inches
Height: 8 feet, 6 inches
Span: 38 feet, 7 inches
Other
Fuel capacity: 108 gallons
Rate of climb: 1340 fpm (217 single)
Seats: Four

Make: Piper
Model: PA-44-180 Seminole
Year: 1989– (none built from 1991 to 1995)
Engines
Make: Lycoming
Model: O-360-A1H6
Horsepower: 180
TBO: 2000 hours
Speeds
Maximum: 193 mph

Fig. 11-19 . *New Piper Seminole.*

Cruise: 186 mph
Stall: 63 mph
Transitions
    Takeoff over 50-foot obstacle: 2200 feet
    Ground run: NA
    Landing over 50-foot obstacle: 1490 feet
    Ground roll: NA
Weights
    Gross: 3800 pounds
    Empty: 2600 pounds
Dimensions
    Length: 27 feet, 7 inches
    Height: 8 feet, 6 inches
    Span: 38 feet, 7 inches
Other
    Fuel capacity: 108 gallons
    Rate of climb: 1330 fpm
    Seats: Four

Part 3

Alternative Airplanes, the Rumor Mill, and AD Facts

12

Alternative Aircraft

THERE IS NO TYPICAL AIRPLANE OWNER. In fact, most are individuals with different flying interests, varied financial backgrounds, and assorted pilot skills. Fortunately, there is considerable flexibility in flying and airplane ownership, allowing for specialized pursuits.

FLOATPLANES

Ever wish you could go somewhere to really get away from it all? Perhaps the answer to your dream is a floatplane. Water-based airplanes offer a means of transportation to otherwise inaccessible locations (Fig. 12-1).

To fly a water-based plane, you need a seaplane rating, which is relatively easy to obtain. Many flying schools around the country offer seaplane

Fig. 12-1. *Float-equipped planes open new vistas of enjoyment and utility.*

training (see *Trade-A-Plane* or other similar publications). Training for a seaplane rating is practical in nature, involving only flying. There is no written test and quite often the price for the training is fixed and the rating guaranteed.

Insurance rates for water-flight planes are high unless you have extensive floatplane experience with no accident history. Even then, the rates are higher for a float-equipped airplane than a comparable land-only plane. The reasoning behind the higher rates is the higher loss ratio with floatplanes compared to land planes. For example, a typical ground loop in a land-based aircraft can result in several hundred dollars of damage, and the damaged aircraft can often be taxied to a repair facility. The same type of mishap on the water could mean a sunken airplane, resulting in difficult or impossible recovery and thousands of dollars in damages.

As compared to land-only airplanes, more maintenance is required on floatplanes, mainly for corrosion control and hull repair.

Areas of floatplane activity

Floatplane flying is found in every state of the union, although some have more activity than others. The following states offer excellent float flying: Alaska, California, Florida, Louisiana, Maine, Massachusetts, Minnesota, Washington, and Wisconsin. Canada offers some of the finest floatplane flying in the world, as many desirable locations are accessible only by air.

For more information

AOPA sponsors the Seaplane Pilots Association (SPA). The group was formed in 1972 and now claims several thousand members. SPA's objectives are to assist seaplane pilots with technical problems, provide a national lobbying effort, and more. Membership includes the quarterly magazine *Water Flying*, *Water Flying Annual*, and other written communications that include timely tips and safety measures. SPA sponsors numerous fly-ins. Contact:

Seaplane Pilots Association
4315 Highland Park Boulevard, Suite C
Lakeland, FL 33813-1639
(888) 772-8923, (863) 701-7979
spa@seaplanes.org
www.seaplanes.org

HOMEBUILTS

Airplane homebuilding has been around since the Wright brothers, and is really the grassroots of general aviation. Today, the homebuilt airplane field accounts for a large percentage of all newly registered airplanes (Figs. 12-2 through 12-4). In fact, it is the only segment of the U.S. fleet of airplanes that is actually growing.

Courtesy of James Koepnick, EAA

Fig. 12-2. *Homebuilt tube-and-fabric aircraft.*

Courtesy of NEICO Aviation

Fig. 12-3. *Homebuilt composite airplane, the Lancair.*

Homebuilders construct an airplane most suitable to their needs and tastes: utility, speed, maneuverability, or individuality. Airplanes built by their owners can be constructed of tube and fabric, all metal, or composite methods and materials (a method of construction using materials similar to those found in fiberglass powerboat construction). Homebuilders generally labor to make their airplanes as custom as possible.

Homebuilding is not cheap, but the rewards are plentiful. No one is prouder than the homebuilder pointing to an airplane and saying, "I built that." That simple statement makes up for the years spent constructing the plane (Fig. 12-5).

Contact:

Experimental Aircraft Association
EAA Aviation Center
3000 Poberezny Road
Oshkosh, WI 54902
(920) 426-4800
www.eaa.org

Fig. 12-4. *The Avid Flyer comes as a complete kit, all parts included. No welding is needed, either!*

Fig. 12-5. *Thousands come to see the homebuilt airplanes each year at Oshkosh.*

GLIDERS

Most pilots have heard at one time or another that obtaining a glider rating makes a better pilot. No doubt there is some truth to that statement. After all, glider pilots get only one shot at a landing; powered airplane pilots can always go around and do it again (Figs. 12-6 and 12-7).

Power pilots tend to think only of getting there in a hurry. Glider flying is more back to the basics, more attuned to the surroundings. However,

Fig. 12-6. *This entry-level glider offers loads of performance.*

Fig. 12-7. *High-performance gliders set world-class records.*

make no mistake: Gliders are very sophisticated aircraft. For information about gliding, contact:

Soaring Society of America, Inc.
P.O. 2100
Hobbs, MN 88241
(505) 392-1177, fax (505) 392-8154
www.ssa.org

POWERED GLIDERS

The powered glider, first brought to popularity in Europe, is perhaps one of the most intriguing types of aircraft ever built. It offers reasonably good glider flight characteristics, yet can be flown under power over long distances. Powered gliders are new to the United States and the costs, as in all of aviation, are quite high (Fig. 12-8).

Fig. 12-8. *This powered glider offers interest for the worlds of both sports and aviation.*

WAR BIRDS

World War II saw the greatest use of air power the world has ever seen. Many fine examples of the war's historic aircraft have been restored to like-new condition.

The fighters, bombers, and patrol airplanes of World War II (and other eras) can be seen at many air shows around the country. Some of the planes are small, such as the Army L-3 (Aeronca 7AC) and L-4 (Piper J3). Others are big and powerful like the famous Mustang, Bearcat, and B17 (Fig. 12-9).

Some modern small airplanes, such as the Cessna T-41 (military training version of the model 172) and O-2 (military observation version of the model 337), can also be claimed as war birds. They were used for training, logistical, and forward-fire control duties.

Unfortunately, most war birds are not for the average pilot. They drink fuel at a rate only OPEC could appreciate, although the direct operating costs are quite nominal when compared to the very costly maintenance required to keep them flying—to say nothing of purchase and insurance costs. Further, the larger fighters and bombers require pilot skills few general aviation pilots possess. Many ex-military airplanes are owned by

Fig. 12-9. *P-51.*

the Commemorative Air Force (CAF), formerly known as the Confederate Air Force:

Commemorative Air Force, Inc.
P.O. Box 62000
Midland, TX 79711-2000
(432) 563-1000, fax (432) 563-8046
www.cafhq.org

or the Warbirds of America (a division of the EAA):

Warbirds of America, Inc.
P.O. Box 3086
Oshkosh, WI 54903-3086
(920) 426-4800
www.warbirds-eaa.org

Both organizations are excellent sources of information relative to their particular types of airplanes.

13

Hangar Flying
Used Airplanes

THIS CHAPTER CONTAINS comments, statements, rumors, and facts often heard about used airplanes (attributions are in parentheses).

OLDER AIRPLANES

The conventional-gear tube and fabric airplanes built prior to 1960.

"Buy the airplane for the engine, the rest is cheap to fix." (mechanic)

"The old steel tubing is very prone to destructive corrosion, rust, like around the tailwheel." (rebuilder)

"Most wing struts have been subject of one AD or another—be sure the work was really done." (owner)

"You ought to recover every ten years. This gives you a chance to see what's going on underneath. Not like a metal airplane, where you have to strain to see everything. These get stripped to the bones." (rebuilder)

"Parts are getting hard to get for the smaller engines. More so each year." (mechanic)

"I had to STC an O-200 into my plane 'cause I couldn't get another engine cheaper." (owner)

"Don't the polished aluminum look great? Should have seen all the corrosion that was inside the rafters and tail. Should of seen all the money he sunk in getting it fixed." (owner's wife)

"Heel brakes on the older birds don't do much, but you won't nose it over either." (instructor)

NEWER AIRPLANES

The all-metal airplanes built after 1955.

"Inspect the cables and pulleys carefully. They often freeze, break, warp, or bind. Also, the cables can fray and kink. Unfortunately, they can be very hard to inspect easily." (mechanic)

"Fiberglass parts are usually easy to fix. Not so for some of the miracle plastics used on some models. Stuff just won't hold a patch." (owner)

"Check the skin near the landing gear attach points for signs of warping or twisting. It's caused by hard landings." (mechanic)

"Corrosion hides well in dark places where you can't see." (owner)

AERONCA

"I learned to fly in a 7AC. Now I have a 7ECA and love it." (owner)

"The older Aeroncas never got as popular as the Pipers. That makes cost a little less. The old Chiefs are about the cheapest. (owner)

"My company uses the Scout for pipeline patrol. I can put it down anyplace and get it out again." (pilot)

"I suggest any pilot from a tri-gear plane get some good instruction before attempting a taildragger." (instructor)

BEECHCRAFT

"Beechcraft 19, 23, Sport, and Musketeer airplanes have speed-to-horsepower ratios that are low when compared to other planes in their class. The 150-hp versions take off and climb marginally on hot days and at high altitudes. Stalls are abrupt and complete, unlike the Cessna and Piper stalls. Resale prices are poor on these planes and there is little demand for them. Additionally, Beech parts are notoriously expensive." (broker)

"My Bonanza Vee-tail has never worried me. I know there was a speed limit, but I never worried. When I had the AD modification done, cracks were found. Next time some problem indicates a need to slow down, you can bet I'll do what I am told." (owner)

"We like our Sundowner. It's not fast, but it sure is comfortable." (owners)

"I prefer the three-piece tail over the Vee. It seems more stable laterally." (owner)

"Resale on lower models of Beech, such as the 19 and 23 series, are poor." (dealer)

"Rebuild costs for the IO-346 are out of sight. There are no parts being made." (mechanic)

"The Musketeers have wonderfully large cabins. Much bigger than Piper or Cessna." (spouse of pilot)

"My Musketeer has two doors; many don't." (owner)

"I like the visibility from my Skipper, but it doesn't climb too well." (owner)

"Many model 33, 35, and 36 planes are not equipped with dual controls." (broker)

"The Skipper has two doors; Piper should have done the same on their Tomahawk." (Tomahawk owner)

"Those V-tails are the classiest planes out there." (Sunday driver)

"If you need AD work on D-18 wings, you might as well junk it." (mechanic)

"The gear on the model 23 is about the stoutest in the industry." (mechanic)

"No such thing as a cheap Beech part." (mechanic)

"I can still get new parts from a real airplane builder." (owner)

BELLANCA

"My Viking will do nearly any aerobatic stuff you would want." (owner)

"The wood wing worries me; I am afraid it will deteriorate and cost money." (owner)

"A lot of plane for the money, but hangar it and save the fabric cover." (owner)

"We bought a Cruisemaster in a basket and made it into a family project. Now we go all over and love it. Cheap for me to maintain, too!" (owner)

CESSNA

"The Cessna 150 is the best two-place airplane for the money in today's market. The 172 is the same for four-placers. Either will take care of you, as long as you take care of it." (author)

"The Lycoming O-320-H2AD engine is the worst thing that ever happened to general aviation. It is found only in 1977 through 1980 Cessna 172s. It is an engine capable of self-destruction and has been the subject of many very expensive ADs." (mechanic)

"The Cessna 172 is so good that one even got past the Russian Air Defense system in May 1987 when a 19-year-old German pilot flew hundreds of miles over the USSR and landed in Red Square." (history)

"Before purchasing a used 182, check the firewall for damage from hard landings. The nosegear can cause the firewall to buckle if the airplane is landed wheelbarrow fashion." (broker)

"My 175 has an O-360 engine. Common sense told me that the GO-300 engine running at 3,000 rpms would wear out faster than one operating at 2,400 rpms, and the gear box needed regular maintenance." (owner)

"The flaps on a 140 are a joke!" (owner)

"The 182 is a true four-place plane. It will carry four people and all their luggage." (owner)

"Insuring the Cessna 172 is easy, with no real secrets to them. They are reliable, easy to fly, and replacement parts are available." (underwriter)

"Cessna gives very poor support to small airplane owners." (owner)

"I just paid over $600 for a set of seat tracks for my Cessna 182. I feel ripped off; they couldn't possibly be worth more than $20 or $30." (owner)

"The 150-hp Cardinal is an underpowered dog!" (instructor)

"I see all kinds of planes, and some are real expensive, but I really like to see an old 172 that has been all fixed up. The pilot is usually the guy who fixed it, and he's real proud of it." (line boy)

"I could afford to buy a retractable if I wanted it, but why pay for all the extra maintenance for a little extra speed? Slow down and enjoy life." (owner)

"The 172 flies like a 150, just bigger." (owner)

"I just had a 180-hp engine installed in my 172. Wow! What get-up-and-go! This really helps, as the ranch strip is kind of bad in the spring." (rancher)

"The visibility in a busy traffic pattern is poor, but that's the same for all high-wingers." (instructor)

"Love the barn-door flaps because they can really save a high approach." (rancher)

"I have a '63 model with manual flaps, and plan to keep it. I don't like the electric flaps because there's too much that can break on them." (owner)

"I have Deemers wing tips on the plane, and they allow me to make it into my farm strip easier. The strip is only 990 feet long, but clear on both ends." (farmer)

"I use mogas in my '64 Hawk. Seems okay, and it saves me money. I wish the FBO would pump it because the gas in the trunk of the car scares me." (owner)

"172s sell themselves. They are roomy, look good, and are reasonable in price. They're just a real good value, something you don't often see these days." (broker)

"I have rarely seen a bargain 172; generally, you get what you pay for." (mechanic)

"I wish I had bought a barn full of 172s back in about 1980; they'd be worth a fortune today!" (author)

CIRRUS

"That panel is just like a 767—all glass!" (student)

COMMANDER

"The factory support on the 112's AD problems was great—after the legal wrangling." (owner)

"My 114 is quite choppy in rough air." (owner)

"The current Commander Aircraft Company supports the older models for parts." (owner)

"If all the ADs and mods are taken care of, you'll have a tough bird." (owner)

ERCOUPE

"Cute, but glides like a piano." (instructor)

"Watch for corrosion in the wingroot." (mechanic)

"My legs got messed up in the war. If it weren't for the 'Coupe I'd be sitting on the ground instead of flying!" (owner)

"Our A model was cheap to buy, but loads of work has gone into it. I don't think I would recommend them as cheap airplanes." (owner)

"Some of the 'Coupes are fast nearing the half-century mark and require real tender loving care." (mechanic)

"I like to fly with the window open." (owner)

"Sure am glad Univair is around to supply parts." (owner)

GULFSTREAM

"The nosewheel is a swivel affair. You have no control of it except by differential brake steering. It is very good in close quarters." (owner)

"Watch for delamination of the control and wing surfaces. There is an AD about this problem." (mechanic)

"The laminated fiberglass landing gear will save most botched landings. They are great mistake absorbers." (instructor)

"The AA-1s glide like bricks." (owner)

"You need miles for takeoff on a real hot day." (instructor)

"Glad to see them back in production." (owner in 1993)

"Too bad they went under again." (owner in 1996)

HELIO

"Just because it is slow and STOL, don't forget that it's a taildragger and just waiting to eat your lunch." (former owner)

"The crosswind gears work great." (owner)

"The geared engines are expensive to rebuild, but it's a low price to pay for the STOL work I can do." (patrol pilot)

LUSCOMBE

"A pilot's airplane. Keeps you in shape on crosswind landings." (instructor)

"Luscombes have a poor reputation for ground loops." (FBO)

"Too bad they didn't keep going. The Sedan was way ahead of everyone else." (owner)

"Wax and polish is all I ever do, but it looks good." (owner)

MAULE

"Most pilots will tell you they are good performers for the market they are built for: that is, utility use. They are noisy and drafty." (patrol pilot)

"The specification numbers given for Maule airplanes are somewhat higher than what I see." (owner)

"Nothing wrong with tube and fabric!" (FBO)

MEYERS

"One tough airplane and no ADs to prove it!" (owner)

"Goes fast and is just great for two." (owner)

"Real quality—Just real quality." (owner)

PIPER

"The PA-22 Tri-Pacer is probably the last of the affordable four-place airplanes. Just remember it is fabric-covered. Like a Maule, the PA-22 is noisy and drafty." (FBO)

"Watch those crosswinds when you taxi." (owner)

"Some PA-28s are not true four-place planes; they are a two-place with the rear seat added." (instructor)

"The warrior is very stable for instrument work." (owner)

"Any Cub is an investment if you take care of it." (FBO)

"Watch the lift struts on all older Pipers. They tend to rust and crack. There is an AD out about this problem." (mechanic)

"The PA-28 180 is a true four-place airplane." (owner)

"That great big throttle quadrant, and only 112 horses." (Tomahawk owner)

"The first Apaches flew like a rock on one engine." (FBO)

"I bought a Super Cub that had been a sprayer. Had to rebuild most of the airframe. Those chemicals ate it all up. Think what they do to people." (owner)

"It is very easy to overload the Seneca. It's just so big!" (rancher)

"The tail shakes like it will fall off when a stall breaks." (Tomahawk student)

"The Tee-tailed Lance doesn't fly as well as the straight-tailed version. It lacks authority on takeoff and requires a faster and longer roll." (owner)

"Wish Piper had put two doors on [the Tomahawk] like the Skipper." (instructor)

"I had a bug get caught in the gear sensor and couldn't get the gear up on my Arrow." (owner)

"The automatically operated gear on the Arrow leads to complacency in retractable pilots." (insurance carrier)

"I have had nothing but trouble with my Warrior's (161) cabin door. It never wants to stay closed." (owner)

"You get more Piper for the dollar than Cessna." (broker)

PITTS

"Super airplane that can do anything!" (instructor)

"A little dicey on landings from size and poor visibility, but okay after you get used to it." (owner)

"Many world-class aerobats have cut their teeth on the Pitts." (aerobatics judge)

"Mine is an older homebuild, single-seat version, but I still enjoy it and do all my own servicing." (owner)

"Climbs like a rocket ship!" (new owner)

"Made me into a new pilot. It's a scream when compared to a 150 Airbat!" (owner)

SOCATA

"A class act; all the models are." (looker)

"They are well-built with very little plastic frills and decoration. Solid would be the word." (broker)

"Has a stall bell instead of a stall horn—different, but does the job." (owner)

"Socata offers pretty good parts and technical support." (owner)

STINSON

"The barn-door fin on the Stinson will become a weathercock in heavy winds." (owner)

"My wagon is metallized. Nice, but I wonder what's going on under there." (owner)

"On my next major I'm going to get an STC for a 200-hp Lycoming or Continental to replace the Franklin. It'll give better performance and be cheaper to maintain in the long run." (owner)

"Thank the Lord for Univair and all the Stinson stuff they sell." (owner)

"The old Franks [Franklin engines] are getting mighty expensive to fix." (mechanic)

SWIFT

"There is room for only two people in my Swift. The luggage compartment is a joke. You can take your toothbrush if you pack it carefully." (owner)

"The landing felt super smooth as I flared, then I realized the gear wasn't down." (past owner)

"The best thing I ever did was put 150 seats in mine." (owner)

TAYLORCRAFT

"T-Crafts are good performers; just remember they are fabric-covered." (FBO)

"Kind of light in heavy winds, but enough control to handle it though." (owner)

"The O-200 engine is very reliable." (patrol pilot)

"I bought mine for $2500 from a fellow not wanting to finish rebuilding it. It was all in boxes except for the airframe." (owner)

"Not sophisticated, but cheap!" (owner)

HOMEBUILTS

"The new kit tube-and-fabrics are great. Easy to build and you can take them home at night." (builder)

"I am concerned about how long the two-cycle engines will last." (mechanic)

"Some of them are works of art." (photographer)

"Many of the homebuilders would rather build than fly. As soon as they finish an airplane, they are ready to start on the next one." (EAA member)

"I cannot see putting 50 or 60 thousand dollars into a high-performance homebuilt. I'd rather buy something that will retain its value." (Bonanza owner)

"If I built one, I would have to burn it when I got finished with it— afraid of product liability, even at my level." (Piper owner)

"You built it—you fix it." (builder)

"I feel safer in the plane I built than in most other planes. Why? Because I maintain it myself." (builder)

14

Airworthiness Directives

THE SUBJECT OF AIRWORTHINESS DIRECTIVES (ADS) is complicated and frustrating. It requires attention to detail and is the word of the law.

AC 39-7C

SUBJECT: Airworthiness Directives

Date: 11/16/95

Initiated by: AFS-340

1. PURPOSE.
This advisory circular (AC) provides guidance and information to owners and operators of aircraft concerning their responsibility for complying with airworthiness directives (AD's) and recording AD compliance in the appropriate maintenance records.

2. CANCELLATION.
AC 39-7B, Airworthiness Directives for General Aviation Aircraft, dated April 8, 1987, is canceled.

3. PRINCIPAL CHANGES.
References to specific Federal Aviation Regulations have been updated and text reworded for clarification throughout this document.

4. RELATED FEDERAL AVIATION REGULATIONS.
14 Code of Federal Regulations (CFR) part 39; part 43, §§ 43.9 and 43.11; part 91, §§ 91.403, 91.417, and 91.419.

5. BACKGROUND.
The authority for the roles of the Federal Aviation Administration (FAA) regarding the promotion of safe flight for civil aircraft may be found generally at Title 49 of the United States Code (USC) § 44701 et. seq. (formerly, Title VI of the Federal Aviation Act of 1958 and related statutes). One of the

ways the FAA has implemented its authority is through 14 CFR part 39, Airworthiness Directives. Pursuant to its authority, the FAA issues AD's when an unsafe condition is found to exist in a product (aircraft, aircraft engine, propeller, or appliance) of a particular type design. AD's are used by the FAA to notify aircraft owners and operators of unsafe conditions and to require their correction. AD's prescribe the conditions and limitations, including inspection, repair, or alteration under which the product may continue to be operated. AD's are authorized under part 39 and issued in accordance with the public rulemaking procedures of the Administrative Procedure Act, 5USC553, and FAA procedures in part 11.

6. AD CATEGORIES.

AD's are published in the Federal Register as amendments to part 39. Depending on the urgency, AD's are issued as follows:

a. Normally a notice of proposed rulemaking (NPRM) for an AD is issued and published in the Federal Register when an unsafe condition is found to exist in a product. Interested persons are invited to comment on the NPRM by submitting such written data, views, contained in the notice may be changed or withdrawn in light of comments received. When the final rule, resulting from the NPRM, is adopted, it is published in the Federal Register, printed and distributed by first class mail to the registered owners and certain known operators of the product(s) affected.

b. Emergency AD's. ADs of an urgent nature are adopted without prior notice (without an NRPM) under emergency procedures as immediately adopted rules. The AD's normally become effective in less than 30 days after publication in the Federal Register and are distributed by first class mail, telegram, or other electronic methods to the registered owners and certain known operators of the product affected. In addition, notification is also provided to special interest groups, other government agencies, and Civil Aviation Authorities of certain foreign countries.

7. AD'S WHICH APPLY TO PRODUCTS OTHER THAN AIRCRAFT.

AD's may be issued which apply to aircraft engines, propellers, or appliances installed on multiple makes or models of aircraft. When the product can be identified as being installed on a specific make or model aircraft, the AD is distributed by first class mail to the registered owners of those aircraft. However, there are times when such a determination cannot be made, and direct distribution to registered owners is impossible. For this reason, aircraft owners and operators are urged to subscribe to the Summary of Airworthiness Directives which contains all previously published AD's and a biweekly supplemental service.

Advisory Circular 39-6, Announcement of Availability–Summary of Airworthiness Directives, provides ordering information and subscription prices on these publications. The most recent copy of AC 39-6 may be obtained, without cost, from the U.S. Department of Transportation, General Services Section, M-4831, Washington, D.C. 20590.

Information concerning the Summary of Airworthiness Directives may also be obtained by contacting the FAA, Manufacturing Standards Section (AFS-613), P.O. Box 26460, Oklahoma City, Oklahoma 73125-0460. Telephone (405) 954-4103, FAX (405) 954-4104.

8. APPLICABILITY OF AD'S.

Each AD contains an applicability statement specifying the product (aircraft, aircraft engine, propeller, or appliance) to which it applies. Some aircraft owners and operators mistakenly assume that AD's do not apply to aircraft with other than standard airworthiness certificates, i.e., special airworthiness certificates in the restricted, limited, or experimental category. Unless specifically stated, AD's apply to the make and model set forth in the applicability statement regardless of the classification or category of the airworthiness certificate issued for the aircraft. Type certificate and airworthiness certification information are used to identify the product affected. Limitations may be placed on applicability by specifying the serial number or number series to which the AD is applicable. When there is no reference to serial numbers, all serial numbers are affected. The following are examples of AD applicability statements:

a. "Applies to Smith (Formerly Robin Aero) RA-15-150 series airplanes certificated in any category." This statement, or one similarly worded, makes the AD applicable to all airplanes of the model listed, regardless of type of airworthiness certificate issued to the aircraft.

b. "Applies to Smith (Formerly Robin Aero) RA-15-150 Serial Numbers 15-1081 through 15-1098." This statement, or one similarly worded, specifies certain aircraft by serial number within a specific model and series regardless of the type of airworthiness certificate issued to the aircraft.

c. "Applies to Smith (Formerly Robin Aero) RA-15-150 series aircraft certificated in all categories excluding experimental aircraft." This statement, or one similarly worded, makes the AD applicable to all airplanes except those issued experimental airworthiness certificates.

d. "Applicability: Smith (Formerly Robin Aero) RA-15-150 series airplanes; Cessna Models 150, 170, 172, and 175 series airplanes; and Piper PA-28-140 airplanes; certificated in any category, that have been modified in accordance with STC SA807NM using ABLE INDUSTRIES, Inc. (Part No. 1234) muffler kits." This statement, or one similarly worded, makes the AD applicable to all airplanes listed when altered by the supplemental type certificate listed, regardless of the type of airworthiness certificate issued to the aircraft.

e. Every AD applies to each product identified in the applicability statement, regardless of whether it has been modified, altered, or repaired in the area subject to the requirements of the AD. For products that have been modified, altered, or repaired so that performance of the requirements of the AD is affected, the owner/operator must use the authority provided in the alternative methods of compliance provision of the AD (see paragraph 12) to request approval from the FAA. This approval may address either no action, if the current configuration eliminates the unsafe condition; or different actions necessary to address the unsafe condition described in the AD. In no case, does the presence of any alteration, modification, or repair remove any product from the applicability of this AD. Performance of the requirements of the AD is "affected" if an operator is unable to perform those requirements in the manner described in the AD. In short, either the requirements of the AD can be performed as specified in the AD and the specified results can be achieved, or they cannot.

9. AD COMPLIANCE.

AD's are regulations issued under part 39. Therefore, no person may operate a product to which an AD applies, except in accordance with the requirements of that AD. Owners and operators should understand that to "operate" not only means piloting the aircraft, but also causing or authorizing the product to be used for the purpose of air navigation, with or without the right of legal control as owner, lessee, or otherwise. Compliance with emergency AD's can be a problem for operators of leased aircraft because the FAA has no legal requirement for notification of other than registered owners. Therefore, it is important that registered owner(s) of leased aircraft make the AD information available to the operators leasing their aircraft as expeditiously as possible, otherwise the lessee may not be aware of the AD and safety may be jeopardized.

10. COMPLIANCE TIME OR DATE.

a. The belief that AD compliance is only required at the time of a required inspection, e.g., at a 100-hours or annual inspection, is not correct. The required compliance time is specified in each AD, and no person may operate the affected product after expiration of that stated compliance time.

b. Compliance requirements specified in AD's are established for safety reasons and may be stated in various ways. Some AD's are of such a serious nature they require compliance before further flight, for example: "To prevent uncommanded engine shutdown with the inability to restart the engine, prior to further flight, inspect. . . ." Other AD's express compliance time in terms of a specific number of hours in operation, for example, "Compliance is required within the next 50-hours time in service after the effective date of this AD." Compliance times may also be expressed in operational terms such as: "Within the next 10 landings after the effective date of this AD. . . ." For turbine engines, compliance times are often expressed in terms of cycles. A cycle normally consists of an engine start, takeoff operation, landing, and engine shutdown.

c. When a direct relationship between airworthiness and calender time is identified, compliance time may be expressed as a calendar date. For example, if the compliance time is specified as "within 12 months after the effective date of this AD..." with an effective date of July 15, 1995, the deadline for compliance is July 15, 1996.

d. In some instances, the AD may authorize flight after the compliance date has passed, provided that a special flight permit is obtained. Special flight authorization may be granted only when the AD specifically permits such operation.

Another aspect of compliance times to be emphasized is that not all AD's have a one-time compliance requirement. Repetitive inspections at specified intervals after initial compliance may be required in lieu of, or until a permanent solution for the unsafe condition is developed.

11. ADJUSTMENTS IN COMPLIANCE REQUIREMENTS.

In some instances, a compliance time other than the compliance time specified in the AD may be advantageous to an aircraft owner or operator. In recognition of this need, and when an acceptable level of safety can be shown, flexibility may be provided by a statement in the AD allowing adjustment of the specified interval. When adjustment authority is provided in an AD, owners or operators desiring to make an adjustment are required to submit data substantiating their proposed adjustment of their local FAA Flight Standards District Office or other FAA office for consideration as specified in the AD. The FAA office or person authorized to approve adjustments in compliance requirements is normally identified in the AD.

12. ALTERNATIVE METHODS OF COMPLIANCE.

Many AD's indicate the acceptability of one or more alternative methods of compliance. Any alternative method of compliance or adjustment of compliance time other than that listed in the AD must be substantiated and approved by the FAA before it may be used. Normally the office or person authorized to approve an alternate method of compliance is indicated in the AD.

13. RESPONSIBILITY FOR AD COMPLIANCE AND RECORDATION.

The owner or operator of an aircraft is primarily responsible for maintaining that aircraft in an airworthy condition, including compliance with AD's.

a. This responsibility may be met by ensuring that properly certificated and appropriately rated maintenance person(s) accomplish the requirements of the AD and properly record this action in the maintenance records. This action must be accomplished within the compliance time specified in the AD or the aircraft may not be operated.

b. Maintenance persons may also have direct responsibility for AD compliance, aside from the times when AD compliance is the specific work contracted for by the owner or operator. When a 100-hour, annual, progressive, or any other inspection required under parts 91, 121, 125 or 135 is accomplished, § 43.15(a) requires the person performing the inspection to determine that all applicable airworthiness requirements are met, including compliance with AD's.

c. Maintenance persons should note that even though an inspection of the complete aircraft is not made, if the inspection conducted is a progressive inspection, determination of AD compliance is required for those portions of the aircraft inspected.

d. For aircraft being inspected in accordance with a continuous inspection program (§ 91.409), the person performing the inspection must ensure that an AD is complied with only when the portion of the inspection program being handled by that person involves an area covered by a particular AD. The program may require a determination of AD compliance for the entire aircraft by a general statement, or compliance with AD's applicable only to portions of the aircraft being inspected, or it may not require compliance at all. This does not mean AD compliance is not required at the compliance time or date specified in the AD. It only means that the owner or operator has elected to handle AD compliance apart from the inspection program. The owner or operator remains fully responsible for AD compliance.

e. The person accomplishing the AD is required by § 43.9 to record AD compliance. The entry must include those items specified in § 43.9(a)(1) through (a)(4). The owner or operator is required by § 91.405 to ensure that maintenance personnel make appropriate entries and, by § 91.417, to maintain those records. Owners and operators should note that there is a difference between the records required to be kept by the owner under § 91.417 and those § 43.9 requires maintenance personnel to make. In either case, the owner or operator is responsible for maintaining proper records.

f. Pilot Performed AD Checks. Certain AD's permit pilots to perform checks of some items under specific conditions. AD's allowing this action will include specific direction regarding recording requirements. However, if the AD does not include recording requirements for the pilot, § 43.9 requires persons complying with an AD to make an entry in the maintenance record of that product. § 91.417(a) and (b) requires the owner or operator to keep and retain certain minimum records for a specific time. The person who accomplished the action, the person who returned the aircraft to service, and the status of AD compliance are the items of information required to be kept in those records.

14. RECURRING/PERIODIC AD'S.
Some AD's require repetitive or periodic inspection. In order to provide for flexibility in administering such AD's, an AD may provide for adjustment of the inspection interval to coincide with inspections required by part 91, or other regulations. The conditions and approval requirements under which adjustments may be allowed are stated in the AD. If the AD does not contain such provisions, adjustments are not permitted. However, amendment, modification, or adjustment of the terms of the AD may be requested by contacting the office that issued the AD or by following the petition procedures provided in part 11.

15. DETERMINING REVISION DATES.
The revision date required by § 91.417(a)(2)(v) is the effective date of the latest amendment to the AD and may be found in the last sentence of the body of each AD. For example: "This amendment becomes effective on July 10, 1995." Similarly, the revision date for an emergency AD distributed by

telegram or priority mail is the date it was issued. For example: "Priority Letter AD95-11-09, issued May 25, 1995, becomes effective upon receipt." Each emergency AD is normally followed by a final rule version that will reflect the final status and amendment number of the regulation including any changes in the effective date.

16. SUMMARY.

The registered owner or operator of an aircraft is responsible for compliance with AD's applicable to airframe, engine, propeller, appliances, and parts and components thereof for all aircraft it owns or operates. Maintenance personnel are responsible for determining that all applicable airworthiness requirements are met when they accomplish an inspection in accordance with part 43.

SECTION 39

Airworthiness Directives are described in Section 39 of the FARs as applying to aircraft, aircraft engines, propellers, or appliances when an unsafe condition exists in a product and when that condition is likely to exist or develop in other products of the same type design.

CIVILITIES OF ADS

Airworthiness Directives have been around nearly since there were airplanes, but the attitude of creating them has been changing lately. In the past, airplane manufacturers were proud when they could point to how few ADs had been issued against their products. Recently, however, the manufacturers themselves are requesting ADs to be issued. The manufacturers, however, won't explain this change in attitude.

There are airplane owners who feel that the manufacturers, by creating ADs, are attempting to protect themselves from litigation in product liability suits. Some owners also feel that the ADs are being issued to create a market for expensive and limited, perhaps artificially so, supplies of parts required to comply with the ADs.

Unlike automobile recalls, where the manufacturer, through the local dealer, repairs the problem at no charge to the owner, ADs are generally at the expense of the airplane owner—with a few exceptions, usually when the AD applies to an airplane under warranty.

Issued for whatever reason, Airworthiness Directives are the word of law and must be complied with.

THE AD LISTS

Warning: This listing of ADs was complete at the time of writing, but it is unofficial and incomplete for all maintenance or inspection purposes.

It is intended to serve only as a guide for prospective purchasers, as an aid in determining values and choices. Some ADs might not apply to every airplane of a particular make and model, and some are excluded from the list. Serial number checks must be made for specific aircraft. Consult with an A&P mechanic for additional information. Airworthiness Directives can be issued against airframes, engines, accessories, avionics, etc.

Note that some manufacturers appear to have many more ADs listed than others. This is not a factor of quality, but rather of quantity; the more airplanes built, the more ADs issued. The list is divided under airframes and engines, then grouped by manufacturer. The AD numbers are listed in the left column in numerical order; the right column lists the required procedure and the models affected.

Airframes

AERO COMMANDER/GULFSTREAM/MEYERS/ROCKWELL

59-04-01	Stabilizer rivets: 500
59-06-01	Elevator front spar: 500
59-16-01	Fuel line routing: 500
60-16-04	Cast guide collars: 500
61-14-01	Engine mount inspection: 500
62-08-01	Fuel pressure gauge drain line: 500
63-22-03	Marvel Schebler carburetors: 500
64-02-01	Main landing gear: 500
64-24-04	Cracks in the blade shank: 200
66-28-01	Elevator trim system: 200
67-22-01	Main landing gear: 500
67-23-01	Main landing gear: 200
68-12-05	Modify the rudder control arms: 100
68-19-04	Blade shank cracks: 500
68-21-02	Inspect/replace the aileron cable assembly: 100
73-01-01	Inspect and repair the exhaust system: 100
73-06-02	Fuel siphoning: 500
75-12-09	Inadvertent structural failure: 500
77-16-01	Blade actuating pin failures: 200
86-13-04	Engine cylinder assembly: 200
87-21-07	Fuel filler openings: 500
91-15-04	Possible blade separation: 200
94-04-13	Wing structure cracks: 500
94-04-15	Wing spar cap: 500
97-18-02	Hartzell propeller inspection: 500
98-07-17	Flap system cable: 500
98-08-19	Inspect wing brackets: 500

| 98-08-25 | Nose launching gear bolt: 500 |
| 02-09-08 | Hartzell propeller: 200 |

AERONCA/AMERICAN CHAMPION/CHAMPION/BELLANCA

47-20-01	Gascolator bowl cleaning: 7AC/11 series
47-20-02	Oleopiston: 7AC/11 series
47-30-01	Lift strut wing attach fittings: 7AC/11 series
47-30-05	Exhaust stack inspection: 7AC/11 series
48-04-02	Wing rib rework: 7AC/11 series
48-08-02	Cleveland wheels: 7AC/11 series
48-13-07	Turnbuckle fork replacement: 11 series
48-38-01	Oil cooler installation: 15 Sedan
48-39-01	Control stick rework: 7AC series
48-43-02	Continental C-145-2 engines: 15 Sedan
49-11-02	Wing attach fitting: 7AC/11 series
49-15-01	Seat anchorage rework: 11 series
52-28-01	Fuel transfer placard: 11 series
61-16-01	Lift strut fittings: 15 Sedan
67-03-02	Fuel shutoff valve bracket: 7 EC/GC
68-20-05	Fuel tank inspection: 15 Sedan
71-20-04	Lower fuselage longeron: 7 EC/GC/KCAB
72-18-03	Rudder and elevator cables: 7 EC/GC/KCAB
72-20-06	Aerobatic flight placard: 7 EC/GC/KCAB & 8 KCAB
74-15-07	Propeller mounting: 8 GCBC
74-23-04	Rib flanges: 8 KCAB
75-04-12	Markings on tachometer: 8 GCBC
75-17-16	Engine power loss: all except Aeronca
76-22-01	Adjustable front seat: all except Aeronca
77-22-05	Replace wing lift struts: 7 EC/GC/KCAB
79-22-01	Exhaust system: 7 EC/8 GCBC & KCAB
80-21-06	Muffler core, body—inspect each 100 h: all except Aeronca
81-16-04	Competition harness installation: all except Aeronca
87-18-09	Wing spars: 8 GCBC
89-18-06	Front folding seats: all except Aeronca
90-15-15	Wing front spar strut fittings: 8 KCAB
93-10-02	Cylinder valve retainer: 15 Sedan
93-11-03	Connecting rod: 15 Sedan
96-09-06	Air filter gasket: 15 Sedan
96-18-02	Wing front strut attach fittings: all except Aeronca
97-18-02	Hartzell propeller inspection: 7GC series
98-05-04	Inspect wood spars: 8GCBC
99-24-10	Standby vacuum system: all
00-25-02	Wood spars: all Champion
04-04-10	Propeller mounting bolts: Aeronca

AVIAT

90-20-05	Seat back: A-1
91-23-02	Carb air intake box: A-1
96-09-06	Air filter gasket: A-1

BEECH/RAYTHEON

47-33-05	Horizontal stabilizer spar: 18 S
47-33-06	Stabilizer spar replacement: 18 S
47-33-07	Alternate static source: 18 S
47-34-01	Expansion type filler cap: 18 S
47-34-02	Tail cone drain holes: 18 S
47-34-03	Fuel cell seal: 18 S
47-47-07	Engine identification plate: 35 series
47-47-08	Fuel line chafing: 35 series
48-08-01	Starter rework: 35 series
48-13-01	Generator control box: 18 S
48-34-01	Stabilizer attachment fittings: 18 S
49-04-01	Aileron chain: 35 series
49-26-01	Trailing antenna rework: 35 series
49-29-02	Rudder spring bushings: 18 S
49-31-01	Emergency fuel pump "O" rings: 35 series
49-48-01	Thompson fuel pump: 35 series
50-42-01	Elevator tab replacement: 35 series
51-14-01	Fuel booster pump removal: 35 series
53-01-02	Fuel selector valve rework: 35 series
55-22-01	Oil outlet check valve: 35 series
58-13-02	Tail light assembly: 50 series
59-08-01	Landing gear limit switches: 35 series
59-18-02	Fuel line modification: 18 S
62-08-03	Control wheels: 35/50/95
63-06-01	Cabin door hinge: 23
63-25-01	Fuselage modification: 35
64-06-01	Cabin heater/muffler weld assembly: 23
64-24-02	Wing attach bolts: 50
64-27-01	Control wheel column aft stop: 35
65-03-01	Gap seal strips: 18 S
65-11-01	Instrument static system: 23
65-14-01	Alternator support bracket: 35
65-19-02	Aileron control column links: 23
66-30-01	Static pressure line: 18 S
68-13-02	400 hours inspect Hartzell prop: S35/V35/V35A/C33A/E33A
68-14-01	Pneumatic de-icer system: 55/95

68-17-06	Modify the master brake cylinder: 23
68-23-06	Volpar tri-gear: 18 S
68-26-06	Placard installation: 95/55
69-18-01	Fuel system modification: 35
70-03-05	Turning type takeoffs: 33/35/36
70-12-02	Modify the seat tracks: 36
70-15-03	Fuel selector valve: 19/23
71-24-10	Security of control wheel: 33/35/36/55/95
71-25-03	Fuel and oil restrictor: 19/23/24
72-01-02	Oil drain tubes: 55
72-11-02	Engine fuel interruption: 33/35/36/95
72-16-02	Volpar nose gear fork: 18 S
72-18-01	Fuel selector valve light: 35
72-22-01	Landing gear uplock rollers: 33/35/36/55/95
72-25-06	Nose gear extension: 18 S
73-12-11	Carburetor air box valve: 19/23
73-14-03	Landing gear strut piston: 33/35/36/55
73-20-07	Wing attach bolts/brackets: 19/23/24
73-21-02	Failure of flex hose assemblies: 55
73-23-03	Flexible induction air ducts: 19/23/24
73-23-06	Inspect/replace throttle control cable: 19/23
73-25-04	Maximum design weight placard: 19
73-26-08	Roll and pitch servoclamps: 55
74-03-02	Landing gear shock piston: 33/35/36/55
74-05-01	Restricted fuel vent tank lines: 19/23/24
74-12-02	Aft facing seats: 36
74-14-02	Carburetor and air controls: 19/23
74-14-05	Placard: spins: 19/23
74-23-09	Inflight situations: 19/23
74-24-03	Strobe light system: 33/35/36/55
75-01-04	Fuel selector valve: 19/23/24
75-01-07	Exhaust system failures: 35
75-05-02	Engine oil filter: 33/35/36
75-13-02	Propeller blade vibration: 95
75-15-08	Engine lubrication: 35
75-16-10	Elevator trim tab: 55/95
75-17-37	Mixture and/or heat control: 19/23/24
75-20-08	Propeller governor: 24
75-27-09	Wing structure: 18 S
76-03-05	Housing cable and battery relay: 33/35
76-05-04	Inspect stabilizer attach fitting each 1000 hours: 35
76-10-09	Wing flaps: 19/23/24
76-13-05	Electrical power: 55
76-22-02	"Y" type shoulder harness: 33/35/36/55

76-25-05	Aileron control: 19/23/24
77-02-04	Wing tip strobe lights: 55
77-03-05	Dye inspect the main landing gear housing: 19/23/24
77-11-03	Dye check/replace the prop pitch control: 35
77-19-07	Control column assembly: 18 S
78-04-01	Replace the flap control weld assembly: 19/23/24
78-10-01	Lower rudder torque tube: 18 S
78-11-01	Powerplant fires: 50
78-16-06	Structural integrity trim tab: 19/23/24
78-20-08	Trim tabs push rods: 76
78-22-05	Elevator control push rods: 50
79-01-01	Screws in control system: 33/35/36/55/56/58
79-01-02	In-flight fires: 50
79-09-04	Engine starter relay failures: 76
79-13-01	Powerplant fire detection: 50
79-17-06	Main landing gear: 76
79-23-05	Attachment bolt threads and nut: 77
79-23-06	Replacing stop bolt nuts: 76
80-07-06	Rudder and rudder trim tab: 76/77
80-07-07	Aileron control system: 77
80-19-12	Engine mount structure: 76
80-21-10	Rework the engine mount and bolts: 77
82-02-03	Elevator control cable: 76
82-09-03	Shoulder harness protection: 77
83-17-05	Changes AFM maneuvers: 33
83-23-03	Reinforce the rudder balance weight bracket: 77
84-08-04	Oil return line check valve: 36
84-09-01	Emergency egress provisions: 33/35/36/55/56/58/95
85-05-02	Fuel selector guard mod: 19/23/24
85-22-05	Nut and bolt replacement: T34/50
87-02-08	Stabilator hinge assemblies: 19/23/24
87-08-05	Modify fuel system: 36
87-18-06	Seat recline actuator handle: 33/35/36/55/58
88-10-01	Fuel boost pump: 23/24
88-20-01	Remove sound deadener material from firewall: 33/36
88-21-02	Fill holes in seat track: 33/35/36/55/58
89-05-02	Elevator control fittings: 33/T34/35/36/55/56/95
89-24-09	Install inspection openings in wings: 19/23/24
89-26-08	Inspect/modify propeller: 33/35/36
90-08-14	Inspect wing spar structure: 55/56/58/95
91-15-20	Cracked engine mounts: 55/58/95
91-17-01	Elevator trim tab actuators: 33/T34/36/55/56/58/95
91-18-19	Shoulder harnesses: 33/35/36/55/58

91-20-08	Fresh air blower housing: 33/35/36
91-23-01	Nose landing gear fork: 77
92-17-01	Fuel metering unit: 35/36
92-18-11	Landing gear circuit: 36
92-24-01	Elevator arm separation: T34
93-03-02	Loran deviation signal: 33/36/58
93-24-03	Rudder spar cracks: 33/36
94-04-18	Elevator balance arm: T34
94-20-04	V-tail structural failure: 35
95-04-03	Spar failure: 33/35/36
95-04-10	Wing attach nut: T34
96-18-14	Replace propeller hub: T34/36
97-06-10	Inspect/replace main gear assembly: 76
97-06-11	Replace tail control rod assembly: 35
97-09-09	Inspect cabin longeron: 58
97-14-15	Door handle locks: 33/36/50/55/56/58/95
98-04-24	Revise flight manual about icing: 55/58
98-13-02	Fabricate V_{ne} card: 35
00-18-02	Inspect elevator skin assembly: 55/56
00-22-01	Rudder bellcrank tube: 58
00-24-04	Rudder control cable: 58
01-08-08	Replace exhaust clamp: 33/35/36
01-11-03	Modify cooling blower: F33A/A36/B36TC/58
01-23-10	Flap actuator assembly: A36/B36TC/58/35/33/95/55
03-01-01	Inspect for missing rivets: 36/58
04-25-51	Wing structure inspection: T34

BELLANCA

46-41-01	Rudder bellcrank replacement: 14-13 series
46-41-02	Aileron control bushing: 14-13 series
46-41-03	Control wheel universal joints: 14-13 series
47-07-01	Fuel selector valve indexing: 14-13 series
47-14-01	Flap hinge bracket: 14-13 series
47-25-09	Fin and stabilizer fittings: 14-13 series
47-32-17	Landing gear inspection covers: 14-13 series
47-32-18	Sprocket cotter pin: 14-13 series
47-50-13	Kopper aeromatic props: 14-13/19 series
47-51-13	Cabin heat control valve: 14-13 series
48-05-03	Trim tab bracket bolts: 14-13 series
48-13-06	Engine cowl brackets: 14-13 series
48-21-01	Accelerator pump linkage: 14-13 series
49-13-02	Landing gear drag strut: 14-13 series
51-16-01	Trim tab modification: 14-13/19 series
52-17-01	Landing gear chain guard: 14-13 series

52-28-02	Fuel pump relief valve: 14-13 series
53-16-01	Elevator trim tab looseness: 14-19 series
63-06-02	Rudder bellcrank: 14-13/19
63-17-01	Main landing gear: 14-13
63-20-01	Brake cylinder support bracket: 14-13
68-23-08	Horizontal tail surface: 14-19/17-30
69-12-04	Stabilizer strut clevis: 14-19/17-30
71-10-01	Fuel leakage: 17-30
72-01-01	Engine flex assemblies: 14-19/17-30
72-01-03	Engine fuel hose: 17-30/31
72-13-03	Fuel boost pump circuit: 17-30
73-03-03	Engine induction box bolts: 17-30
73-05-02	Rudder control system: 17-30/31
75-11-06	Check valve: 17-31
75-20-06	Vertical tubes: 14-19 and 17-30/31
76-08-04	Wood deterioration: all
76-20-07	Wood decayed wing spars: 14-13
76-23-03	Exhaust system: 17-30/31
77-22-02	Nose landing gear engine mount: 14-19/17-30
79-19-05	Aileron control system: 14-13
86-25-06	Wing main and auxilary fuel tanks: 17-30/31
87-11-01	Fuel caps and drains: 17-30/31
90-02-17	Landing gear fitting assemblies: all except 14-13
96-18-07	Inspect nose gear: 17-30/31
97-18-02	Hartzell propeller inspection: 14-13/19

CESSNA

46-44-01	Rudder stop bolts: 120/140 series
46-44-02	Safety belt bracket: 120/140 series
46-44-03	Windshield restraining channel: 120/140 series
46-44-04	Carburetor hot air ducts: 120/140 series
46-44-05	Engine mounting bolts: 120/140 series
47-06-10	Aileron carry-through bar: 120/140 series
47-06-11	Forward doorpost cracks: 120/140 series
47-26-02	Wing leading edge rework: 120/140 series
47-43-01	Primer line relocation: 120/140 series
47-43-02	Fuel selector valve handle: 120/140 series
47-43-03	Seaplane spreader struts: 120/140 series
47-43-04	Rudder control cable horns: 120/140 series
47-43-05	Elevator spar web reinforcement: 120/140 series
47-43-06	Aileron support ribs: 120/140 series
47-43-08	Beech R003-201 propeller blades: 120/140 series
47-50-02	Fuselage bulkhead: 120/140 series
48-05-04	Operator limitations placard: 120/140 series

48-07-01	Stabilizer attaching bolts: 120/140 series
48-25-02	Welded exhaust muffler: 120/140 series
48-25-03	Wing drag wire system: 120/140 series
48-43-02	Continental C-145-2 engines: 170 series
50-31-01	Fin spar reinforcement: 120/140 series
51-21-01	Rudder rib flanges: 120/140 series
57-04-01	Aileron hinge: 310
59-10-03	Flasher switch: 172/180/182
61-25-01	Met-Co-Aire landing gear: 120/140
62-22-01	Vacuum pump modification: 150/175
62-24-03	Cabin heat system: 120/140
66-19-02	Fuel system: 210
67-03-01	Exhaust gas heat exchangers: 150
67-31-04	Removal of glove compartment: 150
68-07-09	Longitudinal control system: 177
68-17-04	Stall warning system: most 100 models
68-18-02	Oil pressure line: 177
68-19-05	Utility category placard: 172
69-08-11	Fuel boost pump: most 200 models
69-12-03	Fuel crossfeed line: 310
69-14-01	Fuel starvation: 310/320
69-15-03	Muffler assembly: 170/172/175
69-15-09	Fuel starvation: 310/320
70-01-02	Fuel quantity transmitter float: 177
70-03-04	Turbosupercharger turbine: 310/320
70-10-06	Oil pressure instrument line: 172
70-15-02	Fuel quantity indicators: 336/337
70-24-04	Fuel shutoff valve: 177
71-01-03	Stabilizer attach angles: 177
71-09-01	Nacelle electrical failures: 320
71-09-07	Exhaust manifold heat exchanger: 206/207/210
71-17-08	Rear engine stoppage: 337
71-18-01	Fuel selector valve placard: 172
71-22-02	Cracks in nose gear fork: 150/172/175/182/210
71-24-04	Hoses in engine compartment: 177/206/210/336/337
72-03-03	Wing flap jack screw: most 100/200 series
72-03-07	Landing gear failure: 310
72-07-02	Selector valve placard: 172
72-07-09	Cracks/loose bolts in fin and rudder: 182/205/206/207/210
72-14-08	Flexible hose assemblies: 310/320
72-23-03	Wing flap actuator: 336/337
73-01-02	Alternate air system: 310
73-07-07	Fuel hoses and electrical wire: 310/320

73-17-01	Fuel transfer pump placard: 170/172/175/177/180/182
73-20-04	Fuel boost electrical circuit: 172
73-23-07	Defective spar attach fittings: 150/172/180/182/185
73-25-01	Heater fuel pump relay: 310
74-04-01	Aft fuselage bulkhead assembly: 172
74-06-02	AVCON mufflers: 150/170/172/175/177
74-16-06	Oil pressure gage line: 177
74-20-01	Main landing gear trunion: 310
74-20-08	Fire procedures checklist: 310
75-05-02	Engine oil filter: 182
75-07-02	Air filter seal: 177
75-09-05	Pitot system: 337
75-09-06	Engine induction air duct: 185
75-11-01	Fuel lines: 320
75-15-08	Engine lubrication: 150
75-16-01	Wing fuel tanks: 180/182/185/206/207
75-23-07	Emergency exit window: 310
75-23-08	Engine exhaust system: 310/320
76-01-01	High speed flutter: 150
76-04-01	Main gear extension: 205/210
76-04-03	ARC PA-500A actuators: most except 150/152/320
76-08-02	Wing tip nose cap: 310/320
76-14-07	Landing gear saddles: 205/210
76-14-08	Trim tab actuator: 177
76-21-06	Loss of engine oil: 172/177
77-02-09	Wing flap system: 150/172/182/206/207
77-04-05	Carburetor air intake: 180/182/185
77-08-05	Single engine takeoffs: 337
77-12-08	External electrical ground power: most except 150/152/310/320
77-14-09	Aircraft controllability: 182
77-16-05	Fuel selector valve: 205/206/207/210
77-23-11	ELT installation: 182
78-01-14	ELT installation: 182
78-07-01	Engine oil pump: 206/207/210
78-09-05	Wing spar caps and web doublers: 336/337
78-11-05	Autopilot actuator: 210/310
78-16-04	Weight and balance data: 336
78-25-07	Vertical fin attach brackets: 150/152
78-26-12	Fuel quantity system: 210
79-02-06	Exhaust system: 152
79-03-03	Engine oil pressure: 206/207/210
79-08-03	Electrical system: 100/200 series
79-10-14	Fuel tank venting: most except 310/320

79-15-01	Fuel flow distribution: 200 series
79-19-06	Fuel flow system per STC: 210
79-25-07	Alternator ground: 180/182/185/200 series
80-01-06	Flap actuator assembly: 152/172
80-04-08	Fuel lines: 172
80-06-03	Wing flap cable clamp: 150/152/172
80-07-01	Oil pressure: 172/200 series
80-10-01	Airglas Engineering ski installation: 180/185
80-11-04	Cracked nutplates: 150/152
80-13-14	Fuel flow transducer: 310
80-19-08	Fuel/air mixture: 172 RG
80-21-03	Roll axis flight control: 210
81-05-01	Fuel tank capacity: 152/172
81-14-06	Rudder trim/nose gear bungee: 172 RG
81-15-05	Engine and airplane mod: 210
81-16-09	Elevator control system: 172
81-23-03	Engine exhaust system: 210
82-06-10	Vacuum-driven instruments: 210
82-07-02	Engine crankcase breather: 170/172/175
83-10-03	Control wheel yoke guide mod.: 172
83-13-01	Placard about fuel cap sealing: 182
83-14-04	Cabin heater shroud mod.: 150/172/182/185
83-17-06	Aileron balance weights: 182 RG (Robertson STOL)
83-22-06	Aileron hinge pin: most 100 series
84-10-01	Bladder fuel cells: 182/185/200 series
85-02-07	Fuel selector: 200 series
85-03-01	Throttle and mixture controls: 200 series
85-10-02	Engine induction airbox: 200 series
85-11-07	Turbocharger oil reservoir: 210
85-17-07	Wing rear spar: 206/207
85-20-01	Cabin heat deactivated: 172 RG
86-11-07	STC SA2353NM (volt. regulator): 206
86-15-07	Modification—larger engine: 150
86-19-11	Contaminated fuel: 172/177/185/200 series
86-24-07	Engine controls installation: most 100/200 series
87-20-03	Seat tracks: most except 310/320
87-20-04	Fuel system: 185
87-21-02	Fuel filler openings: 310/320
87-21-05	Placard—spins: 150/152
88-12-12	Fuel system: 177 RG
88-15-06	Battery location: 150
88-22-07	Hose assemblies: 200 series
88-25-04	Before flight—check for magneto moisture: 210
90-02-13	Main landing gear: 310/320

90-06-03	Heater/muffler: 172
90-21-08	Fuel tank bladder: 180
91-22-01	Fuel line chafing: 205/210
93-13-09	Intercooler installation: 210
93-24-15	Instrument panel rheostat: 150/172/180/182/185
94-12-08	Preflight procedures: 210
96-12-23	Wing stall fence: 150/152 (Bush STOL)
97-01-13	Replace hoses: all
97-06-16	Inspect propeller: 152
97-12-06	Check cowling for rubbing: 172
97-18-02	Hartzell propeller inspection: most
98-01-01	Inspect static air source: 172/182
98-01-14	Replace mufflers: 182
98-02-05	Replace mufflers: 172
98-04-28	Revise flight manual about icing: 310
98-05-14	Revise flight manual about icing: 210
98-13-10	Fabricate placard: 182S
98-13-41	Inspect control cables: 172R
98-14-07	Modify forward door post: 172R
98-16-04	Inspect for missing wing stiffener: 180/182/185
98-20-33	Revise flight manual about icing: 210
98-23-02	Modify ski system: 180/185
98-25-03	Inspect aileron control cable: 172R
99-13-04	Correct aileron stop bolts: 206 series
99-18-04	Inspect aileron control pivot bolt: 172R
99-27-02	Inspect fuel selector: 170/172/175/177
00-02-14	Fabricate inspection placard: 182S
00-04-01	Replace oil pressure switch: 172R/182S/206H/T206H
00-06-01	Measure fuel standpipe: 100 series
01-06-06	Inspect main landing gear: 172RG
01-09-06	Inspect stabilizer attachment: 206/T206
01-23-03	Modify door post: 127N/P/RG
02-07-01	Replace stabilizer brackets: 206, 207, 210 series
02-22-17	Inspect flap bellcrank
03-21-04	Inspect right flap bellcrank: 208 series
03-24-13	Install update autopilot software: 1997 and up 100 series and 206
04-19-01	Shoulder harness adjuster: most except 310/320
01-25-03	Inspect rivets on elevator and rudder: SR20/22
02-24-08	Modify parachute system: SR20/22

COMMANDER

| 73-14-04 | Engine and prop controls: 112 |
| 73-14-06 | Cracks in prop spinner bulkhead: 112 |

73-24-01	Control surface hinge failure: 112
75-22-09	Ailerons: 112
76-23-02	Cabin vents and air ducts: 114
77-01-08	Engine failure: 112
85-03-04	Front seat and belt attach: 112/114
88-05-06	Inspect vertical stabilizer for cracks: 112/114
90-04-07	Wing spars: 112/114

ERCOUPE/ALON/FORNEY/MOONEY (415 ALL/A2/F1/M10)

46-23-01	Muffler replacement
46-23-02	Engine breather line hose
46-23-03	Aileron control column fitting
46-38-02	Aileron control stop
46-38-03	Fuel system elbow fitting
46-46-01	Fuselage gas tank overflow line
46-49-01	Nose wheel replacement
47-20-03	Fuel pump line alternation
47-20-04	Baggage compartment zipper
47-20-05	Belly skin reinforcement
47-20-06	Aileron reinforcement
47-20-08	Battery box drain
47-20-09	Voltage regulator check
47-42-20	Control column shaft
50-07-01	Elevator trim tab stop
52-02-02	Aileron inspection
52-25-02	Federal nose ski
54-26-02	Control cable fraying
55-22-02	Template fuel tank
57-02-01	Rudder horn attachment
59-05-04	Rear spar reinforcement
59-25-05	Rudder reinforcement
60-09-02	Nose gear bolts
67-06-03	Rudder bellcrank
70-17-02	Rudder bellcrank tie rod screws: M-10
86-22-09	Fuel line nipple
94-18-04	Wing panels
02-26-02	Inspect wing center section for corrosion: all
03-21-01	Install wing inspection panels: M-10

GULFSTREAM/AMERICAN GENERAL

70-05-05	Muffler inspection: AA-1 series
70-25-05	Bungee mounting plate: AA-1 series
71-04-03	Nose gear strut: AA-1 series
72-06-02	Rudder, aileron, elevator cables: AA-1/AA-5 series

72-07-10	Wear on elevator bungee housing: AA-1 series
73-13-07	Placard installation: AA-1 series
75-07-04	Rudder control bar assemblies: AA-1/AA-5 series
75-09-07	Mixture control wires: AA-1 series
76-01-02	Engine cowl: AA-5 series
76-04-05	Mixture control wire: AA-5 series
76-17-03	Delaminations in bonded skin: AA-1/AA-5 series
76-22-09	Oil coolers: AA-5 series
77-07-04	Carburetor heat valve assembly: AA-5 series
77-08-03	Static system: AA-5 series
78-13-04	Fuel measurement gauge: AA-1 series
78-20-09	Fuel selector valves: GA-7
78-22-06	Elevator control: GA-7
79-03-06	Rudder assembly: GA-7
79-05-06	Chafing of fuel line: GA-7
79-05-07	Fuel leakage in tank ribs: GA-7
79-22-04	Aileron system: AA-5 series
81-24-03	Engine power loss prevention: AA-5 series
85-21-02	Seat belt attachment bracket: AA-1 series
85-26-05	Rudder torque tube: GA-7
89-18-08	Fuel tank/fuel system: AA-5/GA-7
92-06-08	Prevent elevator binding: GA-7
95-19-15	Wing attach. shoulder bolt: AA-5 series

HELIO H-395/295

58-25-04	Fin hinge attachment
59-17-05	Tail spar fitting
59-17-06	Pulley guards
63-18-02	Engine breather
82-16-08	L and R main spar carry-thru
97-18-02	Hartzell propeller inspection

LAKE

64-17-05	Engine breather tube
65-15-03	Nose gear drag strut bolt
67-11-01	Fuel pressure gage line
72-07-11	Flap pushrod end bearings
74-09-06	Fuel restrictor fitting
74-26-02	Rudder control
76-02-01	Engine power
76-12-11	Engine power loss
76-24-02	Engine oil coolers
78-14-05	Main beam fittings
79-06-01	Engine mount straps

79-21-09	Electrical system
86-23-05	Fuel shutoff valve plate
94-22-02	Autoflight control system
98-10-13	Inspect stabilizer fitting
00-10-22	Inspect spar doublers
02-21-05	Inspect wing spar doublers

LUSCOMBE

46-30-01	Control stick horn adjustment: 8 series
47-10-40	Rudder control arm: 8 series
47-22-01	Bulkhead reinforcement: 8 series
48-08-02	Cleveland wheels: 8 series
48-09-03	Kollsman airspeed baffle: 8 series
48-49-01	Vertical stabilizer spar: 8 series
49-40-01	Trim tab horn attachment: 11A Sedan
49-43-02	Stabilizer spar inspection: 8 series
49-45-01	Landing gear bulkhead: 11A Sedan
50-37-01	Fuel system modifications: 8 series
51-10-02	Control cable inspection: 8 series
51-21-03	Rudder bellcrank: 11A Sedan
55-24-01	Corrosion inspection: 8 series
61-03-05	Fuel line interference: 8 series
62-24-03	Cabin heat system: 8 series
79-25-05	Stabilizer forward attach: 8 series
94-16-02	Vertical stabilizer: 8 series
96-24-17	Wing corrosion inspection: 8 series
97-18-02	Hartzell propeller inspection: 11A

MAULE

65-18-01	Fuselage fabric: M-4
65-28-04	Fuel system modifications: M-4
68-07-08	Rudder trim tab control: M-4
69-20-02	Aileron control system pulley: M-4
71-06-06	Seat track and guide: M-4
71-17-01	Fuel line blockage: M-4
73-11-06	Engine overheat: M-4
75-11-02	Fuel lines: M-4/5
78-13-08	Engine fuel supply: M-5
79-12-01	Tail-to-fuselage attach tube: M-4/5
81-14-02	Rudder pedal V-bar: M-4/5
82-03-05	Engine stoppage prevention: M-6
84-09-07	Plug type drain fittings: M-4/5
86-17-11	Fuel crossover supply line: M-5
95-26-18	Wing lift strut: most

96-10-05	Engine fire: M-4
98-15-18	Inspect wing lift struts: all
00-09-06	Inspect sleeve terminals: all

MOONEY

57-10-01	Landing gear bellcrank: M-20
58-19-03	Carb air box: M-20A
58-23-03	Aileron counterbalance: M-20
59-06-03	Rudder hinge bearing: M-20/A
59-11-03	Heater jacket bulkheads: M-20/A
59-14-02	Fuel selector valve: M-20/A
59-25-06	Carb alternate air source: M-20/A
61-22-07	Fuel/oil gage lines: M-20/A
63-10-05	Empennage attach bracket: M-20/A
65-12-03	RPM range limitation: M-20E
65-22-03	Tail truss: M-20/A
66-07-05	Aural gear warning: M-20B/C/D/E
67-11-05	Tail truss: M-20/A
67-22-07	Rudder assembly: M-20F
67-23-06	Dukes fuel pumps: M-20B/C/D/E
73-21-01	Rod end bearing lubrication: most M-20 series
75-04-09	Landing gear actuator: M-20B/C/E/F
75-09-08	Engine mount: most M-20 series
75-23-04	Dukes gear actuator: most M-20 series
76-10-05	Fuel pressure switch diaphragm: M-20B/C/E/F
77-06-01	Alternate static source valve: M-20J
77-08-06	Oil cooler: M-20E/F/J
77-17-04	Control wheel shaft: most M-20 series
77-18-01	Stewart-Warner oil cooler: M-20E/F/J
78-15-02	Main landing gear side brace bolts: M-20F/J
78-17-01	Pacific Comm. ELT "Alert 50": M-20B/C/F/J
79-06-04	Edo-Aire Mitchell pitch servo: M-20F
80-02-12	Fuselage tubular structure: M-20K
80-13-03	Fuel filter installation: M-20E/F/J
85-24-03	Water entrapment in fuel tanks: most models
86-19-10	Deteriorated wooden structures: M-20/A
88-25-11	Baggage door: M-20J/K
91-03-15	Tailpipe coupling: M-20M
92-08-15	Rudder imbalance: M-20J
95-17-06	Exhaust system cracks: M-20K
95-18-12	Fuel selector valve: M-20K
95-26-16	Alternate air door: M-20J
97-18-02	Hartzell propeller inspection: M-20 series

97-26-08	Filler cap modification: M-20F/J/L
98-24-11	Inspect aileron control links: M20/M20M
99-11-07	Install air conditioner placard: M20R
04-25-04	Fuel bladder drain valve inspection: M-20 series

NAVION

47-11-01	Rudder-nose gear bellcrank: Navion/A/B
47-11-02	Hartzell propeller blade rework: Navion/A/B
47-21-04	Fuel scupper drain line: Navion/A/B
47-21-05	Propeller control friction lock: Navion/A/B
47-21-07	Carburetor air intake scoop: Navion/A/B
47-21-08	Generator stud insulator: Navion/A/B
47-21-09	Hydraulic system modifications: Navion/A/B
47-21-10	Propeller guide pin dowel: Navion/A/B
47-31-01	Hydraulic cylinder lines: Navion/A/B
47-31-02	Carburetor vapor return line: Navion/A/B
47-31-03	Fuel drain cock: Navion/A/B
48-06-03	Hartzell propeller blade: Navion/A/B
48-08-03	Fuel valve control support clip: Navion/A/B
48-29-01	Fuel pump drain: Navion/A/B
49-05-03	Continental engine bearings: Navion/A/B
49-09-02	Booster pump rework: Navion/A/B
49-11-01	Carter fuel pump rework: Navion/A/B
49-12-02	Remove fuel pump vent plugs: Navion/A/B
49-28-01	Unsatisfactory spinner: Navion/A/B
50-10-02	Air intake hose: Navion/A/B
51-07-01	Throttle housing: Navion/A/B
52-08-02	Carburetor ducting: Navion/A/B
52-24-01	Aileron control chain guard: Navion/A/B
52-26-01	Stabilizer spar web: Navion/A/B
53-03-01	Hartzell propeller vibration: Navion/A/B
53-08-01	Horizontal stabilizer fitting: Navion/A/B
54-18-01	Wing-to-fuselage attachment: Navion/A/B
55-01-01	Fuselage frame cracks: Navion/A/B
56-03-03	Wing rear spar inspection: Navion/A/B
56-27-04	Engine cooling kit removal: Navion/A/B
59-06-04	Hydraulic actuating cylinders: Navion/A/B
62-03-03	Landing gear selector valve: most models
63-21-05	Main and nose retraction links: most models
64-04-05	Outer wing panel: D/E/F
68-20-06	Rudder horn inspection: most models
75-05-02	Engine oil filter: Navion/A/B
97-18-02	Hartzell propeller inspection: all

PIPER

46-36-02	Airscoop rework: PA-12
46-36-03	Muffler brace: PA-12
46-37-01	Fuel strainer gasket: J3/PA-12
46-37-02	Fuel strainer position: J3
47-22-03	Landing gear tie strap: PA-12
47-47-01	Landing gear reinforcement: PA-12
47-47-02	Battery box insulating spacer: PA-12
47-47-04	Starter solenoid replacement: PA-12
47-50-03	Canvas seat inspection: J3
47-50-06	Shock strut cracks: J3/PA-11
48-01-02	Cowl support braces: PA-12
48-03-03	Header tank installation: PA-11
48-13-03	Battery hold down brackets: PA-12
48-13-04	Fuse-clip attachment: PA-12
48-14-01	Fuel line elbow: PA-12
49-14-01	Elevator connector tube fitting: J3/PA-11/12
49-27-02	Aileron bellcrank castings: PA-12/14
50-05-01	Nicropress sleeve rework: J3/PA-11/12
50-23-01	Shock strut fittings: PA-15/17
51-19-02	Oil radiator hose: PA-18/20/22
51-21-02	Aileron hinge brackets: PA-12
51-23-03	Battery box insulation: PA-16/20/22
51-27-03	Nose wheel drain: PA-22 Tri-Pacer
52-07-03	Lift strut rework: J3/PA-11
53-04-01	Control stick attachment: PA-18
53-24-04	Fuel-hydraulic lines: PA-18/20/22
54-19-01	Brake line cover plate: PA-18
55-07-02	Landing gear tube reinforcement: PA-22 Tri-Pacer
55-08-04	Ignition filter replacement: PA-22 Tri-Pacer
55-11-02	Thermal relief valves: PA-23 Apache
55-13-01	Idler bellcrank installation: PA-23 Apache
55-14-02	Hydraulic cylinder replacement: PA-23 Apache
55-21-02	Front wing spar attachment: PA-23 Apache
55-22-03	Fuel tank cap: PA-22 Tri-Pacer
56-27-03	Aileron balance weight brackets: PA-23 Apache
57-01-02	Stabilizer attachment bolts: PA-23 Apache
57-05-03	Elevator push-pull tube: PA-23 Apache
57-10-02	Oil pressure gage line: PA-23 Apache
57-13-08	Anti-retraction device: PA-23 Apache
57-19-01	Elevator rod end bearing: PA-23 Apache
57-21-01	Rudder trim tab pin: PA-23 Apache
57-22-01	Fire prevention rework: PA-16/20/22
58-01-06	Stabilizer fitting: PA-23 Apache

58-01-07	Control system turnbuckles: J3/5
58-04-03	Rudder trim tab adjustment: PA-23 Apache
58-12-02	Aileron hinge brackets: J3/PA-15/16/17
58-16-01	Cigar lighter fuse: PA-20/22
58-22-03	Goodrich G-3-787 wheel: PA-23
58-25-05	Door latch modification: PA-24
59-06-05	Nose gear bungee: PA-24
59-07-05	Oil cooler lines clearance: PA-24
59-08-03	Operating limitations placard: PA-23 Apache
59-10-08	Gas tank cap vent: PA-18/20/22
59-12-09	Control wheel sprocket stud: PA-24
59-13-02	Aileron balance weight: PA-24
59-26-02	Fuel cell vent tubes: PA-24
60-01-07	Tail brace wires: PA-12/14/20/22
60-03-07	Engine starter solenoid: PA-23 Apache
60-03-08	Fuel control cable and wire: PA-23 Apache
60-05-03	Safety belt: PA-22 Tri-Pacer
60-10-08	Fuel selector valve: PA-18/20/22
60-24-03	Fuel vent tubes: PA-24
61-16-05	Control cable replacement: PA-18/22
61-16-06	Fuel selector valve handles: PA-24
61-20-02	Exhaust stack reinforcement: PA-24
62-02-05	Rudder cable attachment lug: PA-18
62-10-03	Aileron counterweights: PA-24
62-19-03	Propeller attach bolts: most PA-28 series
62-26-05	Exhaust system: PA-24
62-26-06	Exhaust system: PA-28-150/160
63-07-03	Rudder trim tabs: PA-23 Apache
63-12-02	Elevator rib cracks: PA-23 Apache
63-24-03	Rudder trim tab control rod: PA-23 Apache
63-26-03	Tail area inspection: PA-23 Apache
63-27-03	Landing gear retraction motor: PA-24
64-05-04	Upper main oleobearings: PA-22 Tri-Pacer/Colt
64-06-06	Control wheels: most PA-28 series
64-09-05	Induction system alternate air: PA-30
64-10-04	Carburetor air box deflector: PA-24
64-16-06	Nose gear retraction tubes: PA-30
64-21-05	Hartzell propeller governors: PA-23/30
64-22-03	Landing gear safety switch: PA-24
64-28-03	Heavy walled torque tube: PA-30
65-06-06	Fuel quantity gauge sender: most early PA-28 series
65-11-04	Stabilator control system: PA-24/30
65-25-03	Nose landing gear drag clevis: PA-24
66-18-03	Induction system icing: PA-23 Aztec

66-18-04	Baggage door latch: PA-24/30
66-20-05	Propeller spinner: PA-28-150/160/180
66-28-06	Stabilator system: PA-30
66-30-07	Fuel purge valve hose assembly: PA-24/32
67-12-06	Inspect tubes for corrosion: most early PA-28/32 series
67-19-05	Oxygen cylinder channels: PA-30
67-20-04	Main landing gear torque link: most PA-28/32 series
67-24-02	Engine fuel starvation: PA-22
67-26-02	Fuel selector valve: most PA-28/32 series
67-26-03	Fuel system screens and filters: most early PA-28/32 series
68-01-03	Wing-to-fuselage fittings: PA-32
68-05-01	Exhaust mufflers: most tube & fabric models
68-07-04	Engine mount: PA-23 Aztec
68-12-04	Gear retraction fittings: PA-28R
68-13-03	Fuel cell collapse: PA-24
68-21-03	Exhaust tailpipe: PA-23 Aztec
69-12-01	Air induction inlet hose: PA-28R
69-13-03	Heater exhaust tube: PA-23/30
69-15-01	Control wheel pin: some PA-28/32 series
69-22-02	Failure of control wheel: most early PA-28/32 series
69-23-03	Lower longeron inspection: PA-18
69-24-04	Minimum control speed: PA-30
70-09-02	Propeller spinner: PA-28R
70-15-17	Operating limitation placard: PA-30
70-16-05	Cracks in muffler: PA-28-140/150/160/180
70-18-05	Landing gear torque link bolts: most early PA-28/32
70-22-05	Electrical system modification: PA-23/30
70-26-04	Stabilizer balance weight tube: most early PA-28/32
71-09-05	Front seat belt hardware: PA-32
71-12-01	Engine control support bracket: PA-23
71-12-05	Electric trim switch mod.: PA-23/24/30/39
71-14-06	Prevent grounding of magnetos: most early PA-28 series
71-18-03	Left nose wheel door hinge: PA-23
71-21-06	Engine fire wall flex duct: PA-23 Aztec
71-21-08	Binding of fuel selector handle: PA-28-140/180/R
72-01-07	Failure of engine mount: PA-18
72-08-06	Main landing gear torque links: most early PA-28/32 series
72-11-01	Possible fire from fuel vapor: PA-23 Aztec
72-14-05	Engine exhaust system cracks: PA-23 Aztec
72-14-07	Stabilator hinge fittings: most early PA-28/32 series
72-17-01	Induction air box valve: PA-34
72-17-05	Electric trim switch: PA-28/180/235
72-18-06	Stabilator tip balance weights: PA-34
72-21-03	Fuel line failure: PA-20/22

72-21-07	Longitudinal stability: PA-23 Aztec
72-22-05	Operation limitation placard: PA-24
72-24-02	Throttle movement placard: PA-28-140/150/160
73-09-06	Throttle hangup: PA-22 Tri-Pacer
73-11-02	Main landing gear support: PA-34
73-13-01	Rudder trim tab: PA-34
73-14-02	Cracks in exhaust system: PA-34
74-06-01	Air Research turbochargers: PA-23 Aztec
74-09-02	Stall warning light: PA-32
74-09-04	Rear seat belt installation: many early PA-28 series
74-10-01	Outboard flap hinge: PA-23 Aztec
74-13-01	Stabilator torque tube: PA-23/24/39
74-13-03	Stabilator attach belts: PA-23/24/30/39
74-13-04	Throttle control cable: PA-28-140
74-14-04	Reduced weight placard: PA-28-151
74-16-08	Aft bulkhead assembly: PA-30/39
74-17-04	Fabric reinforcement: PA series
74-17-08	Fuel line clamps: PA-34
74-18-06	Fuel quantity placards: PA-28-235
74-18-13	Nose gear vibration: PA-32
74-19-01	Outer wing spars: PA-28-235/32/34
74-22-05	Full valve stem and cap nut: PA-23
74-24-12	Aileron centering assembly: PA-28-151
74-26-07	Incorrect placard: PA-28-180
75-02-03	Nose wheel fork assembly: most early PA-28/32 series
75-05-02	Engine oil filter: PA-24
75-08-03	Fuel drain valves: most early PA-28 series
75-10-03	Baggage door: PA-32
75-11-05	Glare shield panel: PA-23 Aztec
75-12-06	Fin forward spar: PA-24
75-16-04	Carburetor air box valve: PA-28-151
75-20-03	Usable fuel data: PA-34
75-24-02	Loose seats: some PA-28/32/34
75-25-03	Engine oil: PA-34
75-27-08	Torque tube bearing fittings: PA-24/39
76-06-06	PA-14 wing flaps: PA-12
76-11-07	Magnetic compass deviations: PA-23 Aztec
76-11-09	Fuel leakage: PA-32
76-15-08	Nose gear trunion: PA-28R/32
76-18-04	Fuel valve: PA-28/32
76-18-05	Forward fin attachment: PA-30/39
76-19-07	Stabilator weight assembly: PA-24
76-25-06	Oil hose rupture: PA-28-140
77-01-01	Fuel quantity gage: most PA-28 series/PA-32

77-01-03	Carburetor air filter boxes: PA-28-151
77-01-05	Strobe NAV light wire: PA-23 Aztec
77-08-01	Aileron spar cracks: PA-24/30/39
77-09-10	Electric trim switch: PA-23/24/30/39
77-12-01	Fuel system: PA-28/235 & PA-32
77-13-21	Prevent landing gear collapse: PA-24/30/39
77-23-03	Rod end bearings: most PA-28/32 series
78-02-03	Stabilator: PA-23 Aztec
78-08-03	Hinge bracket assembly: PA-23 Apache
78-10-03	Fuel venting: most tube & fabric models
78-12-07	Fuel selector valve: PA-30/39
78-16-08	Loss of engine oil: PA-32
78-21-03	Fuel lines: PA-34
78-22-01	Rudder or elevator binding: PA-38
78-22-07	Control column travel: PA-32
78-23-01	Fuel drain assembly: PA-23/32
78-23-04	Wing rear spar: PA-38
78-23-09	Control wheels: PA-38
78-26-06	Vertical fin attach. plate: PA-38
79-02-02	Fuel supply lines: PA-44
79-02-05	Fuel flow: PA-28-161/R
79-03-02	Rudder hinge bearing: PA-38
79-05-11	Engine power control loss: PA-44
79-08-02	Stabilizer bracket fitting: PA-38
79-11-06	Landing gear selector lever: PA-23
79-12-08	Fuel selector valves: PA-24/30/39
79-12-11	Fuel primer mod kits: PA-44
79-13-03	Fire hazard prevention: most late PA-28 series
79-13-04	Fuel leakage prevention: PA-32
79-17-05	Fuel gauges and instruments: PA-38
79-20-10	Incorporation of Piper Kit: PA-24/30/39
79-22-02	Fuel leakage: most late PA-28-series
79-26-01	Stabilator skins and frame: PA-23 Aztec
79-26-04	Rudder skin cracks: PA-32
80-09-04	Vertical fin and stabilator: PA-34
80-12-03	Nose landing gear: PA-44
80-14-01	Fuel tank vent system: PA-32
80-14-02	Throttle cable: PA-28/32 series
80-14-03	Bendix, King, or Narco transmitters: most late models
80-18-10	Fuel selector valves: PA-23
80-19-01	Carbon monoxide in cabin: PA-28R
80-20-05	Turbocharger exhaust system: PA-32
80-21-11	Left and right ailerons: PA-44

80-22-13	Rudder upper hinge: PA-38
80-26-04	Rudder pulley at cabin step: PA-23
81-04-05	Flap-spar hinge attachment: PA-23
81-06-08	Lower fuselage attachment tab: PA-18
81-10-01	Nose cone spars: PA-44
81-10-04	Aileron modification: PA-44
81-11-02	Oil drain valves: PA-28R
81-16-10	Outboard leading edge skin: PA-44
81-23-05	Seat springs and battery box: PA-28-151/161
81-23-07	Engine mount assembly: PA-38
81-24-07	NLG modification: PA-32
81-25-05	Wing lift strut forks: most tube & fabric models
82-02-01	Aileron balance weights: PA-38
82-04-08	Landing gear retraction: PA-34
82-06-11	Inspect/modify nose gear: PA-28R
82-23-01	Placard near flap actuator: PA-24/30/39
82-27-08	Forward fin spar web: PA-38
83-10-01	Water and fuel system inspection: PA-24/30/39
83-14-05	Heat exchanger/tailpipe assemblies: PA-34
83-14-08	Stall characteristics: PA-38
83-19-01	Forward and aft fin spars mod.: PA-38
83-19-03	Lower spar cap inspection: PA-24/30/39
83-22-01	Fuel weight limitations: PA-23 Aztec
85-02-05	Piper P/N 81090-02 placard: most PA-28 and up
85-06-04	Fuel tank: most tube & fabric models
86-17-01	Ammeter replacement: most PA-28/32 series
86-17-07	Hydraulic hoses: PA-23
86-20-11	Ammeter replacement: PA-44
87-04-01	Air valve linkage: PA-46
88-04-05	Baggage compartment door: PA-34
88-21-07	Fuel filler compartment: PA-23
88-25-08	Engine cooling system: PA-46
90-19-03	Main landing gear system: PA-38
91-21-09	Alternate air heat: PA-24
92-08-04	Loss of rudder control: PA-34
92-13-04	Water contamination in the fuel: PA-23 Apache
92-13-05	Inability to extend nose gear: PA-34
92-13-06	Jammed trim tab: PA-46
92-13-07	Loose empennage rivets: PA-46
92-15-14	Undetected low vacuum: PA-46
93-05-10	Nose landing gear: PA-32
93-06-02	Possible fuel leakage: PA-23
93-10-06	Wing strut corrosion: most tube & fabric models

93-24-14	Nose landing gear: PA-34
94-13-10	Hi-Shear rivets: PA-24/30/39
94-13-11	Main landing gear: PA-34/44
94-14-14	Nose gear: PA-28R/44
95-20-07	Landing gear inspection: most retractable models
95-26-13	Oil cooler hoses: most PA-28/32 series
96-10-01	Landing light: most older PA-28 series
96-10-03	Flap handle: most PA-28/32/34/44 models
97-01-01	Inspect/replace main gear brace: PA-28R/32R/34
97-18-02	Hartzell propeller inspection: most
98-03-16	Replace rudder hinge bracket: PA-38
98-04-27	Revise flight manual about icing: PA23/32
99-01-05	Inspect wing struts/forks: all high-wing models
99-05-09	Replace induction air filter: PA-23/28/33/34
99-16-06	Reinforce wing attackment brackets: PA-46
01-12-01	Inspect flap bellcranks: PA-46
01-23-08	Eddy current propeller hub inspection: PA-32
03-09-13	Inspect flap control tube: PA-23 (160/235/250)
04-03-32	Replace circuit breakers: PA-46

PITTS S2-B AND S2-S

96-09-08	Longeron failure
96-09-08	Inspection of the rear cabane struts
96-10-12	Flight control stick
96-12-03	Wing attach fitting

REPUBLIC R-3 SEABEE

47-21-11	Firewall stud bushings
47-21-12	"No Smoking" placard
47-21-13	Elevator push-pull tube rivets
47-21-14	Elevator control cable guide
47-21-15	Radifilters
47-21-16	Fuel strainer drain
47-21-17	Backfire screen
47-21-18	Mixture control support bracket
47-21-19	Control clamps or ferrules
47-21-20	Oil pressure gage line
47-21-21	Tip float struts
47-21-22	Engine mounting bolt washers
47-21-23	Engine cooling fan
47-47-10	Float strut rework
47-47-11	Propeller reverse control
47-47-12	Carburetor antiswirl vanes

47-47-13	Hartzell propeller hub
47-47-14	Oil screen inspection
47-51-08	Tailwheel horns
48-01-03	Elevator trim tab bushing
48-11-04	Hydraulic pump handle
49-03-01	AC fuel pump
49-31-02	Fuel placard
53-23-03	Lift strut fittings
97-18-02	Hartzell propeller inspection

SOCATA TB-9/10/20/21TC

86-17-03	Elevator attachments: TB-20
86-21-08	Battery tray: TB-10/20
87-03-11	Stabilizer rod ends: TB-10/20/21
87-12-09	Aileron weight attach rivets: TB-20/21
87-22-02	Inspect for cracks: TB-10/20/21
88-02-05	Main landing gear: TB-20/21
90-02-18	Fuel system: TB20/21
90-25-17	Engine oil cooler: all
91-05-02	Inspect for cracks: TB-20/21
91-15-10	Horizontal stab balance weight: all
98-04-03	Inspect safety belt attachment: all
98-08-21	Inspect rear wing attachment: TB-10
98-16-03	Inspect wing attachments: TB-9/10
98-18-13	Dye inspect main landing gear: TB-20/21
01-23-04	Inspect rudder hinge: TB-9/10/20/21
01-23-05	Modify front seats: TB-9/10/20/21
02-20-04	Install exhaust extension: TB-21
03-04-03	Inspect aileron control cable: all

STINSON 108 SERIES

47-50-11	Ash tray modification
47-50-12	Stabilizer attachment fitting
49-16-02	Wing fabric inspection
50-17-02	Rudder cable inspection
50-25-01	Fuel drip strip

SWIFT GC1B

46-23-04	Addition of rivets in wing skin
46-33-02	Fuselage bulkhead stiffener
46-42-01	Cabin heater valve replacement
47-06-01	Landing gear adjustment
47-06-02	Landing gear retraction placard

47-06-03	Elevator cable collars
47-06-04	Landing gear washers
47-06-05	Battery vent plugs
47-06-06	Engine breather line
47-25-06	Carburetor flexible air duct
47-25-07	Oil radiator outlet sleeve
48-28-01	Fuselage bulkhead rework
51-02-02	Asbestos cloth removal
51-08-03	Stabilizer spar attachment
51-10-05	Fuselage reinforcement rework
51-11-04	Landing gear rework
56-16-04	Landing gear torque knees
58-10-03	Landing gear stop ring
64-05-06	Engine mount
97-18-02	Hartzell propeller inspection

TAYLORCRAFT

47-13-02	Fuel hose: BC-12D
47-16-03	Wing strut fittings: BC-12D
50-41-01	Elevator horn bolt: BC-12D
51-09-03	Fuel shutoff valve clip: BC-12D/19/21
75-18-05	Engine mount bolts: 19
78-20-11	Aileron control malfunction: BC-12D/19
79-04-04	Charging circuit fire hazard: 19
87-03-08	Oil pressure gauge hose: BC-12D/19/21

VARGA 2150/2180

80-02-08	Rudder balance weight bolts
80-13-08	Throttle stop installation
81-20-02	Fuel vent tube
82-08-04	Elevator horn assembly

MOST SMALL AIRPLANES

80-06-05	Test the magneto impulse coupling: Slick
81-15-03	Replace inlet air filter
81-16-05	Inspect the magneto coil: Slick
82-11-05	Comply with service bulletin (magnetos)
82-20-01	Inspect impulse couplers
84-26-02	Each 500 hours replace paper air-filter element
86-01-06	Replace dry vacuum pumps
96-09-06	Inspect/replace air filter assembly
98-23-01	Replace dry air pump coupling
00-04-10	Replace propeller mounting bolts
01-07-03	Replace BASCO-serviced Hartzell propeller

Engines

CONTINENTAL AND TELEDYNE CONTINENTAL

46-36-01	Piston pins: A-65/75
47-40-02	Piston pins: C-75/85/90/125
49-50-01	Generator drive coupling disc: C-75 thru 145
50-20-01	Crankcase cracks: C-145
50-32-01	Oil pump failure: C-165/E-165
51-26-03	Hydraulic valve lifters: E-185
56-06-01	Piston pin assemblies: 470/E-185/E-225
60-06-04	Generator gear retaining nut: C-165/E-185/E-225
60-12-01	Piston pins: E-185/225
63-15-01	Exhaust valves: C-165/E-/185/225/470
69-18-05	Turbocharger oil scavenge pumps: 520
70-13-03	Balance tube assembly: 520
70-14-07	Fuel injection pump needle: 360/470/520
71-09-03	Fuel leakage: 360/520
72-20-02	Cylinder assemblies: 470
72-24-03	Oil filter adaptor: IO-346/520
73-13-04	Carburetor bowl 1/4" drain plug: 470
74-18-07	Cylinder failure: 360
74-18-08	Fuel nozzle line leakage: 360
75-09-13	Turbocharger oil inlet adapters: 470/520
76-13-09	Exhaust flange and elbow: 520
77-05-04	Crankshaft failure: 470/520
77-13-03	Cylinder head cracking: O-200
77-13-22	Crankcase cracks: 520
78-06-01	Flexible intake elbow: 520
79-05-09	Oil pressure relief valve: IO-346/470/520
80-01-04	Cylinder holddown flanges: 520
80-06-05	Test magneto impulse coupler: Lyc & Cont
80-22-05	Crankshaft front oil seal: 520
81-07-06	AC fuel pumps: most four cylinder engines
81-13-10	Engine oil pump drive shaft: 360
81-16-05	Inspect magneto coil: Lyc & Cont
81-23-02	Oil filter adapter gasket: 360
81-24-06	Fuel pump: 520
82-09-01	Fuel pressure regulator leaks: 520
84-19-04	Engine driven fuel pump: 520
84-25-05	Turbocharger oil scavenge reservoir: 520
85-08-02	Cylinder assemblies: 470
86-08-07	Engine driven fuel pump: 520
86-13-04	Cylinder assemblies: 520
87-14-02	Starter adaptor shaftgear: 520

87-26-08	Crankshaft and piston pins: 520
88-03-06	Oil filter: 360/470/520
88-17-03	Starter adapter shaft gear: 520
89-14-01	Crankshaft end play check: 520
89-24-01	Scavenge oil pump gears: 520
89-24-03	Turbocharger inlet assembly: 520
91-19-03	Champion oil filter: IO-346/360/470/520
92-04-09	Rocker shaft: 360
93-08-17	Incorrect oil pick-up tube: 470/520
93-10-02	Cylinder valve retainer: 200/300/360/470/520
93-11-03	Connecting rod: O-200/300
93-16-15	Fuel pumps: 520
93-18-03	Inspect/replace venturi: Lyc & Cont
93-19-04	Replace specified floats: thru O-300
93-22-05	Carb intake housing: C-75/85/90/O-200
94-01-03	Replace specified magneto coils: most
94-05-05	Rocker shaft bosses: 300
94-14-12	Low octane detonation: as listed in AD by N-number
95-03-14	Engine mount cracks: IO-346/520
95-08-10	Turbocharger check valves: 360
95-21-15	Engine teardown and analytical inspection: as listed in AD by N-number
96-12-04	PMA approved pistons: 470
96-12-06	Cylinder cracking: O-200
97-21-02	Replace cylinders and cylinder pins: E-165/185/225 and O-470
97-26-17	Inspect crankshaft: IO-360/520
98-01-08	Inspect roller rocker arms: IO-520/TSIO-520/O-470/IO-470/IO-550
98-17-11	Check crankshaft: all with Nelson Balancing Service repairs
98-19-02	Replace piston pins manufactured by Superior Air Parts: IO-360 series
99-19-01	Inspect crankshaft: 470/520 series
00-11-51	Replace magneto: O-300/360/520 series
00-23-21	Crankshaft core sample—replace crankshaft: IO/TIO-360/470/520/550
02-13-04	Inspect/replace magneto: C125 models and up

FRANKLIN

48-50-01	Cylinder base flange failure: 6A4
51-15-02	Crankcase cracks: 6A4
94-14-11	Low octane detonation: 6A4
96-02-04	Detonation damage: 6A4

LYCOMING AND TEXTRON LYCOMING

51-24-01	Exhaust valve seats: 435
54-02-01	Magnet timing: 290
55-02-02	Drive adaptor gasket: 320
59-10-07	Cylinder baffle clamps: 320/360/480/540
60-11-06	Crankshaft counterweights: 480
62-23-05	Prop shaft oil seal rings: 435/480
63-14-03	Oil pump drive shaft: 540
63-23-02	Exhaust valve stem: 320
64-16-05	Oil seal failure: 320/360/540
65-03-03	Crankshaft flange failure: 320
66-06-03	Connecting rod assemblies: 360
66-14-03	Crankshaft idler shaft: 540
66-20-04	Oil filter adaptor gasket: 320/360/540
67-22-06	Fuel diaphragm: 320/360/540
69-08-09	Manifold pressure gage placard: 540
69-25-08	Reduction gear assembly: 435/480
71-05-02	Crankshaft main bearings: 360
71-11-02	Hydraulic tappet plunger: 360
71-13-01	Fuel injector manifold: 540
73-23-01	Piston pin failures: 320/360/540/720
75-08-09	Oil pump: 235/290/320/360/540
75-09-15	Fuel flow divider gasket: 320/360/540/720
77-07-07	Oil filler extension: 320
77-20-07	Rocker arm retaining studs: 320
78-12-08	Oil pump driving impeller: 320
78-12-09	Crankshaft assembly: 320
78-23-08	Fuel injector tube: 540
78-23-10	Center body bellows seal assembly: 320/360/540/720
78-25-01	Slick magnetos: 235
79-04-05	Fuel diaphragm: 320/360/540/720
79-10-03	Engine mounting bolts: 320/360
79-15-02	Economizer channel plug: 360
80-02-13	Turbocharger oil drain flange: 360
80-04-03	Push rods: 320/360
80-06-05	Test magneto impulse coupler: Lyc & Cont
80-14-07	Exhaust valve spring seats: 320/360
80-17-10	Seized throttle movement: 540
80-25-02	Engine push rods: 235
81-03-05	Mixture control shaft: 540
81-16-05	Inspect magneto coil: Lyc & Cont
81-18-04	Oil pump: 235/290/320/360/540
83-22-04	Injector fuel diaphragm stem: 540
84-13-05	Crankshaft flange: 360

84-19-03	Engine cylinder assembly: 540
87-10-06	Rocker arm assemblies: 320/360/540
89-15-10	Fuel leak check: 540
90-04-06	Prop governor oil line: 235/290/320/360
91-08-07	Fuel pump vent hose: 540
91-10-04	Exhaust transition flange bolts: 540
91-14-22	Crankshaft gear retaining bolt: most
91-21-01	Engine exhaust system: 540
92-12-05	Piston pin failure: 320/360/480/540/720
92-12-10	Fuel injector line failure: 540
93-02-05	Fuel injection lines: 320/360/540/720
93-05-21	AC, Textron, Rajay fuel pumps: 320/360/540
93-05-22	Fuel injector lines: 540
93-11-11	AC, Textron, Rajay fuel pumps: 320/360/540
93-14-15	Operation placard: 360
93-18-03	Inspect/replace venturi: Lyc & Cont
93-19-04	Replace specified floats: thru O-320
94-14-13	Low octane detonation: as listed in AD by N-number
95-03-10	Push rod inspection: 235
95-07-01	Connecting rod bolts: 360/540
95-26-02	Low octane detonation: as listed in AD by N-number
96-09-10	Oil pumps: 235/290/320/360/540
96-23-03	Replace fuel pumps: IO 320/360/540
97-01-04	Inspect/replace cylinder heads: IO-540
97-15-11	Replace piston pins: as listed by serial number
98-02-08	Crankshaft corrosion inspection: 320 (160-hp) series/360 series
98-17-11	Check crankshaft: all with Nelson Balancing Service repairs
98-18-12	Torque inspection of relief valve: IO 320/360/540/541/720
98-19-02	Replace Superior Air Parts piston pins: IO-360 series
99-03-05	Replace crankshaft gear bolt: O-540 series
99-04-04	Inspect/replace magneto coupling O-540 series
02-12-07	Replace oil filter plate/gasket: most engine models
03-14-03	Torque check pump relief valve: 320/360/540/720 series
04-05-24	Replace gear retaining bold: various models

PORSCHE

| 99-04-15 | Replace valve springs: 3200 series |

15

Conversions and Modifications

NOT EVERYONE IS SATISFIED WITH THEIR AIRPLANE as it came from the factory—they want to "improve" upon the design. Whether this improvement is to meet a need created by a work requirement, the owner's aesthetic point of view, or purely a personal point of pride, airplanes can be and are modified.

In general, most airplane modifications (mods) are made with a view towards improved performance. This may be a speed improvement for getting there in a hurry or a reduction of the stall speed to allow for improved slow-flight.

Gap seals, speed fairings, wheel pants, curved windshields, and larger engines (more horsepower) are all modifications generally related to speed improvements. However, speed aside, STOL (short takeoff and landing) improvements are probably the most popular modifications. STOL mods can include changes to the shape of the wing, stall dams, gap seals, vortex generators, and/or the installation of a larger engine. To be sure, a good STOL modification is expensive.

The expenses of airplane modifications bring up a very important point: Money spent on a airplane mod must be considered an expenditure and not an investment. It is rare when the expense of a modification is reflected completely in an airplane's overall value. Generally, there will be an increase. However, while the value may increase, it will not increase an amount equal to the cost of the modification. Further, modifications usually reduce the market standing of an airplane. In other words, not everyone wants a modified airplane.

Some simple mods, such as wing tips or vortex generators, are relatively inexpensive, give near instant gratification, and can be considered an adjunct to safety. Others modifications, such as a full-blown STOL or engine horsepower change, are not inexpensive. In fact, it is not unheard of to spend more on the modification than the initial purchase price of the airplane.

Before undertaking an expensive modification, consider changing airplanes—to a more suitable airplane for the job—whether for speed, range, or STOL performance reasons. If you are in real need of a fast, long-range, four-place airplane, consider moving up to a retractable, such as a Mooney. If you are flying from unimproved short strips, consider a Maule or a Cessna 180, against converting a trigear airplane to a taildragger.

You will probably spend less purchasing a different airplane than you would pay for modifying the current one, to say nothing of the downtime involved with modification work. And, as mentioned in Chap. 3, the unmodified Maule or 180 will be much easier to sell than a highly modified 172.

MOD DIRECTORY

The following is a listing of popular modifications and suppliers available for most airplanes noted in this book. Points of contact are given (address and telephone and fax numbers). In some cases modifications consist only of the FAA required paperwork (in the form of a Supplemental Type Certificate and drawings) for your local mechanic to work from. Others are complete installations done at the installer's location.

A complete listing of Supplemental Type Certificates (STCs) for older airplanes may be found in *The Pilot's Guide to Affordable Classics* by Bill Clarke (McGraw-Hill, 2d ed., 1993).

Aero Twin Inc.: cargo/baggage/folding seats (Cessna 208)

> 2404 Merrill Field Drive
> Anchorage, AK 99501
> (907) 274-6166, fax (907) 274-4285
> www.aerotwin.com

Aerospace Systems & Technologies, Inc.: icing kits for many complex airplanes

> 3213 Arnold Avenue
> Salina, KS 67401
> (785) 493-0946, fax (785) 493-0950

Air Plains Services: horsepower (Cessna 172/182/185)

> Wellington Airport
> P.O. Box 541
> Wellington, KS 67152
> (800) 752-8481, fax (620) 326-5346
> www.airplains.com

Aircraft Conversion Technologies: taildragger/fuel tanks/horsepower (Cessna 150/152/172)

6245 Aerodrome Way
Georgetown, CA 95634
(530) 333-4807, fax (530) 333-0627

Altruair: molded speed cowlings and baffles (most 150- to 210-hp engines)

1780 Joe Crosson Drive
1405 N. Johnson Avenue
El Cajon, CA 92020
(619) 449-1570

Aviation Performance Products: cowlings/exhaust systems (Piper PA-24)

1710 Williamsburg Way
Melbourne, FL 32934
(407) 254-2880
www.aviationperformanceproducts.com

Beryl D'Shannon Aviation: horsepower/vortex generators (Beech 33/35/36)

P.O. Box 548
Wayzata, MN 55391
(800) 328-4629, fax (800) 546-4217
www.beryldshannon.com

Boundry Layer Research, Inc.: vortex generators (Cessna 100 series)

9730 29th Avenue W., C-106
Everett, WA 98204
(800) 257-4847, fax (425) 355-3046
www.blrvgs.com

Colemill Enterprises: horsepower (Cessna 310)

2640 Airpark Drive
Nashville, TN 37206
(615) 226-4256, fax (615) 226-4702
www.colemill.com

Cub Crafters, Inc.: STOL/baggage/struts/recover (Piper PA-18)

P.O. Box 9823
Yakima, WA 98909
(509) 248-9491, fax (509) 248-1421

Davis Aviation Services: horsepower/propellers
(Cessna 180/182/185/206/207/210)

P.O. Box 192
Bristol, TN 37621
(423) 652-1113, fax (423) 652-2503
www.davisaviationservices.com

Flint Aero, Inc.: fuel tanks (Cessna most)

1935 North Marshall Avenue
El Cajon, CA 92020
(619) 448-1551, fax (619) 448-1571
www.flintaero.com

Globe Fiberglass Ltd.: wingtips (Piper PA-28/32)

3470 Aircraft Drive
Lakewood, FL 33811
(800) 899-2707, fax (941) 646-6919
www.globefiberglass.com

Horton STOL Craft: STOL/speed (Cessna 100, 200)(336/337)
(Piper PA-28)

421 North West Road
Wellington, KS 67152
(620) 326-2241, fax (620) 326-2244

Kenmore Air Harbor: horsepower (Cessna 180 and up)

P.O. Box 82064
Kenmore, WA 98028
(800) 543-9595, fax (206) 486-5471
www.kenmoreair.com

Lake Aero Styling & Repair: horsepower/speed (Mooney M20 series)

4725 Island Springs Road
Lakeport, CA 95453
(707) 263-0412, fax (707) 263-0420
www.lasar.com

Lo Presti Speed Merchants: cowlings/props/spinners
(Piper PA-24/32/30/39)

2620 Airport North Drive
Vero Beach, FL 32960
(772) 562-4757, fax (772) 563-0446
www.speedmods.com

Maple Leaf Aviation Ltd.: wheel fairings (Cessna 150/177/182/206)

455 Agnew Drive
Brandon, MB Canada R7A5Y8
(204) 728-7618

Met-Co-Aire: wing tips/fuel tanks (Beech 35) (Cessna most 100/210)
(Piper PA-23/24/28)

P.O. Box 2216
Fullerton, CA 92837

(714) 870-4610
www.metcoaire.com

Micro Aerodynamics, Inc.: vortex generators (most small planes)

4000 Airport Road, Suite D
Anacortes, WA 98221
(800) 677-2370, fax (360) 293-5499
www.microaero.com

Mod Works: fuel tanks/windows/fairings/speed brakes (Mooney M20)

27256 Mooney Avenue
Punta Gorda, FL 33982
(941) 637-6770
www.modworks.com

O&N Aircraft Modifications, Inc.: fuel tanks/baggage (Cessna 150/172/182/210) (Grumman AA-5)(mooney M20)

210 Windsock Lane
Seaman's Airport
Factoryville, PA 18419
(570) 945-3769, fax (570) 945-7282
www.onaircraft.com

Penn Yan Aero Services, Inc. : horsepower (Cessna 172) (Piper PA-18)

2499 Bath Road
Penn Yan, NY 14527
(315) 536-2333, fax (315) 536-2335
www.pennyanaero.com

Peterson's Performance Plus: STOL/horsepower (Cessna 182)

1465 SE 30th
El Dorado Airport
El Dorado, KS 67042
(316) 320-1080, fax (316) 321-3842
www.260se.com

Precise Flight, Inc.: speed brakes (most models)

63354 Powell Butte Road
Bend, OR 97701
(800) 547-2558, (541) 388-1105
www.preciseflight.com

Rocket Engineering Corporation: horsepower (Mooney M20)

East 6427 Rutter Avenue
Spokane, WA 99212
(509) 535-4401, fax (509) 534-2025
www.rocketengineering.com

Sierra Industries: fuel tanks/STOL (Cessna most) (Beech 35/36)
(Piper PA-23/24/28/39)

> Garner Municipal Airport
> 122 Howard Langford Drive
> Uvalde, TX 78801
> (830)278-4381, fax (830) 278-7649

Southwest Texas Aviation, Inc.: speed/cooling (Mooney M20)

> 1815 Airport Drive
> San Marcos Municipal Airport
> San Marcos, TX 78666
> (512) 353-3455

Univair Aircraft Corporation: taildragger/flaps/horsepower: (most classics)

> 2500 Himalaya Road
> Aurora, CO 80011
> (888) 433-5433, fax (303) 375-8888
> www.univair.com

Wipaire, Inc.: floats, skis, wingtips/horsepower (Cessna 206/208)

> 1700 Henry Avenue
> South St. Paul, MN 55075
> (651) 451-1205, fax (651) 451-1786
> www.wipaire.com

THE FAA AND MODIFICATIONS

Before undertaking any modifications to an airplane, talk over your ideas with the FAA at your local FSDO. Do this early in the thinking stage, before any planning or spending of money. In most cases, they will have direct experience with similar airplanes and modifications.

Its important to remember that the FAA is there for safety reasons and sometimes must function to protect us from ourselves. Seeking advice from the FAA can save loads of frustration and expense later on.

One last comment about the local FSDOs: For every negative story I hear about them, whether it be about enforcement actions, an unapproved Form 337, ramp check, etc., I hear a hundred good ones. Unfortunately, the latter are never publicized. The local FSDO is a great source of information and experience—use them.

16

Light-Sport Aircraft

ON JULY 20, 2004, THE FEDERAL AVIATION ADMINISTRATION issued new certification requirements for light-sport aircraft, pilots, and repairmen that, "will make recreational flying safer while keeping it affordable and fun."

With this new rule, the FAA created two new aircraft airworthiness certificates: one for special light-sport aircraft (LSA), which may be used for personal purposes as well as for compensation while conducting flight training, rental, or towing; and the other a separate airworthiness certificate for experimental light-sport aircraft, which may be used only for personal use. The new rule also established the requirements for light-sport aircraft maintenance, inspections, pilot training, and certification. A welcome point of this new rule is pilot medical self-certification.

Simply stated, the new rules provide for a new entry level of pilot—with less training and requirements—to fly lightweight simple airplanes.

THE AIRPLANES

The basics of the new rule's aircraft requirements (as it applies to fixed wing airplanes):

Maximum gross takeoff weight: 1320 lb (1430 lb for seaplanes)

Maximum stall speed: 51 mph (45 kts)

Maximum speed in level flight with maximum continuous power: 138 mph (120 kts)

Two-place maximum (pilot and one passenger)

Single nonturbine engine

Fixed or ground-adjustable propeller

Fixed landing gear (repositional for seaplanes allowed)

Unpressurized cabin

A light-sport aircraft can be manufactured and sold ready-to-fly under a new special light-sport aircraft certification without FAR Part 23 compliance. Aircraft must meet consensus standards. LSA may be used for sport and recreation, flight training, and aircraft rental.

Aircraft of foreign manufacture meeting the specific requirements of the LSA rule are acceptable for use.

Experimental LSA

Homebuilt airplanes can be licensed as experimental light-sport aircraft (E-LSA), provided that the airplane meets the basic LSA rules stated above. Airplanes certified under the E-LSA rules may be used only by the owner of the aircraft for recreational, sport, and instructional flight.

An aircraft previously operated as an ultralight trainer must be transitioned to the E-LSA category no later than January 31, 2008.

Standard category aircraft

Aircraft with a standard airworthiness certificate that meet the LSA specifications may be flown by sport pilots. However, the original airworthiness certification category for that aircraft will not be changed to LSA. This means that a sport pilot can fly an aircraft with a standard airworthiness certificate if the aircraft meets the definition of a light-sport aircraft.

Night flight

An LSA can be operated at night if it is equipped according to FAR 91.209 and the pilot holds at least a private pilot certificate and a minimum of a third-class medical.

SPORT PILOTS

The sport pilot rule creates a new layer of pilot certificates (student, pilot, and instructor) for operating any aircraft that meets the definition of a light-sport aircraft. Rule specifics:

Requires FAA knowledge (written) and practical (flight) test.

Credits ultralight training and experience toward a sport pilot certificate.

Credits sport pilot flight time toward more advanced pilot ratings.

Requires either a third-class FAA medical certificate or a current and valid U.S. driver's license as evidence of medical eligibility via medical self-certification (see notes 1 and 2).

Does not allow carrying passengers for compensation or hire.

Allows sharing ("pro rata") operating expenses with another pilot.

Allows daylight/VFR flight only.

Allows sport pilots to fly vintage and production aircraft (standard airworthiness certificate) that meet the definition of a light-sport aircraft.

Note 1: Medical self-certification means that a pilot holding a license of sport pilot or higher who flies a light-sport aircraft does not need an FAA medical examination. Rather, the pilot self-certifies by holding a state automobile driver's license and agreeing to fly only when physically and mentally capable.

Note 2: A pilot having had a medical denied or revoked by the FAA will be required to obtain a special-issuance medical before being allowed to base his or her medical fitness solely on driver's license requirements.

Sport pilot restrictions

The flight restrictions of a sport pilot certificate are as follows:

No flights into Class A airspace

No flights into Class B, C, or D airspace unless you receive training and a logbook endorsement

No flights outside the United States without advance permission

No sightseeing flights with passengers for charity

No flights above 10,000 feet MSL

Daytime/VFR flight only

No commercial flights

As a sport pilot, you can fly any aircraft that meets the definition of a light-sport aircraft in any one of the FAA certification categories, including experimental aircraft (including home-built aircraft), standard category aircraft (type certificated in accordance with FAR Part 43), primary category aircraft, special light-sport aircraft, or experimental light-sport aircraft.

LSA MAINTENANCE AND REPAIR

As with normally certified aircraft, maintenance and repair of light-sport aircraft have specific requirements and levels of required competency (of the person doing the maintenance/repair).

The annual condition inspection on special light-sport airworthiness certificated aircraft can be completed by

An appropriately rated A&P mechanic

An appropriately rated repair station

A light-sport repairman with a maintenance rating

Preventive maintenance can be performed by a certificated pilot. No rating is required to perform maintenance on experimental light-sport airworthiness certificated aircraft.

The sport pilot/light-sport aircraft rule creates two new levels of repairman certificates: light-sport aircraft repairman with a maintenance rating and light-sport aircraft repairman with an inspection rating. To earn an FAA repairman certificate of any type, you must

Be at least 18 years of age.

Speak, read, and understand English.

Demonstrate the requisite skill to determine whether an E-LSA or S-LSA is in a condition for safe operation.

Be a U.S. citizen or legal permanent resident.

The inspection rating requires the completion of a 16-hour course on the inspection requirements of the particular class of light-sport aircraft. The maintenance rating requires the completion of a 120-hour course on the maintenance requirements of the class of light-sport aircraft.

PRIVATE PILOTS AND LSA

Private pilots, or higher, may fly as sport pilots and operate any sport-pilot-eligible aircraft (must meet LSA definition) in the categories or classes in which they are rated using their valid driver's license or third-class medical as their medical certification. The pilot must have a current flight review.

The primary advantage of operating as a sport pilot is that you can avoid the problems associated with maintaining a third-class medical. You can use your valid state driver's license as your medical as long as your most recent medical application was not denied, suspended, or revoked. If you use your driver's license to establish medical fitness, you must carry it with you when you fly.

To operate as a sport pilot, you are restricted to

Daylight/VFR flight

Aircraft that meet the definition of a light-sport aircraft

Flying only your approved category/class of LSA

Having a current flight review

The main disadvantages to flying as a sport pilot are that your flying is limited to light-sport planes only, daylight/VFR flight only, and carrying only a single passenger.

As a private pilot, you are restricted to flying the category and class of aircraft shown on your pilot certificate. Therefore, if you are rated as airplane–single-engine land, you are allowed to fly any single-engine airplane that meets the definition of a light-sport aircraft. *Note:* If you are going to fly a taildragger, you must have a tailwheel endorsement in your logbook.

A private pilot, flying as a sport pilot, does not need an additional endorsement to fly into Class D, C, or B airspace. Private pilots have already been trained and tested for this activity.

LISTINGS OF LIGHT-SPORT AIRCRAFT

According to the stated requirements for an LSA (weight, speed, etc.), the following standard-category make/model aircraft are allowed under the new LSA rule (the list is not all-inclusive):

Aeronca: 7 series, 11 series, and most older models prior to 1940

Ercoupe: 415C, 415CD

Interstate: SA1, SA1A

Luscombe: 8 series except E and F

Piper: J3, J4, L4, PA11, PA15, PA17, and most models prior to 1940

Porterfield: 40 through 65

Taylorcraft: BC series (including BC12D)

The specifications and photos of many of the preceding aircraft can be found in Chap. 7 of this book.

Should your interests lean toward a new airplane, you have two options: buy or build. As of this writing, several aircraft companies have plans to introduce light-sport aircraft in the near future. Some of these aircraft are currently being flown under ultralight rules, have their roots in ultralight flight, or are among the growing number of homebuilt aircraft available (see Figs. 16-1 through 16-5).

Courtesy of Zenith Aircraft Company, www.zenithair.com

Fig. 16-1. *All-metal Zodiac kit.*

Fig. 16-2. *RS-80 Tiger Moth kit.*

Fig. 16-3. *CH701 STOL kit.*

Fig. 16-4. *EuroFox factory-built.*

Fig. 16-5. *Fisher Dakota Hawk.*

Appendixes

A

NTSB Accident Ranking of Small Airplanes

THE NATIONAL TRANSPORTATION SAFETY BOARD (NTSB) and FAA investigate all airplane accidents. Based on thousands of investigations, the NTSB has amassed a tremendous amount of numerical data. In the late 1970s, they reduced this information to chart form by comparing specific types of aircraft accidents with makes and models of small airplanes.

The placement of an aircraft make and model on NTSB charts is determined by the frequency of accidents per 100,000 hours of operation, compared to other aircraft listed on the same chart; aircraft with poor accident records are at the top, and aircraft with better records are at the bottom.

Note: Although the initial study was completed in the late 1970s, trends have remained consistent during the ensuing years.

ACCIDENTS CAUSED BY ENGINE FAILURE

Globe GC-1	12.36
Stinson 108	10.65
Ercoupe	9.50
Grumman AA-1	8.71
Navion	7.84
Piper J-3	7.61
Luscombe 8	7.58
Cessna 120/140	6.73

Piper PA-12	6.54
Bellanca 14-19	5.98
Piper PA-22	5.67
Cessna 195	4.69
Piper PA-32	4.39
Cessna 210/205	4.25
Aeronca 7	4.23
Aeronca 11	4.10
Taylorcraft	3.81
Piper PA-24	3.61
Beech 23	3.58
Cessna 175	3.48
Mooney M-20	3.42
Piper PA-18	3.37
Cessna 177	3.33
Cessna 206	3.30
Cessna 180	3.24
Cessna 170	2.88
Cessna 185	2.73
Cessna 150	2.48
Piper PA-28	2.37
Beech 33/35/36	2.22
Grumman AA-5	2.20
Cessna 182	2.08
Cessna 172	1.41

ACCIDENTS CAUSED BY IN-FLIGHT AIRFRAME FAILURE

Bellanca 14-19	1.49
Globe GC-1	1.03
Ercoupe	0.97
Cessna 195	0.94
Navion	0.90
Aeronca 11	0.59
Beech 33/35/36	0.58

Luscombe 8	0.54
Piper PA-24	0.42
Cessna 170	0.36
Cessna 210/205	0.34
Cessna 180	0.31
Piper PA-22	0.30
Aeronca 7	0.27
Beech 23	0.27
Cessna 120/140	0.27
Piper PA-32	0.24
Taylorcraft	0.24
Piper J-3	0.23
Mooney M-20	0.18
Piper PA-28	0.16
Cessna 177	0.16
Cessna 182	0.12
Cessna 206	0.11
Grumman AA-1	0.09
Cessna 172	0.03
Cessna 150	0.02

ACCIDENTS RESULTING FROM A STALL

Aeronca 7	22.47
Aeronca 1	18.21
Taylorcraft	6.44
Piper J-3	5.88
Luscombe 8	5.78
Piper PA-18	5.49
Globe GC-1	5.15
Cessna 170	4.38
Grumman AA-1	4.23
Piper PA-12	3.27
Cessna 120/140	2.51
Stinson 108	2.09
Navion	1.81

Piper PA-22	1.78
Cessna 177	1.77
Grumman AA-5	1.76
Cessna 185	1.47
Cessna 150	1.42
Beech 23	1.41
Ercoupe	1.29
Cessna 180	1.08
Piper PA-24	0.98
Beech 33/35/36	0.94
Cessna 175	0.83
Piper PA-28	0.80
Mooney M-20	0.80
Cessna 172	0.77
Cessna 210/205	0.71
Bellanca 14-19	0.60
Piper PA-32	0.57
Cessna 206	0.54
Cessna 195	0.47
Cessna 182	0.36

ACCIDENTS CAUSED BY HARD LANDINGS

Beech 23	3.50
Grumman AA-1	3.02
Ercoupe	2.90
Cessna 177	2.60
Globe GC-1	2.58
Luscombe 8	2.35
Cessna 182	2.17
Cessna 170	1.89
Beech 33/35/36	1.45
Cessna 150	1.37
Cessna 120/140	1.35
Cessna 206	1.30

Piper PA-24	1.29
Aeronca 7	1.20
Piper J-3	1.04
Grumman AA-5	1.03
Cessna 175	1.00
Cessna 180	0.93
Cessna 210/205	0.82
Piper PA-28	0.81
Cessna 172	0.71
Piper PA-22	0.69
Taylorcraft	0.48
Cessna 195	0.47
Piper PA-18	0.43
Piper PA-32	0.42
Cessna 185	0.42
Navion	0.36
Mooney M-20	0.31
Piper PA-12	0.23
Stinson 108	0.19

ACCIDENTS RESULTING FROM A GROUND LOOP

Cessna 195	22.06
Stinson 108	13.50
Luscombe 8	13.00
Cessna 170	9.91
Cessna 120/140	8.99
Aeronca 11	7.86
Aeronca 7	7.48
Cessna 180	6.49
Cessna 185	4.72
Piper PA-12	4.67
Piper PA-18	3.90
Taylorcraft	3.58
Globe GC-1	3.09

Grumman AA-1	2.85
Piper PA-22	2.76
Ercoupe	2.74
Beech 23	2.33
Bellanca 14-19	2.10
Piper J-3	2.07
Cessna 206	1.73
Cessna 177	1.61
Grumman AA-5	1.47
Piper PA-32	1.42
Cessna 150	1.37
Piper PA-28	1.36
Piper PA-24	1.29
Cessna 210/205	1.08
Cessna 182	1.06
Cessna 172	1.00
Mooney M-20	0.65
Beech 33/35/36	0.55
Navion	0.36
Cessna 175	0.17

ACCIDENTS CAUSED BY UNDERSHOT LANDINGS

Ercoupe	2.41
Luscombe 8	1.62
Piper PA-12	1.40
Globe GC-1	1.03
Cessna 175	0.99
Grumman AA-1	0.95
Taylorcraft	0.95
Piper PA-22	0.83
Piper PA-32	0.70
Bellanca 14-19	0.60
Aeronca 11AC	0.59
Piper PA-28	0.59

Aeronca 7	0.59
Piper PA-24	0.57
Piper J-3	0.57
Stinson 108	0.57
Cessna 120/140	0.53
Cessna 195	0.47
Grumman AA-5	0.44
Piper PA-18	0.43
Beech 23	0.43
Cessna 185	0.41
Mooney M-20	0.37
Cessna 170	0.36
Navion	0.36
Cessna 150	0.35
Cessna 210/205	0.33
Cessna 206	0.32
Cessna 172	0.26
Cessna 182	0.24
Beech 33/35/36	0.21
Cessna 180	0.15
Cessna 177	0.10

ACCIDENTS RESULTING FROM LANDING OVERSHOOT

Grumman AA-5	2.35
Cessna 195	2.34
Beech 23	1.95
Piper PA-24	1.61
Piper PA-22	1.33
Cessna 175	1.33
Stinson 108	1.33
Cessna 182	1.21
Aeronca 11	1.17
Luscombe 8	1.08
Piper PA-32	1.03

Globe GC-1	1.03
Mooney M-20	1.01
Cessna 172	1.00
Cessna 170	0.99
Grumman AA-1	0.95
Piper PA-12	0.93
Cessna 210/205	0.89
Cessna 177	0.88
Piper PA-18	0.81
Cessna 206	0.81
Piper PA-28	0.80
Cessna 120/140	0.71
Ercoupe	0.64
Bellanca 14-19	0.60
Cessna 180	0.56
Navion	0.54
Aeronca 7	0.48
Cessna 150	0.35
Piper J-3	0.34
Cessna 185	0.31
Beech 33/35/36	0.23

PERSONAL FLYING

The accident figures for 1993, as stated in the NALL Report 1996 (published by the AOPA), show that personal flying represents 38 percent of all general aviation flight, 65 percent of all the accidents, and 68 percent of all the fatal accidents. In round numbers, that means that owners and pilots account for about 40 percent of all noncommercial and nonmilitary flying. Yet those same pilots and owners have nearly 70 percent of the accidents. These numbers should be cause for deep reflection!

B

Government Aviation Offices and Contacts

U.S. GOVERNMENT AVIATION TELEPHONE NUMBERS

Whenever you need a question about general aviation answered, or you have a question about general aviation aircraft, you can always turn to the Federal Aviation Administration. They have many offices (Flight Standards District Offices and Regional Offices) spread around the country with experts to serve the public. Because of current security concerns, you *must* call and make an appointment before visiting any FSDO.

FAA HQ (Washington, DC)	(202) 267-3484
FAA HQ web site	www.faa.gov
FAA Consumer Hotline	(800) 322-7873
FAA Safety Hotline	(800) 255-1111
FAA Mike Monroney Aeronautical Center (aircraft records; Oklahoma City, OK)	(405) 954-3116
FAA Technical Center (Atlantic City, NJ)	(609) 485-6641

FAA Regional Offices

Alaska Region (Anchorage, AK)	(907) 271-5514
Central Region (Kansas City, MO)	(816) 329-3050
Eastern Region (Jamiaca, NY)	(718) 553-3240

Great Lakes Region (Des Plaines, IL)	(847) 294-7252
New England Region (Burlington, MA)	(781) 238-7200
Northwest Mountain Region (Renton, WA)	(425) 227-2001
Southern Region (College Park, GA)	(404) 305-6000
Southwest Region (Fort Worth, TX)	(817) 222-5001
Western-Pacific Region (Los Angeles, CA)	(310) 725-3608

Flight Standards District Offices in the United States

ALABAMA

| Birmingham | (205) 731-1557 | fax (205) 731-0939 |

ALASKA

Anchorage	(907) 271-2000	fax (907) 271-4777
Fairbanks	(907) 474-0276	fax (907) 479-9650
Juneau	(907) 586-7532	fax (907) 586-8833

ARIZONA

| Scottsdale | (480) 419-0111 | fax (480) 419-0800 |

ARKANSAS

| Little Rock | (501) 918-4400 | fax (501) 918-4403 |

CALIFORNIA

Fresno	(559) 487-5306	fax (559) 454-8808
Long Beach	(562) 420-1755	fax (562) 420-6765
Los Angeles	(310) 215-2150	fax (310) 645-3768
Oakland	(510) 748-0122	fax (510) 748-9559
Riverside	(909) 276-6701	fax (909) 689-4309
Sacramento	(916) 422-0272	fax (916) 422-0462
San Diego	(619) 557-5281	fax (619) 557-7156
San Jose	(408) 291-7681	fax (408) 279-5448
Van Nuys	(818) 904-6291	fax (818) 786-9732

COLORADO

| Denver | (303) 342-1100 | fax (303) 342-1176 |

CONNECTICUT

Windsor Locks   (860) 654-1000

DISTRICT OF COLUMBIA

Washington   (703) 661-8160   fax (703) 661-8744

FLORIDA

Ft. Lauderdale   (954) 635-1300
Miami   (305) 716-3400
Orlando   (407) 812-7700   fax (407) 812-7710
Tampa   (813) 287-4900

GEORGIA

Atlanta   (404) 305-7200   fax (404) 305-7215

HAWAII

Honolulu   (808) 837-8300   fax (808) 837-8399

IDAHO

Boise   (208) 387-4000   fax (208) 387-4020

ILLINOIS

Chicago   (847) 928-8000   fax (847) 928-8002
Springfield   (217) 744-1910   fax (217) 744-1947
West Chicago   (630) 443-3100   fax (630) 443-3155

INDIANA

Indianapolis   (317) 487-2400   fax (317) 487-2429
South Bend   (574) 245-4600   fax (574) 233-9387

IOWA

Des Moines   (515) 289-3840   fax (515) 289-3855

KANSAS

Wichita   (316) 941-1200   fax (316) 941-1276

KENTUCKY

Louisville   (502) 753-4200   fax (502) 753-4205

LOUISIANA

Baton Rouge   (225) 932-5900

MAINE

Portland   (207) 780-3263   fax (207) 780-3296

MARYLAND

Baltimore   (410) 787-0040   fax (410) 787-8708

MASSACHUSETTS

Boston   (781) 274-7130   fax (781) 274-6725

MICHIGAN

Belleville   (734) 487-7222   fax (734) 487-7221
Grand Rapids   (616) 954-6657   fax (616) 940-3140

MINNESOTA

Minneapolis   (612) 713-4211   fax (612) 713-4290

MISSISSIPPI

Jackson   (601) 664-9800   fax (601) 664-9910

MISSOURI

Kansas City   (816) 891-2100   fax (816) 891-2155
St. Louis   (314) 429-1006   fax (314) 429-6367

MONTANA

Helena   (406) 449-5270   fax (406) 449-5275

NEBRASKA

Lincoln   (402) 475-1738   fax (402) 474-7013

NEW JERSEY

Teterboro   (201) 556-6600   fax (201) 556-6623

NEW MEXICO

Albuquerque   (505) 764-1200   fax (505) 764-1233

NEVADA

Las Vegas   (702) 269-1445   fax (702) 269-8013
Reno   (775) 858-7700   fax (775) 858-7737

NEW YORK

Albany	(518) 785-5660	fax (518) 785-7165
Farmingdale	(631) 755-1300	fax (631) 694-5516
Garden City	(516) 228-8029	fax (516) 228-8827
Rochester	(585) 436-3880	fax (585) 436-2322

NORTH CAROLINA

Charlotte	(704) 319-7020	fax (704) 319-7022
Greensboro	(336) 662-1000	fax (336) 662-1080

NORTH DAKOTA

Fargo	(701) 232-8949	fax (701) 235-2863

OHIO

Cincinnati	(513) 979-6400	fax (513) 979-6420
Cleveland	(440) 686-2001	fax (440) 686-2080
Columbus	(614) 255-3120	fax (614) 255-3159

OKLAHOMA

Oklahoma City	(405) 951-4200	fax (405) 951-4282

OREGON

Portland	(503) 681-5500	fax (503) 681-5555

PENNSYLVANIA

Allegheny	(412) 466-5357	
Allentown	(610) 264-2888	
Harrisburg	(717) 774-8271	fax (717) 774-8327
Philadelphia	(610) 595-1500	
Pittsburgh	(412) 262-9034	fax (412) 264-9302

SOUTH CAROLINA

West Columbia	(803) 765-5931	fax (803) 253-3999

SOUTH DAKOTA

Rapid City	(605) 737-3050	fax (605) 737-3069

TENNESSEE

Memphis	(901) 322-8600	fax (901) 322-8601

Nashville	(615) 324-1300	fax (615) 324-1360

TEXAS

Dallas (Love Field)	(214) 902-1800	
Dallas (DFW)	(817) 684-6700	
Ft. Worth	(817) 491-5000	fax (817) 491-5014
Houston	(281) 929-7000	fax (281) 929-7059
Lubbock	(806) 740-3800	fax (806) 740-3809
San Antonio	(210) 308-3300	fax (210) 308-3399

UTAH

Salt Lake City	(801) 257-5020	fax (801) 257-5066

VIRGINIA

Richmond	(804) 222-7494	

WASHINGTON

Seattle	(425) 917-6604	fax (425) 917-6638
Spokane	(509) 353-2434	fax (509) 353-2122

WISCONSIN

Milwaukee	(414) 486-2920	fax (414) 486-2921

WEST VIRGINIA

Charleston	(304) 347-5199	fax (304) 343-2011

WYOMING

Casper	(307) 261-5425	fax (307) 261-5424

CANADIAN GOVERNMENT AVIATION TELEPHONE NUMBERS

In Canada, the ready source for your aviation related questions and problems is Transport Canada. Similar to the FAA, Transport Canada has many offices scattered around the country. Their staff is always friendly and knowledgeable.

Transport Canada Civil Aviation

National Headquarters (Ottawa, ON)	(613) 990-2309

TRANSPORT CANADA REGIONAL OFFICES

Atlantic Regional Office (Moncton, NB)	(800) 387-4999
Ontario Regional Office (Toronto, ON)	(416) 952-0230
Pacific Regional Office (Vancouver, BC)	(604) 666-3518
Prairie and Northern Regional Office (Edmonton, AB)	(204) 983-3152
Quebec Regional Office (Dorval, PQ)	(514) 633-2714

Transport Canada Service Centers

ALBERTA

Calgary	(403) 292-5227
Edmonton	(780) 495-2764

BRITISH COLUMBIA

Abbotsford	(604) 504-4650	fax (604) 504-4651
Kelowna	(250) 491-3700	fax (250) 491-3710
Prince George	(250) 561-5294	fax (250) 561-5145
Richmond	(604) 666-8777	fax (604) 666-8877
Sidney	(250) 363-6627	fax (250) 363-2729
Vancouver	(604) 666-5851	

MANITOBA

Winnipeg	(204) 983-3152

NEW BRUNSWICK

Moncton	(506) 851-7314

NEWFOUNDLAND AND LABRADOR

Goose Bay (Labrador)	(709) 896-6190
St. John's	(709) 772-6197

NORTHWEST TERRITORIES

Yellowknife	(867) 920-6684

ONTARIO

Bottonville	(905) 477-7061	fax (905) 477-1314
Hamilton	(905) 679-3477	fax (905) 679-0033

London	(519) 452-4032	fax (519) 452-3598
Ottawa	(613) 952-1637	fax (613) 998-7382
Sudbury	(705) 693-5804	fax (705) 693-1088
Thunder Bay	(807) 474-2570	fax (807) 475-5816
Toronto	(905) 405-5180	fax (905) 405-3305

QUEBEC

Almo	(418) 669-0529	fax (418) 669-0530
Dorval	(514) 633-2717	fax (514) 633-2720
Sept-Iles	(418) 961-2008	fax (418) 961-2009
Ste-Foy	(418) 640-2690	fax (418) 640-2680
Val d'Or	(819) 825-9250	

SASKATCHEWAN

Saskatoon	(306) 975-8941

YUKON

Whitehorse	(867) 393-6850

WEB SITES

The FAA operates many web sites, some are considered official, while others are unofficial information only. Most can be reached by pointer or hyperlink from the main FAA web site of www.faa.gov. FAA Region Offices sponsor web sites, as do many local FSDOs.

Transport Canada operates a web site at www.tc.gc.ca which hyperlinks with other government and aviation sites. It is an excellent site, worthy of a visit even if you are not involved with a Canadian registered airplane.

C

State Aviation Agencies

WHEN IT COMES TO DAILY LIVING it seems that everyone wants to make a regulation or have an official say in our lives. Aviation is no exception— just look at the Federal Aviation Regulations (FARs). And if the Federal regulations are not enough, most states have an aviation agency of one type or another that has additional regulations for us to follow. Add to that the various state and local level insurance and taxation departments.

Some state aviation agencies offer assistance, up-to-date news, and information to their pilots/owners. Unfortunately, a few states use their aviation agencies only as taxing houses.

Check with your state and local governments for information relative to the registration and taxation of your airplane. Be aware that a sales or transfer tax is collected by most states when an airplane's ownership is bought, sold, or otherwise transferred. Some states also require state aircraft registration and collect use taxes on an annual basis. Check with the appropriate departments to prevent later tax difficulties.

Alabama

Alabama Department of Transportation
Aeronautics Bureau
1409 Coliseum Boulevard
Montgomery, AL 36130
(334) 242-6820, fax (334) 353-6540
www.dot.state.al.us/bureau/aeronautics
State insurance department: (334) 269-3550
State tax department: (334) 242-1170

Alaska

Alaska Department of Transportation and Public Facilities
Statewide Aviation
4111 Aviation Avenue
Anchorage, AK 99502
(907) 269-0730, fax (907) 269-0489
www.dot.state.ak.us/stwdav/stwdav.html
State insurance department: (907) 465-2515
State tax department: (907) 269-6620

Arizona

Arizona Department of Transportation
Division of Aeronautics
255 East Osborn Street, Suite 101
Phoenix, AZ 85012-2373
(602) 294-9144, fax (602) 294-9141
www.dot.state.az.us/aero
State insurance department: (602) 914-8444
State tax department: (602) 255-3381

Arkansas

Arkansas Department of Aeronautics
National Airport Terminal
One Airport Drive, 3d Floor
Little Rock, AR 72202
(501) 376-6781, fax (501) 378-0820
State insurance department: (800) 852-5494
State tax department: (501) 682-2242

California

California Department of Transportation
Division of Aeronautics
1120 N Street
Sacramento, CA 95814
(916) 654-4959, fax (916) 653-9531
www.dot.ca.gov/aeronautics
State insurance department: (213) 897-8921
State tax department: (800) 400-7115

Colorado

Colorado Department of Transportation
Division of Aeronautics

5126 Front Range Parkway
Watkins, CO 80137
(303) 261-4418, fax (303) 261-9608
www.colorado-aeronautics.org
State insurance department: (303) 894-7490
State tax department: (303) 238-7378

Connecticut

Connecticut Department of Transportation
Bureau of Aviation and Ports
2800 Berlin Turnpike
Newington, CT 06131-7546
(860) 594-2000
State insurance department: (860) 297-3800
State tax department: (860) 297-5962

Delaware

Delaware Department of Transportation
Office of Aeronautics
800 Bay Road
Dover, DE 19901
(302) 760-2149, fax (302) 739-2251
www.deldot.net
State insurance department: (302) 739-4241
State tax department: (302) 577-8460

Florida

Florida Department of Transportation
Aviation Office
605 Suwannee Street
Tallahassee, FL 32399-0450
(850) 414-4500, fax (850) 412-8045
www.dot.state.fl.us/aviation
State insurance department: (850) 413-3100
State tax department: (850) 488-6800

Georgia

Georgia Department of Transportation
Office of Intermodal Programs—Aviation
276 Memorial Drive SW
Atlanta, GA 30303-3743
(404) 651-9221, fax (404) 657-4221

www.dot.state.ga.us
State insurance department: (404) 656-2070
State tax department: (404) 968-0480

Hawaii

Hawaii Department of Transportation
Airports Division
Honolulu International Airport
400 Rodgers Boulevard, Suite 700
Honolulu, HI 96819-1880
(808) 838-8600, fax (808) 838-8734
www.hawaii.gov/dot/airports
State insurance department: (808) 586-2799
State tax department: (808) 974-6321

Idaho

Idaho Transportation Department
Division of Aeronautics
3483 Rickenbacker Street
Boise, ID 83705
(208) 334-8775, fax (208) 334-8789
www.itd.idaho.gov/aero/aerohome.htm
State insurance department: (208) 334-2250
State tax department: (800) 972-7660

Illinois

Illinois Department of Transportation
Division of Aeronautics
Capital Airport
One Langhorne Bond Drive
Springfield, IL 62707-8415
(217) 785-8500, fax (217) 524-1022
State insurance department: (217) 782-4515
State tax department: (217) 782-3336

Indiana

Indiana Department of Transportation
Aeronautics Section
100 N. Senate Avenue, Room N901
Indianapolis, IN 46204-2217
(317) 232-1496, fax (317) 232-1499

www.in.gov/dot/modetrans
State insurance department: (317) 232-2385
State tax department: (317) 232-1497

Iowa

Iowa Department of Transportation
Office of Aviation
800 Lincoln Way
Ames, IA 50010
(515) 239-1659, fax (515) 233-7983
www.iawings.com
State insurance department: (515) 281-5705
State tax department: (515) 281-3114

Kansas

Kansas Department of Transportation
Division of Aviation
700 SW Harrison, 2d Floor NE
Topeka, KS 66603-3754
(785) 296-2553, fax (785) 296-3833
www.ksdot.org/divaviation
State insurance department: (785) 296-3071
State tax department: (785) 368-8222

Kentucky

Kentucky Transportation Cabinet
Division of Aeronautics
200 Mero Street
Frankfort, KY 40622
(502) 564-4480, fax (502) 564-7953
www.kytc.state.ky.us/aeronautics/home.htm
State insurance department: (800) 595-6053
State tax department: (502) 564-3058

Lousiana

Louisiana Department of Transportation and Development
Aviation Division
8900 Jimmy Wedell Dr.
Baton Rouge, LA 70807
(225) 274-4112, fax (225) 274-4181
www.dotd.louisiana.gov/intermodal/aviation

State insurance department: (225) 342-5900
State tax department: (225) 219-2448

Maine

Maine Department of Transportation
Office of Passenger Transportation
16 State House Station
Augusta, ME 04333-0016
(207) 624-3250, fax (207) 624-3251
State insurance department: (207) 624-8475
State tax department: (207) 287-2076

Maryland

Maryland Department of Transportation
Aviation Administration
Terminal Building, 3d Floor
BWI Airport, MD 21240
(410) 859-7111, fax (410) 850-4729
www.marylandaviation.com
State insurance department: (410) 468-2020
State tax department: (410) 767-1300

Massachusetts

Massachusetts Aeronautics Commission
10 Park Plaza, Room 6620
Boston, MA 02116-3933
(617) 973-8881, fax (617) 973-8889
State insurance department: (617) 521-7794
State tax department: (617) 887-6367

Michigan

Michigan Department of Transportation
Multi-Modal Transportation Services Bureau
2700 East Airport Service Drive
Lansing, MI 48906-2160
(517) 335-9568, fax (517) 321-6522
State insurance department: (517) 373-0220
State tax department: (517) 373-3200

Minnesota

Minnesota Department of Transportation
Aeronautics Office
222 East Plato Boulevard
St. Paul, MN 55107-1618
(651) 296-8202, fax (651) 297-5643
www.dot.state.mn.us/plane.html
State insurance department: (651) 296-6848
State tax department: (651) 296-6181

Mississippi

Mississippi Department of Transportation
Aeronautics Division
401 Northwest Street
Jackson, MS 39201
(601) 359-7850, fax (601) 359-7855
www.gomdot.com/aero/default.htm
State insurance department: (601) 359-3569
State tax department: (601) 923-7010

Missouri

Missouri Department of Transportation
Aviation Section
DOT Building
2217 St. Mary's Boulevard
Jefferson City, MO 65102
(573) 526-5570, fax (573) 526-4709
www.modot.state.mo.us/othertransportation
State insurance department: (573) 751-4126
State tax department: (573) 751-3505

Montana

Montana Department of Transportation
Aeronautics Division
2630 Airport Road
Helena, MT 59601
(406) 444-2506, fax (406) 444-2519
www.mdt.state.mt.us/aeronautics
State insurance department: (406) 444-2040
State tax department: (406) 444-6900

Nebraska

Nebraska Department of Aeronautics
3431 Aviation Road, Suite 150
Lincoln, NE 68524
(402) 471-2371, fax (402) 471-2906
www.aero.state.ne.us
State insurance department: (402) 471-2201
State tax department: (402) 471-5729

Nevada

Nevada Department of Transportation
1263 South Stewart Street
Carson City, NV 89712
(775) 888-7002, fax (775) 888-7203
www.nevadadot.com/traveler/aviation
State insurance department: (775) 687-4270
State tax department: (702) 486-2300

New Hampshire

New Hampshire Department of Transportation
Division of Aeronautics
John O. Morton Building
P.O. Box 483 Hazen Drive
Concord, NH 03302-0483
(603) 271-1697, fax (603) 271-3914
www.nh.gov/dot/aeronautics
State insurance department: (800) 852-3416
State tax department: (603) 271-2191

New Jersey

New Jersey Division of Aeronautics
P.O. Box 610
Trenton, NJ 08625-0610
(609) 530-2900, fax (609) 530-4549
State insurance department: (609) 292-5360
State tax department: (609) 292-6400

New Mexico

New Mexico Department of Transportation
Aviation Division
1550 Pacheco Street

Santa Fe, NM 87504
(505) 476-0930, fax (505) 476-0942
State insurance department: (505) 827-4500
State tax department: (505) 827-0700

New York

New York State Department of Transportation
Aviation Services Bureau
50 Wolf Road
Albany, NY 12232
(518) 457-8343, fax (518) 457-9779
www.dot.state.ny.us/pubtrans/airhome.html
State insurance department: (518) 474-6600
State tax department: (800) 225-5829

North Carolina

North Carolina Department of Transportation
Division of Aviation
1050 Meridian Drive
RDU Airport, NC 27623
(919) 840-0112, fax (919) 840-9267
www.dot.state.nc.us/transit/aviation
State insurance department: (919) 733-2032
State tax department: (877) 252-3052

North Dakota

North Dakota Aeronautics Commission
Bismarck Municipal Airport
2301 University Drive, Building 1652-22
Bismarck, ND 58504
(701) 328-9650, fax (701) 328-9656
www.state.nd.us/ndaero
State insurance department: (701) 328-2440
State tax department: (701) 328-2770

Ohio

Ohio Department of Transportation
Office of Aviation
2829 West Dublin-Granville Road
Columbus, OH 43235-2786
(614) 793-5040, fax (614) 793-8972

www.dot.state.oh.us/aviation
State insurance department: (800) 686-1526
State tax department: (800) 282-1780

Oklahoma

Oklahoma Aeronautics Commission
3700 N. Classen Boulevard, Suite 240
Oklahoma City, OK 73118
(405) 604-6900, fax (405) 604-6919
www.aeronautics.state.ok.us
State insurance department: (405) 521-2828
State tax department: (405) 521-3160

Oregon

Oregon Department of Aviation
3040 25th Street SE
Salem, OR 97302-1125
(503) 378-4880, fax (503) 373-1688
www.aviation.state.or.us
State insurance department: (503) 378-4271
State tax department: (503) 378-4988

Pennsylvania

Pennsylvania Department of Transportation
Bureau of Aviation
P.O. Box 3457
Harrisburg, PA 17105-3457
(717) 705-1260, fax (717) 705-1255
www.dot.state.pa.us
State insurance department: (717) 787-2317
State tax department: (717) 787-8201

Rhode Island

Rhode Island Airport Corporation
Theodore Francis Green Airport
2000 Post Road
Warwick, RI 02886-1533
(401) 737-4000, fax (401) 732-3034
State insurance department: (401) 277-2223
State tax department: (401) 222-1040

South Carolina

South Carolina Department of Commerce
Division of Aeronautics
Columbia Metropolitan Airport
2553 Airport Blvd.
West Columbia, SC 29170
(803) 896-6262, fax (803) 896-6266
www.scaeronautics.com
State insurance department: (803) 737-6180
State tax department: (803) 898-5709

South Dakota

South Dakota Department of Transportation
Office of Air, Rail and Transit
Becker-Hansen Building
700 East Broadway Avenue
Pierre, SD 57501-2586
(605) 773-3574, fax (605) 773-3921
www.sddot.com/fpa/aeronautics
State insurance department: (605) 773-3563
State tax department: (605) 773-3311

Tennessee

Tennessee Department of Transportation
Aeronautics Division
607 Hangar Lane, Building 4219
Nashville, TN 37217
(615) 741-3208, fax (615) 741-4959
State insurance department: (615) 741-2241
State tax department: (800) 342-1003

Texas

Texas Department of Transportation
Division of Aviation
150 E. Riverside Drive, 5th Floor, South Tower
Austin, TX 78704
(512) 416-4500, fax (512) 416-4510
www.dot.state.tx.us/AVN/avninfo.htm
State insurance department: (512) 463-6169
State tax department: (800) 434-5464

Utah

Utah Department of Transportation
Aeronautical Operations Division
135 North 2400 West
Salt Lake City, UT 84116
(801) 715-2260, fax (801) 715-2276
State insurance department: (801) 538-3800
State tax department: (801) 297-2200

Vermont

Vermont Agency of Transportation
Operations Division
National Life Building, Drawer 33
Montpelier, VT 05633
(802) 828-2833, fax (802) 828-2848
www.vermontairports.com/air.htm
State insurance department: (802) 828-3301
State tax department: (802) 828-2551

Virginia

Virginia Department of Aviation
5702 Gulfstream Road
Richmond, VA 23250-2422
(804) 236-3624, fax (804) 236-3635
www.doav.virginia.gov
State insurance department: (804) 371-9741
State tax department: (804) 367-8031

Washington

Washington Department of Transportation
Aviation Division
3704 172nd Street NE, Suite K2
Arlington, WA 98223
(360) 651-6300, fax (360) 651-6319
www.wsdot.wa.gov/aviation
State insurance department: (800) 562-6900
State tax department: (800) 647-7706

West Virginia

West Virginia Department of Transportation
Aeronautics Commission

1900 Kanawha Boulevard East, Building 5, Room A-503
Charleston, WV 25305
(304) 558-3436, fax (304) 558-0333
www.wvdot.com/1_airports/1_airports.htm
State insurance department: (304) 558-3354
State tax department: (304) 558-3333

Wisconsin

Wisconsin Department of Transportation
Bureau of Aeronautics
4802 Sheboygan Avenue, Room 701
Madison, WI 53705
(608) 266-3351, fax (608) 267-6748
www.dot.wisconsin.gov/modes/air.htm
State insurance department: (608) 266-3585
State tax department: (608) 266-2772

Wyoming

Wyoming Department of Transportation
Aeronautics Division
5300 Bishop Boulevard
Cheyenne, WY 82009-3340
(307) 777-3952, fax (307) 637-7352
State insurance department: (307) 777-7401
State tax department: (307) 777-7961

D

Resources for the Small Airplane Owner

THE SMALL AIRPLANE OWNER MUST HAVE RESOURCES AVAILABLE for assistance in the areas of ownership, operation, and maintenance. In some instances, contacting the airplane's manufacturer will provide support. In other instances, such as in the case of airplanes no longer produced, contact an airplane type club for support.

GENERAL AVIATION SUPPORT

The Aircraft Owners and Pilots Association (AOPA) is made up of airplane owners and pilots from all over the world. Its purpose is to assist its membership by representing them in large-scale legal issues, fighting for membership rights, and providing many types of services (Fig. D-1).

The AOPA publishes a monthly magazine, *AOPA Pilot*, which helps keep the membership informed about general aviation issues. If you own or fly small airplanes, you should join the AOPA. For more information about the AOPA, contact it at (301) 695-2000 or at www.aopa.org.

MANUFACTURERS AND TYPE CLUBS

Some airplane manufacturers are excellent sources of information, parts, and service for all planes built under their name, while others support only planes that they currently manufacture. For example, Cessna continues to support all airplanes built under their name. Piper, on the other hand, supports only their modern airplanes. Unfortunately, many of the airplanes available on today's used market are without manufacturer-provided owner support.

Fig. D-1. *AOPA Headquarters.*

Most of the popular makes and models of small planes, such as those seen in this book, are represented by a type club. A type club is an association or organization of people interested in a particular make and model of airplane. These organizations generally publish a newsletter, host national or regional gatherings, and can answer questions about their particular type of airplanes. You should join and support the club that applies to your airplane.

The following is a list of manufacturers and type clubs (current manufacturers and/or type certificate holders, if any, are placed at the top of each group):

Airplanes

AERO COMMANDER/ROCKWELL

Twin Commander Aircraft Corporation
9010 59th Drive NE
Arlington, WA 98223
(360) 435-9797, fax (360) 435-1112
www.twincommander.com

Twin Commander Flight Group
601 E. Jefferson, Suite 5
Blue Springs, MO 64014

(816) 224-0346, fax (816) 224-6877
cloudcraft@aol.com

AERONCA, CHAMPION, AMERICAN CHAMPION, BELLANCA CHAMPION

American Champion Aircraft Corporation
P.O. Box 37
Rochester, WI 53167
(262) 534-6315, fax (262) 534-2395
www.amerchampionaircraft.com

Univair Aircraft Corporation
2500 Himalaya Road
Aurora, CO 80011
(303) 375-8882, fax (303) 375-8888
www.univair.com

Aeronca Aviators Club
P.O. Box 66
Coxsackie, NY 12051
www.aeronca.org

Aeronca Sedan Club
115 Wendy Court
Union City, CA 94587
(510) 489-5642

Alexandria Aircraft LLC
2504 Aga Drive
Alexandria, MN 56308
(320)763-4088, fax (320)763-4095
www.alexandriaaircraftllc.com

Bellanca-Champion Club
P.O. Box 100
Coxsackie, NY 12051-0100
518-731-6800
szegor@bellanca-championclub.com
www.bellanca-championclub.com

International Aeronca Association
401 1st Street East
Clark, SD 57225
(605) 532-3862

National Aeronca Association
P.O. Box 2219
Terre Haute, IN 47802
(812) 232-1491

AVIAT/CHRISTIAN/HUSKY

Aviat Aircraft, Inc.
P.O. Box 1240
672 South Washington Street
Afton, WY 83110
(307) 886-3151, fax (307) 886-9674
aviat@aviataircraft.com
www.aviataircraft.com

BEECHCRAFT/RAYTHEON

Raytheon Aircraft Company
P.O. Box 85
Wichita, KS 67201-0085
(316) 676-7072
www.raytheonaircraft.com

American Bonanza Society
Mid-Continent Airport
P.O. Box 12888
Wichita, KS 67277
(316) 945-1700, fax (316) 945-1710
www.bonanza.org

Beechcraft Duke Association
P.O. Box 819
Mesa, AZ 85201
(602) 969-2291

Classic Bonanza Association
P.O. Box 878002
Plano, TX 75086
(972) 227-4741

Duke Flyers Association
P.O. Box 2599
Mansfield, OH 44906-0599
(419) 529-3822

Midwest Bonanza Society
P.O. Box 36782
Groose Point Farms, MI 48236
http://home.earthlink.net/~jwhitehead

North East Bonanza Group
64 Lawrence Road
South Windsor, CT 06074
www.northeastbonanzagroup.org

Rocky Mountain Bonanza Society
32675 Woodside Drive
Evergreen, CO 80439
(303) 670 2244
www.rmbss.org

Staggerwing Club
1885 Millsboro Road
Mansfield, OH 44906
(419) 529-3822

Staggerwing Museum Foundation
P.O. Box 550
Tullahoma, TN 37388
(931) 455-1974

T-34 Association
P.O. Box 925
Champaign, IL 61824
(217) 356-3063

Twin Beech Association, Inc.
P.O. Box 8186
Fountain Valley, CA 92728
(714) 964-4864

Twin Beech 18 Society
c/o Staggerwing Museum Foundation, Inc.
P. O. Box 550
Tullahoma, TN 37388
(931) 455-1974

Twin Bonanza Association
19684 Lakeshore Drive
Three Rivers, MI 49093
(269) 279-2540
www.twinbonanza.com

World Beechcraft Society
500 S.E. Everett Mall Way, Suite A7
Everett, WA 98208-8111
(425) 267-9235
www.worldbeechcraft.com

BELLANCA/VIKING

Alexandria Aircraft LLC
2504 Aga Drive
Alexandria, MN 56308

(320) 763-4088, fax (320) 763-4095
www.alexandriaaircraftllc.com

Bellanca-Champion Club
P.O. Box 100
Coxsackie, NY 12051-0100
(518) 731-6800
szegor@bellanca-championclub.com
www.bellanca-championclub.org

International Viking Owners Group
2640 Ovelisco Place
Carlsbad, CA 92009-6527
(760) 931-1909
http://home.att.net/~ vikingdrvr

CESSNA

Cessna Aircraft Company
P.O. Box 1521
Wichita, KS 67201
(316) 517-6056
www.cessna.textron.com

Univair Aircraft Corporation
2500 Himalaya Road
Aurora, CO 80011
(888) 433-5433
www.univair.com

Cardinal Club
1701 St. Andrew's Drive
Lawrence, KS 66047-1763
(785) 842-7016, fax (785) 842-1777
cardinalclub@juno.com

Cardinal Flyers Online
P.O. Box 532
Hampshire, IL 60140
www.cardinalflyers.com

Cessna 150/152 Club
P.O. Box 1917
Atascadero, CA 93423-1917
(805)461-1958

Cessna 172/182 Club
P.O. Box 22631
Oklahoma City, OK 73123
(405) 495-8664, fax (405) 495-8666

SJONES8166@aol.com
www.cessna172-182club.com

Cessna Owners Organization, Inc.
P.O. Box 5000
Iola, WI 54945
(715) 445-5000, fax (715) 445-4053
jonespub@gglbbs.com
www.cessnaowner.org

Cessna Pilots Association
P.O. Box 5817
Santa Maria, CA 93456
(805) 922-2580, fax (805) 922-7249
info@cessna.org
www.cessna.org

Eastern Cessna 190/195 Association
25575 Butternut Ridge Road
North Olmsted, OH 44070-4505
(440) 777-4025
ccrabs@aol.com

International Cessna 120/140 Association
Box 830092
Richardson, TX 75083-0092
(812) 633-6858
www.cessna120-140.org

International 170 Association
P.O. Box 1667
Lebanon, MO 65536
(417) 532-4847, fax (417) 532-4847
www.cessna170.org

International 180/185 Club
P.O. Box 639
Castlewood, VA 24224
(540) 738-8134

International 195 Club
P.O. Box 737
Merced, CA 95341
(209) 722-6283, fax (209) 722-5124
ewing@elite.net
www.cessna195.org

National 210 Owners Association
P.O. Box 1065
La Canada, CA 91011

(818) 952-6212
avmas98@aol.com

Skymaster Club
2525 NW 71st Place
Ankeny, IA 50021
(515) 289-1439

Straight Tail Cessnas
P.O. Box 234
Darrouzett, TX 79024
(806) 624-3474
doubledr@tgcomp.net

Twin Cessna Flyer
512 Broadway, Suite 102
New Haven, IN 46774
(219) 749-2520, fax (219) 749-6140
twinces@aol.com
www.twinces.org

CIRRUS

Cirrus Owners and Pilots Association
9710 White Blossom Boulevard
Louisville, KY 40241-4178
(502) 412-6500
www.cirruspilots.org

COMMONWEALTH

Rearwin Club
P.O. Box 127
Blakesburg, IA 52536
(515) 938-2773

DIAMOND

Diamond Aircraft Industries, Inc.
1560 Crumlin Sideroad
London, ON N5V 1S2, Canada
(519) 457-4000, fax (519) 457-4021
www.diamondair.com

Diamond Aircraft Pilots and Owners Organization
8716 Skymaster Drive
New Port Richey, FL 34654
www.dapo.org

ERCOUPE/ALON/FORNEY

Univair Aircraft Corporation
2500 Himalaya Road
Aurora, CO 80011
(888) 433-5433, fax (303) 375-8888
www.univair.com

Ercoupe Owners Club
P.O. Box 7117
Ocean Isle Beach, NC 28469-7117
(910) 575-2758
coupecaper@aol.com
www.ercoupe.org

GULFSTREAM AMERICAN/GRUMMAN AMERICAN/
AMERICAN GENERAL

Tiger Aircraft LLC
226 Pilot Way
Martinsburg, WV 25401
(304) 267-1000
www.tigeraircraft.com

FletchAir Company, Inc.
9000 Randolph Street
Houston, TX 77061
(713) 641-2023
fax (713) 643-0070
www.fletchair.com

American Yankee Association
P.O. Box 1531
Cameron Park, CA 95682
(916) 676-4292
www.aya.org

LAKE

Lake Amphibian Flyers Club
7188 Mandarin Drive
Boca Raton, FL 33433
(561) 483-6566
mrodstein@avweb.com

LUSCOMBE

Univair Aircraft Corporation
2500 Himalaya Road

Aurora, CO 80011
(888) 433-5433, fax (303) 375-8888
www.univair.com

Continental Luscombe Association
10251 E. Central Avenue
Del Rey, CA 93616
(559) 888-2745
www.luscombe-cla.org

Luscombe Endowment (Classic Aero Support)
15815 E. Melrose Street
Gilbert, AZ 85296
(480) 650-0883
dcombs@luscombesilvaire.info
www.luscombesilvaire.info

Luscombe Association
1002 Heather Lane
Hartford, WI 53027
(262)966-7627

MAULE

Maule Air, Inc.
2099 Georgia Highway 133 S.
Moultrie, GA 31768
(229) 985-2045, fax (229) 890-2402
www.mauleairinc.com

MEYERS

Meyers Aircraft Owners Association
5852 Bogue Road
Yuba City, CA 95993
(916)673-2724

MOONEY

Mooney Aircraft Corporation
Louis Schreiner Field
Kerrville, TX 78028
(800) 456-3033, (210) 896-6000, fax (210) 896-8180
www.mooney.com

Mooney Aircraft Pilots Association
P.O. Box 460607
San Antonio, TX 78246-0607
(210) 525-8008, fax (210) 525-8085
mapa@mooneypilots.com
www.mooneypilots.com

NAVION

American Navion Society
PMB 335
16420 SE McGillivray, Suite 103
Vancouver, WA 98683-3461
(360) 833-9921
www.navionsociety.org

Navion Skies
P.O. Box 2678
Lodi, CA 95241-2678
(209) 367-9390
www.navionskies.com

PIPER

New Piper Aircraft Company
2926 Piper Drive
Vero Beach, FL 32960
(561) 567-4361
www.newpiper.com

Univair Aircraft Corporation
2500 Himalaya Road
Aurora, CO 80011
(888) 433-5433
fax (303) 375-8888
www.univair.com

Cherokee Pilots Association
P.O. Box 1996
Lutz, FL 33548-1996
(813)948-3616
www.piperowner.com

Cub Club
1002 Heather Lane
Hartford, WI 53027
(262) 966-7627
sskroz@aol.com

Flying Apache Association
6778 Skyline Drive
Delray Beach, FL 33446
(561) 499-1115

International Comanche Society, Inc.
Hangar 3
Wiley Post Airport

Bethany, OK 73008
(405) 491-0321, fax (405) 491-0325
comancheflyer@compuserve.com
www.ics.pxl.net

Malibu and Mirage Owners & Pilots Association
280 N. Bluebird Drive
Green Valley, AZ 85614
(520) 399-1121

Piper Aviation Museum
One Piper Way
Lock Haven, PA 17745
(570) 748-8283
www.pipermuseum.com

Piper Owner Society
P.O. Box 5000
Iola, WI 54945-5000
(715) 445-5000, fax (715) 445-4053
www.piperowner.org

Piper Pilots Association
140 Heimer Road, #560
San Antonio, TX 78232
(210) 525-8008
www.piperpilots.com

Short Wing Piper Club, Inc.
220 Main
Halstead, KS 67056
(316) 835-3307
www.shortwing.com

Super Cub Pilot's Association
P.O. Box 9823
Yakima, WA 98909
(509) 248-9491
www.cubcrafters.com

PITTS

Aviat Aircraft, Inc.
P.O. Box 1240
672 South Washington Street
Afton, WY 83110
(307) 886-3151, fax (307) 886-9674
aviat@aviataircraft.com
www.aviataircraft.com

SEABEE

International Republic SeaBee Owners Club
smestler@pbtcomm.net
www.republicseabee.com

SOCATA

Socata Aircraft
7501 Pembroke Road
Pembroke Pines, FL 33023
(954) 893-1400
www.socataaircraft.com

Socata TB Users Group
14925 North Oaks Drive
Dallas, TX 75240
(972) 960-6408
www.socata.org

STINSON

Univair Aircraft Corporation
2500 Himalaya Road
Aurora, CO 80011
(888) 433-5433, fax (303) 375-8888
www.univair.com

National Stinson Club
14418 Skinner Road
Cypress, TX 77429
(281) 373-0418

Southwest Stinson Club
3005 6th Street
Sacramento, CA 95818
(916) 446-3729

SWIFT

Swift Museum Foundation, Inc.
P.O. Box 644
Athens, TN 37371
(423) 745-9547
www.napanet.net/ ~ arbeau/swift

TAYLORCRAFT

Taylorcraft Aviation, Inc.
4495 W. State Highway 71

La Grange, TX 78945-5150
(800) 217-1399
www.taylorcraft.com/

Univair Aircraft Corporation
2500 Himalaya Road
Aurora, CO 80011
(888) 433-5433, fax (303) 375-8888
www.univair.com

International Taylorcraft Owners Club and Foundation
12809 Greenbower Road
Alliance, OH 44601
(330) 823-9748
www.taylorcraft.org

VARGA/SHINN/MORRISEY

VG-21 Squadron
717 East Brooks Street
Chandler, AZ 85225
(602)786-3578

Engine Manufacturers

CONTINENTAL/TELEDYNE CONTINENTAL

Teledyne Continental Motors
P.O. Box 90
Mobile, AL 36601
(251) 438-3411, fax (251) 432-7352
www.tcmlink.com

FRANKLIN

Franklin Aircraft Engines
P.O. Box 271730
Fort Collins, CO 80527-1730
(970) 224-4404
www.franklinengines.com

LYCOMING/TEXTRON LYCOMING

Textron Lycoming Company
652 Oliver Street
Williamsport, PA 17701
(570) 323-6181
www.lycoming.textron.com

ON THE WORLD WIDE WEB

The World Wide Web (www) can be an interesting place for the airplane owner/buyer to explore. As a starting point, the following web sites are given:

AOPA	www.aopa.com
EAA	www.eaa.org
AvWeb	www.avweb.com
FAA	www.faa.gov
LANDINGS	www.landings.com
NTSB	www.ntsb.gov

Note: Some older web browsers may require that http:// be entered before the preceding web site addresses.

E

Advisory Circulars

THE FOLLOWING AVIATION CIRCULARS, PUBLISHED BY THE FAA, will aid you in understanding the fine points of maintenance records as applicable to small airplane purchasing and ownership.

AC 43-9C

SUBJECT: Maintenance Records

Date: 6/8/98

Initiated by: AFS-340

1. PURPOSE.
This advisory circular (AC) describes methods, procedures and practices that have been determined to be acceptable means of showing compliance with the general aviation maintenance record making and record keeping requirements of Title 14 of the Code of Federal Regulations (14 CFR) parts 43 and 91. This material is not mandatory, nor is it regulatory and acknowledges that the Federal Aviation Administration (FAA) will consider other methods that may be presented. It is issued for guidance purposes and outlines several methods of compliance with the regulations.

Note: The information in this AC does not apply to air carrier maintenance records made and retained in accordance with 14 CFR part 121.

2. CANCELLATION.
AC 43-9B, Maintenance Records, dated January 9, 1984, is canceled.

3. RELATED REGULATIONS.
14 CFR parts 1, 43, 91, and 145.

4. DISCUSSION.
The Code of Federal Regulations state that a U.S. standard airworthiness certificate is effective until it is surrendered, suspended, revoked, or a

termination date is otherwise established by the Administrator. In addition to those terms, a U.S. standard airworthiness certificate is effective only as long as the maintenance, preventive maintenance, and alterations are performed in accordance with parts 43 and 91, and the aircraft are registered in the United States. These terms and conditions are further restated, in block 6, on the front of FAA Form 8100-2, Standard Airworthiness Certificate. Qualified persons, who perform the maintenance, preventive maintenance and alterations, shall make a record entry of this accomplishment, thus maintaining the validity of the certificate of airworthiness. Adequate aircraft records provide tangible evidence that the aircraft complies with the appropriate airworthiness requirements. In accordance with the terms and conditions listed in block 6 of the Standard Airworthiness Certificate, insufficient or non-existent aircraft records may render that standard airworthiness certificate invalid.

5. MAINTENANCE RECORD REQUIREMENTS.

a. Responsibilities. 14 CFR part 91, section 91.417 states that an aircraft owner/operator shall keep and maintain aircraft maintenance records. 14 CFR part 43, sections 43.9 and 43.11 state that maintenance personnel, however, are required to make the record entries.

b. Maintenance Records That Are to Be Retained. Section 91.405 requires each owner or operator to ensure that maintenance personnel make appropriate entries in the maintenance records to indicate that the aircraft has been approved for return to service. Section 91.417(a) sets forth the content requirements and retention requirements for maintenance records. Maintenance records may be kept in any format that provides record continuity; includes required contents; lends itself to the addition of new entries; provides for signature entry; and, is intelligible. Section 91.417(b) requires records of maintenance, alterations, and required or approved inspections to be retained until the work is repeated, superseded by other work, or for one year. It also requires the records, specified in section 91.417(a)(2), to be retained and transferred with the aircraft at the time of sale.

Note: Section 91.417(a) contains an exception regarding work accomplished in accordance with section 91.411. This does not exclude the making of entries for this work, but applies to the retention period of the records for work done in accordance with this section. The exclusion is necessary since the retention period of one year is inconsistent with the 24-month interval of test and inspection specified in section 91.411. Entries for work done per this section are to be retained for 24 months or until the work is repeated or superseded.

c. Section 91.417(a)(1). Requires a record of maintenance, for each aircraft (including the airframe) and each engine, propeller, rotor, and appliance of an aircraft. This does not require separate or individual records for each of these items. It does require the information specified in sections 91.417(a)(1) through 91.417(a)(2)(vi) to be kept for each item as appropriate. As a practical matter, many owners and operators find it advantageous to keep separate or individual records since it facilitates transfer of the record with the item when ownership changes. Section 91.417(a)(1) has no counterpart in section 43.9 or section 43.11.

d. Section 91.417(a)(1)(i). Requires the maintenance record entry to include "a description of the work performed." The description should be in sufficient detail to permit a person unfamiliar with the work to understand what was done, and the methods and procedures used in doing it. When the work is extensive, this results in a voluminous record. To provide for this contingency, the rule permits reference to technical data acceptable to the Administrator in lieu of making the detailed entry. Manufacturer's manuals, service letters, bulletins, work orders, FAA AC's, and others, which accurately describe what was done, or how it was done, may be referenced. Except for the documents mentioned, which are in common usage, referenced documents are to be made a part of the maintenance records and retained in accordance with section 91.417(b).

Note: Certificated repair stations frequently work on components shipped to them without the maintenance records. To provide for this situation, repair stations should supply owners and operators with copies of work orders written for the work, in lieu of maintenance record entries. The work order copy must include the information, required by section 91.417(a)(1) through section 91.417(a)(1)(iii), be made a part of the maintenance record, and retained per section 91.417(b). This procedure is not the same as that for maintenance releases discussed in paragraph 16, and it may not be used when maintenance records are available. Section 91.417(a)(1)(i) is identical to its counterpart, section 43.9(a)(1), which imposes the same requirements on maintenance personnel.

e. Section 91.417(a)(1)(ii). Is identical to section 43.9(a)(2) and requires entries to contain the date the work was completed. This is normally the date upon which the work is approved for return to service. However, when work is accomplished by one person and approved for return to service by another, the dates may differ. Two signatures may also appear under this circumstance; however, a single entry in accordance with section 43.9(a)(3) is acceptable.

f. Section 91.417(a)(1)(iii). Differs slightly from section 43.9(a)(4) in that it requires the entry to indicate only the signature and certificate number of the person approving the work for return to service, and does not require the type of certificate being exercised to be indicated as does section 43.9(a)(4). This is a new requirement of section 43.9(a)(4), which assists owners and operators in meeting their responsibilities. Maintenance personnel may indicate the type of certificate exercised by using airframe (A), powerplant (P), airframe & powerplant (A&P), inspection authorization (IA), or certificated repair station (CRS).

g. Section 91.417(a)(2). Requires six items to be made a part of the maintenance record and maintained as such. Section 43.9 does not require maintenance personnel to enter these items. Section 43.11 requires some of them to be part of entries made for inspections, but they are all the responsibility of the owner or operator. The six items are discussed as follows:

(1) Section 91.417(a)(2)(i). Requires a record of total time-in-service to be kept for the airframe, each engine, and each propeller. Part 1, section 1.1, Definitions, defines time in service, with respect to maintenance time records, as that time from the moment an aircraft leaves the surface of the earth until it touches down at the next point of landing. Section 43.9 does not require this to be part of the entries for maintenance, preventive maintenance, rebuilding, or alterations. However, section 43.11 requires maintenance personnel to make it a part of the entries for inspections made under parts 91, 125, and time-in-service in all entries.

(a) Some circumstances impact the owner's or operator's ability to comply with section 91.417(a)(2)(i). For example, in the case of rebuilt engines, the owner or operator would not have a way of knowing the total time-in-service, since section 91.421 permits the maintenance record to be discontinued and the engine time to be started at zero. In this case, the maintenance record and time-in-service, subsequent to the rebuild, comprise a satisfactory record.

(b) Many components, presently in-service, were put into service before the requirements to keep maintenance records on them. Propellers are probably foremost in this group. In these instances, practicable procedures for compliance with the record requirements must be used. For example, total time-in-service may be derived using the procedures described in paragraph 12; or if records prior to the regulatory requirements are just not available from any source, time-in-service may be kept since last complete overhaul. Neither of these procedures is acceptable when life-limited parts status is involved or when airworthiness directive (AD) compliance is a factor. Only the actual record since new may be used in these instances.

(c) Sometimes engines are assembled from modules (turbojet and some turbopropeller engines) and a true total time-in-service for the total engine is not kept. If owners and operators wish to take advantage of this modular design, then total time-in-service and a maintenance record for each module is to be maintained. The maintenance records specified in section 91.417(a)(2) are to be kept with the module.

(2) Section 91.417(a)(2)(ii). Requires the current status of life-limited parts to be part of the maintenance record. If total time-in-service of the aircraft, engine, propeller, etc., is entered in the record when a life-limited part is installed and the time-in-service of the life-limited part is included, the normal record of time-in-service automatically meets this requirement.

(3) Section 91.417(a)(2)(iii). Requires the maintenance record to indicate the time since last overhaul of all items installed on the aircraft that are required to be overhauled on a specified time basis. The explanation in paragraph 5g (2) also applies to this requirement.

(4) Section 91.417(a)(2)(iv). Deals with the current inspection status and requires it to be reflected in the maintenance record. Again, the explanation in paragraph 5g (2) is appropriate even though section 43.11(a)(2) requires maintenance persons to determine time-in-service of the item being inspected and to include it as part of the inspection entry.

(5) Section 91.417(a)(2)(v). Requires the current status of applicable AD's to be a part of the maintenance record. The record is to include, at minimum, the method used to comply with the AD, the AD number, and revision date; and if the AD has requirements for recurring action, the time-in-service and the date when that action is required. When AD's are accomplished, maintenance persons are required to include the items specified in section 43.9(a)(2), (3), and (4) in addition to those required by section 91.417(a)(2)(v). An example of a maintenance record format for AD compliance is contained in Appendix 1.

(6) Section 91.417(a)(2)(vi). In the past, the owner or operator has been permitted to maintain a list of current major alterations to the airframe, engine(s), propeller(s), rotor(s), or appliances. This procedure did not produce a record of value to the owner/operator or to maintenance persons in determining the continued airworthiness of the alteration since such a record was not sufficient detail. This section of the rule has now been changed. It now prescribes that copies of FAA Form 337, issued for the alteration, be made a part of the maintenance record.

6. PREVENTIVE MAINTENANCE.

a. Preventive maintenance is defined in part 1, section 1.1. Part 43, appendix A, paragraph (c) lists those items which a pilot may accomplish under section 43.3(g). Section 43.7 authorizes appropriately rated repair stations and mechanics, and persons holding at least a private pilot certificate to approve an aircraft for return to service after they have performed preventive maintenance. All of these persons must record preventive maintenance accomplished in accordance with the requirements of section 43.9. AC 43-12, Preventive Maintenance, current edition, contains further information on this subject.

b. The type of certificate exercised when maintenance or preventive maintenance is accomplished must be indicated in the maintenance record. Pilots may use private pilot (PP), commercial pilot (CP), or air transport pilot (ATP) to indicate private, commercial, or airline transport pilot certificate, respectively, in approving preventive maintenance for return to service. Pilots are not authorized by section 43.3(g) to perform preventive maintenance on aircraft when they are operated under part 121, 127, 129, or 135. Pilots may only approve for return to service preventive maintenance that they themselves have accomplished.

7. REBUILT ENGINE MAINTENANCE RECORDS.

a. Section 91.421 provides that zero time may be granted to an engine that has been rebuilt by a manufacturer or an agency approved by the manufacturer. When this is done, the owner/operator may use a new maintenance record without regard to previous operating history.

b. The manufacturer or an agency approved by the manufacturer that rebuilds and grants zero time to an engine is required by section 91.421 to provide a signed statement containing: 1) the date the engine was rebuilt; 2) each change made, as required by an AD; and 3) each change made in compliance with service bulletins, when the service bulletin specifically requests an entry to be made.

c. Section 43.2(b) prohibits the use of the term rebuilt in describing work accomplished in required maintenance records or forms unless the component worked on has had specific work functions accomplished. These functions are listed in section 43.2(b) and, except for testing requirements, are the same as those set forth in section 91.421(c). When terms such as remanufactured, reconditioned, or other terms coined by various aviation enterprises are used in maintenance records, owners and operators cannot assume that the functions outlined in section 43.2(b) have been done.

8. RECORDING TACHOMETERS.

a. Time-in-service recording devices sense such things as electrical power on, oil pressure, wheels on the ground, etc., and from these conditions provide an indication of time-in-service. With the exception of those that sense aircraft lift-off and touchdown, the indications are approximate.

b. Some owners and operators mistakenly believe these devices may be used in lieu of keeping time-in-service in the maintenance record. While they are of great assistance in arriving at the time-in-service, such instruments, alone, do not meet the requirements of section 91.417. For example, when the device fails and requires change, it is necessary to enter time-in-service and the instrument reading at the change. Otherwise, record continuity is lost.

9. MAINTENANCE RECORDS FOR AD COMPLIANCE.

This subject is covered in AC 39-7, Airworthiness Directives for General Aviation Aircraft, current edition. A separate AD record may be kept for the airframe and each engine, propeller, rotor, and appliance, but is not required. This would facilitate record searches when inspection is needed, and when an engine, propeller, rotor, or appliance is removed, the record may be transferred with it. Such records may also be used as a schedule for recurring inspections. The format, shown in Appendix 1, is a suggested one, and adherence is not mandatory. Owners should be aware that they may be responsible for non-compliance with AD's when their aircraft are leased to foreign operators. They should, therefore, ensure that leases should be drafted to deal with this subject.

10. MAINTENANCE RECORDS FOR REQUIRED INSPECTIONS.

a. Section 43.11 contains the requirements for inspection entries. While these requirements are imposed on maintenance personnel, owners and operators should become familiar with them in order to meet their responsibilities under section 91.405.

b. The maintenance record requirements of section 43.11 apply to the 100-hour, annual, and progressive inspections under part 91; inspection programs under parts 91 and 125; approved airplane inspection programs under part 135; and the 100-hour and annual inspections under section 135.411(a) (1).

c. Appropriately rated mechanics are authorized to conduct these inspections and make the required entries. Particular attention should be given to section 43.11(a)(7) in that it now requires a more specific statement than previously required under section 43.9. The entry, in addition to other items, must identify the inspection program used; identify the portion or segment of the inspection program accomplished; and contain a statement that the inspection was performed in accordance with the instructions and procedures for that program.

d. Questions continue regarding multiple entries for 100-hour/annual inspections. As discussed in paragraph 5c, neither part 43 nor part 91 requires separate records to be kept. Section 43.11, however, requires persons approving or disapproving equipment for return to service, after any required inspection, to make an entry in the record of that equipment. Therefore, when an owner maintains a single record, the entry of the 100-hour or annual inspection is made in that record. If the owner maintains separate records for the airframe, powerplants, and propellers, the entry for the 100-hour inspection is entered in each, while the annual inspection is only required to be entered into the airframe record.

11. DISCREPANCY LISTS.

a. Before October 15, 1982, issuance of discrepancy lists (or lists of defects) to owners or operators was appropriate only in connection with annual inspections under part 91; inspections under section 135.411(a)(1); inspection programs under part 125; and inspections under section 91.217. Now, section 43.11 requires that a discrepancy list be prepared by a person performing any inspection required by parts 91, 125, or section 135.411(a)(1).

b. When a discrepancy list is provided to an owner or operator, it says in effect, except for these discrepancies, the item inspected is airworthy. It is imperative, therefore, that inspections be complete and that all discrepancies appear in the list. When circumstances dictate that an inspection be terminated before it is completed, the maintenance record should clearly indicate that the inspection was discontinued. The entry should meet all the other requirements of section 43.11.

c. It is no longer a requirement that copies of discrepancy lists be forwarded to the local Flight Standards District Office (FSDO).

d. Discrepancy lists (or lists of defects) are part of the maintenance record and the owner/operator is responsible to maintain that record in accordance with section 91.417(b)(3). The entry made by maintenance personnel in the maintenance record should reference the discrepancy list when a list is issued.

12. LOST OR DESTROYED RECORDS.

Occasionally, the records for an aircraft are lost or destroyed. In order to re-construct them, it is necessary to establish the total time-in-service of the airframe. This can be done by reference to other records that reflect the time-in-service; research of records maintained by repair facilities; and reference to records maintained by individual mechanics, etc. When these things have been done and the record is still incomplete, the owner/operator may make a notarized statement in the new record describing the loss and establishing the time-in-service based on the research and the best estimate of time-in-service.

a. The current status of applicable AD's may present a more formidable problem. This may require a detailed inspection by maintenance personnel to establish that the applicable AD's have been complied with. It can readily be seen that this could entail considerable time, expense, and in some instances, might require recompliance with the AD.

b. Other items required by section 91.417(a)(2), such as the current status of life-limited parts, time since last overhaul, current inspection status, and current list of major alterations, may present difficult problems. Some items may be easier to reestablish than others, but all are problems. Losing maintenance records can be troublesome, costly, and time consuming. Safekeeping of the records is an integral part of a good record keeping system.

13. COMPUTERIZED RECORDS.

There is a growing trend toward computerized maintenance records. Many of these systems are offered to owners/operators on a commercial basis. While these are excellent scheduling systems, alone they normally do not meet the requirements of sections 43.9 or 91.417. The owner/operator who uses such a system is required to ensure that it provides the information required by section 91.417, including signatures. If not, modification to make them complete is the owner's/operator's responsibility and that responsibility may not be delegated.

14. PUBLIC AIRCRAFT.

Prospective purchasers of aircraft that have been used as public aircraft should be aware that public aircraft may not be subject to the certification and maintenance requirements in Title 14 of the Code of Federal Regulations and may not have records that meet the requirements of section 91.417. Considerable research may be involved in establishing the required records when these aircraft are purchased and brought into civil aviation. The aircraft may not be certificated or used without such records.

15. LIFE-LIMITED PARTS.

a. Present day aircraft and powerplants commonly have life-limited parts installed. These life limits may be referred to as retirement times, service life limitations, parts retirement limitations, retirement life limits, life limitations, or other such terminology and may be expressed in hours, cycles of operation, or calendar time. They are set forth in type certificate data sheets (TCDS), AD's, or the limitations section of FAA-approved airplane or rotorcraft flight manuals. Additionally, instructions for continued airworthiness, which require life-limits be specified, may apply (See CFR 23 Appendix G and CFR 27 Appendix A).

b. Section 91.417(a)(2)(ii) requires the owner or operator of an aircraft with such parts installed to have records containing the current status of these parts. Many owners/operators have found it advantageous to have a separate record for such parts showing the name of the part, part number, serial number, date of installation, total time-in-service, date removed, and signature and certificate number of the person installing or removing the part. A separate record, as described, facilitates transferring the record with the part in the event the part is removed and later reinstalled or installed on another aircraft or engine. If a separate record is not kept, the aircraft record must contain sufficient information to clearly establish the status of the life-limited parts installed.

16. MAINTENANCE RELEASE.

a. In addition to those requirements discussed previously, section 43.9 requires that major repairs and alterations be recorded as indicated in appendix B of part 43 (i.e., on FAA Form 337). An exception is provided in paragraph (b) of that appendix, which allows repair stations certificated under part 145 to use a maintenance release in lieu of the form for major repairs (and only major repairs).

b. The maintenance release must contain the information specified in paragraph (b)(1), (2) and (3), appendix B of part 43, be made a part of the aircraft maintenance record, and retained by the owner/operator as specified in section 91.417. The maintenance release is usually a special document (normally a tag) and is attached to the product when it is approved for return to service. The maintenance release may, however, be on a copy of the work order written for the product. When this is done (for major repairs only) the entry on the work order must meet paragraph (b)(1), (2), and (3) of the appendix. That is to say that the Repair Station is required to give the owner: (1) the customer's work order upon which the repair is recorded; (2) a signed copy of the work order; and (3) a maintenance release which has been signed by an authorized representative of the company. In some cases, a work order and a maintenance release may be a different document. Both must be supplied to the customer.

c. Some repair stations use what they call a maintenance release for other than major repairs. This is sometimes a tag and sometimes information on a work order. When this is done, all of the requirements of section 43.9 must be met (paragraph (b)(3), appendix B, not applicable) and the document is to be made and retained as part of the maintenance records under section 91.417 per discussion in paragraph 5c.

17. FAA FORM 337, MAJOR REPAIR AND ALTERATION.

a. Major repairs and alterations are to be recorded on FAA Form 337, as stated in paragraph 16. This form is executed by the person making the repair or alteration. Provisions are made on the form for a person other than that person performing the work to approve the repair or alteration for return to service.

b. These forms are now required to be made part of the maintenance record of the product repaired or altered and retained in accordance with section 91.417.

c. Detailed instructions for use of this form are contained in AC 43.9-1, Instructions for Completion of FAA Form 337, current edition.

d. Some manufacturers have initiated a policy of indicating, on their service letters and bulletins, and other documents dealing with changes to their aircraft, whether or not the changes constitute major repairs or alterations. Some manufacturers also indicate that the responsibility for completing FAA Form 337 lies with the person accomplishing the repairs or alterations and cannot be delegated. When there is a question, it is advisable to contact the local FSDO for guidance.

18. TESTS AND INSPECTIONS FOR ALTIMETER SYSTEMS, ALTITUDE REPORTING EQUIPMENT, AND AIR TRAFFIC CONTROL (ATC) TRANSPONDERS.

The recordation requirements for these tests and inspections are the same as for other maintenance. There are essentially three tests and inspections (the altimeter system, the transponder system, and the data correspondence test), each of which may be subdivided relative to who may perform specific portions of the test. The basic authorization for performing these tests and inspections, found in section 43.3, is supplemented by sections 91.411 and 91.413. When multiple persons are involved in the performance of tests and inspections, care must be exercised to insure proper authorization under these three sections and compliance with sections 43.9 and 43.9(a)(3) in particular.

19. BEFORE YOU BUY.

This is the proper time to take a close look at the maintenance records of any used aircraft you expect to purchase. A well-kept set of maintenance records, which properly identifies all previously performed maintenance, alterations, and AD compliances, is generally a good indicator of the aircraft condition. This is not always the case, but in any event, before you buy, require the owner to produce the maintenance records for your examination, and require correction of any discrepancies found on the aircraft or in the records. Many prospective owners have found it advantageous to have a reliable unbiased maintenance person examine the maintenance records, as well as the aircraft, before negotiations have progressed too far.

If the aircraft is purchased, take the time to review and learn the system of the previous owner to ensure compliance and continuity when you modify or continue that system.

APPENDIX 1. AIRWORTHINESS DIRECTIVE COMPLIANCE RECORD (SUGGESTED FORMAT) (Fig. E-1)

APPENDIX 1. AIRWORTHINESS DIRECTIVE COMPLIANCE RECORD (SUGGESTED FORMAT)

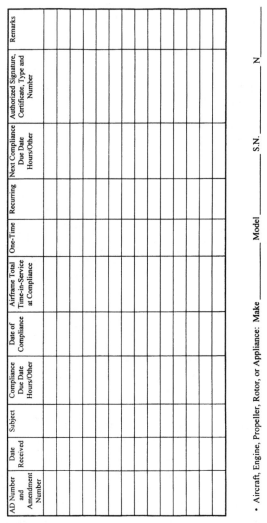

Fig. E-1. *Suggested format for an AD compliance record (from AC 43-9C).*

AC 43-9.1E

SUBJECT: Instructions for Completion of FAA Form 337 (OMB No. 2120-0020), Major Repair and Alteration (Airframe, Powerplant, Propeller, or Appliance)

Date: 5/21/87

Initiated by: AFS-340

1. PURPOSE.
This advisory circular (AC) provides instruction for completing Federal Aviation Administration (FAA) Form 337, Major Repair and Alteration (Airframe, Powerplant, Propeller, or Appliance).

2. CANCELLATION.
AC 43.9-1D, Instructions for Completion of FAA Form 337 (OMB 04-R0060), Major Repair and Alteration (Airframe, Powerplant, Propeller, or Appliance), dated 9/5/79, is canceled.

3. RELATED FEDERAL AVIATION REGULATIONS (FAR) SECTIONS.
FAR Part 43, Sections 43.5, 43.7, 43.9, and Appendix B.

4. INFORMATION.
FAA Form 337 is furnished free of charge and is available at all FAA Air Carrier (ACDO), General Aviation (GADO), Manufacturing Inspection (MIDO), and Flight Standards (FSDO) district offices, and at all International Field Offices (IFO). The form serves two main purposes; one is to provide aircraft owners and operators with a record of major repairs or alterations indicating details and approval, and the other is to provide the FAA with a copy of the form for inclusion in the aircraft records at the FAA Aircraft Registration Branch, Oklahoma City, Oklahoma.

5. INSTRUCTIONS FOR COMPLETING FAA FORM 337.
The person who performs or supervises a major repair or major alteration should prepare FAA Form 337. The form is executed at least in duplicate and is used to record major repairs and major alterations made to an aircraft, an airframe, powerplant, propeller, appliance, or spare part. The following instructions apply to corresponding items 1 through 8 of the form as illustrated in Appendix 1.

    a. Item 1—Aircraft. Information to complete the "Make," "Model," and "Serial Number" blocks will be found on the aircraft manufacturer's identification plate. The "Nationality" and "Registration Mark" is the same as shown on AC Form 8050-3, Certificate of Aircraft Registration.

    b. Item 2—Owner. Enter the aircraft owner's complete name and address as shown on AC Form 8050-3.

    Note: When a major repair or alteration is made to a spare part or appliance, items 1 and 2 will be left blank, and the original and duplicate copy of the form will remain with the part until such time as it is installed on an aircraft. The person installing the part will then enter the required information in blocks 1 and 2, give the original of the form to the aircraft owner/operator, and forward the duplicate copy to the local FAA district office within 48 hours after the work is inspected.

c. Item 3—For FAA Use Only. Approval may be indicated in Item 3 when the FAA determines that data to be used in performing a major alteration or a major repair complies with accepted industry practices and all applicable FAR. Approval is indicated in one of the following methods. (See paragraph 6b for further details.)

(1) Approval by examination of data only—one aircraft only: "The data identified herein complies with the applicable airworthiness requirements and is approved for the above described aircraft, subject to conformity inspection by a person authorized in FAR Part 43, Section 43.7."

(2) Approval by physical inspection, demonstration, testing, etc., of the data and aircraft—one aircraft only: "The alteration (or repair) identified herein complies with the applicable airworthiness requirements and is approved for the above described aircraft, subject to conformity inspection by a person authorized in FAR Part 43, Section 43.7."

(3) Approval by examination of data only—duplication on identical aircraft. "The alteration identified herein complies with the applicable airworthiness requirements and is approved for duplication on identical aircraft make, model, and altered configuration by the original modifier."

d. Item 4—Unit Identification. The information blocks under item 4 are used to identify the airframe, powerplant, propeller, or appliance repaired or altered. It is only necessary to complete the blocks for the unit repaired or altered.

e. Item 5—Type. Enter a checkmark in the appropriate column to indicate if the unit was repaired or altered.

f. Item 6—Conformity Statement.

(1) "A"—Agency's Name and Address. Enter name of the mechanic, repair station, or manufacturer accomplishing the repair or alteration. Mechanics should enter their name and permanent mailing address. Manufacturers and repair stations should enter the name and address under which they do business.

(2) "B"—Kind of Agency. Check the appropriate box to indicate the type of person or organization who performed the work.

(3) "C"—Certificate Number. Mechanics should enter their mechanic certificate number in this block, e.g., 1305888. Repair stations should enter there air agency certificate number and the rating or ratings under which the work was performed, e.g., 1234, Airframe Class 3. Manufacturers should enter their type production or Supplemental Type Certificate (STC) number. Manufacturers of Technical Standard Orders (TSO) appliances altering these appliances should enter the TSO number of the appliance altered.

(4) "D"—Compliance Statement: This space is used to certify that the repair or alteration was made in accordance with the FAR. When work was performed or supervised by certificated mechanics not employed by a manufacturer or repair station, they should enter the date the repair or alteration was completed and sign their full name. Repair stations are permitted to authorize persons in their employ to date and sign this conformity statement.

g. Item 7—Approval for return to service. FAR Part 43 establishes the conditions under which major repairs or alterations to airframes, powerplants, propellers, and/or appliances may be approved for return to service. This portion of the form is used to indicate approval or rejection of the repair or alteration of the unit involved and to identify the person or agency making the airworthiness inspection. Check the "approved" or "rejected" box to indicate the finding. Additionally, check the appropriate box to indicate who made the finding. Use the box labeled "other" to indicate a finding by a person other than those listed. Enter the date the finding was made. The authorized person who made the finding should sign the form and enter the appropriate certificate or designation number.

h. Item 8—Description of work accomplished. A clear, concise, and legible statement describing the work accomplished should be entered in item 8 on the reverse side of FAA Form 337. It is important that the location of the repair or alteration, relative to the aircraft or component, be described. The approved data used as the basis for approving the major repair or alteration for return to service should be identified and described in this area.

(1) For example, if a repair was made to a buckled spar, the description in this part might begin by stating, "Removed wing from aircraft and removed skin from outer 6 feet. Repaired buckled spar 49 inches from tip in accordance with…" and continue with a description of the repair. The description should refer to applicable FAR sections and to the FAA-approved data used to substantiate the airworthiness of the repair or alteration. If the repair or alteration is subject to being covered by skin or other structure, a statement should be made certifying that a precover inspection was made and that covered areas were found satisfactory.

(2) Data used as a basis for approving major repairs or alterations for return to service must be FAA-approved prior to its use for that purpose and includes: FAR (e.g., airworthiness directives), AC's (e.g., AC 43.13-1A under certain circumstances), TSO's parts manufacturing approval (PMA), FAA-approved manufacturer's instructions, kits and service handbooks, type certificate data sheets, and aircraft specifications. Other forms of approved data would be those approved by a designated engineering representative (DER), a manufacturer holding a delegation option authorization (DOA), STC's, and, with certain limitations, previous FAA filed approvals. Supporting data such as stress analyses, test reports, sketches, or photographs should be submitted with the FAA Form 337. These supporting data will be returned to the applicant by the local FAA

district office since only FAA Form 337 is retained as a part of the aircraft records at Oklahoma City.

(3) If additional space is needed to describe the repair or alteration, attach sheets bearing the aircraft nationality and registration mark and the date work was completed.

(4) Showing weight and balance computations under this item is not required; however, it may be done. In all cases where weight and balance of the aircraft are affected, the changes should be entered in the aircraft weight and balance records with the date, signature, and reference to the work performed on the FAA Form 337 that required the changes.

6. ADMINISTRATIVE PROCESSING.

At least an original and one duplicate copy of the FAA Form 337 will be executed. FAA district office processing of the forms and their supporting data will depend upon whether previously approved or non-previously approved data was used as follows:

a. Previously Approved Data. The forms will be completed as instructed in this AC ensuring that item 7, "Approval for Return to Service," has been properly executed. Give the original of the form to the aircraft owner or operator, and send the duplicate copy to the local FAA district office within 48 hours after the work is inspected.

b. Non-previously Approved Data. The forms will be completed as instructed in this AC, leaving item 7, "Approval for Return to Service," blank. Both copies of the form, with supporting data, will be sent to the local FAA district office. When the FAA determines that the major repair or alteration data complies with the applicable regulations and is in conformity with accepted industry practices, data approval will be recorded by entering an appropriate statement in item 3, "For FAA Use Only." Both forms and supporting data will be returned to the applicant who will complete item 7, "Approval for Return to Service." The applicant will give the original of the form, with its supporting data, to the aircraft owner or operator and return the duplicate copy to the local FAA district office who will, in turn, forward it to the FAA Aircraft Registration Branch, Oklahoma City, Oklahoma, for inclusion in the aircraft records.

c. Signatures on FAA Form 337 have limited purposes:

(1) A signature in item 3, "For FAA Use Only," indicates approval of the data described in that section for use in accomplishing the work described under item 8 on the reverse of FAA Form 337.

(2) A signature in item 6, "Conformity Statement," is a certification by the person performing the work that it was accomplished in accordance with applicable FAR and FAA-approved data. The certification is only applicable to that work described under item 8 on the reverse of FAA Form 337.

Note: Neither of these signatures (subparagraph c(1) and c(2)) indicate FAA approval of the work described under item 8 for return to service.

(3) A signature in item 7, "Approval for Return to Service," does not signify FAA approval unless the box to the left of "FAA Flight Standards Inspector" or "FAA Designee" is checked. The other persons listed in item 7 are authorized to "approve for return to service" if the repair or alteration is accomplished using FAA-approved data, is performed in accordance with applicable FAR, and found to conform.

d. FAA Form 337 is not authorized for use on other than U.S.-registered aircraft. If a foreign civil air authority requests the form, as a record of work performed, it may be provided. The form should be executed in accordance with the FAR and this AC. The foreign authority should be notified on the form that it is not an official record and that it will not be recorded by the FAA Aircraft Registration Branch, Oklahoma City, Oklahoma.

e. FAR Part 43, Appendix B, Paragraph (b) authorizes FAA certificated repair stations to use a work order, in lieu of FAA Form 337, for only major repairs. Such work orders should contain all the information provided on the form and in no less detail; that is, the data used as a basis of approval should be identified, a certification that the work was accomplished using that data and in accordance with the FAR, a description of the work performed (as required in item 8 of the FAA Form 337), and approval for return to service must be indicate by an authorized person. Signature, kind of certificate, and certificate number must also appear in the record (reference FAR Section 43.9).

AC 43-10A

SUBJECT: Mechanical Work Performed on U.S. and Canadian Registered Aircraft

Date: 2/25/83

Initiated by: AWS-340

1. PURPOSE.
This circular provides updated information and guidance to aircraft owners/operators and maintenance personnel concerning mechanical work performed on U.S.-registered aircraft by Canadian maintenance personnel and Canadian aircraft by U.S. maintenance personnel.

2. CANCELLATION.
Advisory Circular 43-10 dated January 26, 1976, is canceled.

3. REFERENCES.
Federal Aviation Regulations (FARs) Part 91, Subpart C; Part 43, Section 43.17; and Canadian Ministry of Transport, Engineering and Inspection Manual, Parts I and II.

4. BACKGROUND.
U.S. and Canadian aircraft owners/operators frequently have the need for maintenance to be performed on their aircraft by maintenance persons

of the other country. Civil aviation authorities have recognized the performance of maintenance, repair, and alteration to aircraft of Canadian or United States registry by certificated and appropriately rated maintenance persons of the other country.

5. MECHANICAL WORK PERFORMED ON U.S. AIRCRAFT IN CANADA.
FAR 91, Subpart C, requires owners/operators to have their aircraft maintained in an airworthy condition. Part 43, Section 43.17, mechanical work performed on U.S.-registered aircraft by certain Canadian persons, permits work functions to be performed on U.S. aircraft by the following Canadian persons: an Aircraft Maintenance Engineer (AME) holding a valid mechanic certificate and appropriate ratings, or a person who is an authorized employee (Approved Inspector) performing work for a company whose system of quality control for the inspection and maintenance of aircraft has been approved by the Canadian Ministry of Transport. These persons may perform the following functions on U.S.-registered aircraft:

a. Perform maintenance, preventive maintenance, and alterations if done in accordance with the performance rules of Section 43.13 and maintenance record entries are made in accordance with Section 43.9.

b. Except for an annual inspection, perform any inspection required by Section 91.169, if the inspection is done in accordance with Section 43.15 and the maintenance record entries are made in accordance with Section 43.11.

c. Approve (certify) maintenance, preventive maintenance, and alterations; however, an AME may not approve a major repair or major alteration for return to service. A Canadian Ministry of Transport Airworthiness Inspector or an authorized employee performing work for a company approved by the Canadian Ministry of Transport may approve (certify) this work for return to service, provided that technical data used to accomplish major repairs or major alterations is approved by the FAA Administrator.

6. MECHANICAL WORK PERFORMED ON CANADIAN AIRCRAFT IN THE U.S.
The Canadian requirements approval for return to service of their aircraft in the U.S. are contained in the following excerpts from their Engineering and Inspection Manual.

"2.15.2 Minor modification (alterations) as defined in the U.S. Federal Aviation Regulations (FAR), and periodic routine maintenance shall be completed and documented in accordance with the requirements of FAR 43.

2.15.3 Data relating to major modifications (alterations) shall be approved either by the DOT, or in the case of an aircraft manufactured in the U.S. by the issue of an FAA Supplemental Type Certificate. These modifications shall be certified (approved for return to service) in accordance with the requirements of FAR 43, by a FAA Flight Standards Inspector or by an authorized employee of an appropriately rated Certificated Repair Station.

2.15.4   Repairs shall be certified in accordance with the requirements of FAR 43 by persons appropriately rated and authorized by the Administrator of the FAA. For the purpose of this requirement an engine overhaul shall be deemed to be a Major Repair.

2.15.5   On return of an aircraft to Canada following overhaul, major repair, or major modification, DOT Form 26-0023 shall be completed and a copy forwarded to the appropriate Regional Office together with a copy of FAA Form 337 or equivalent certification document."

7. CANADIAN ENGINES OVERHAULED BY U.S. REPAIR AGENCIES.

The following requirements for Canadian engines are contained in an excerpt from their Engineering and Inspection Manual:

"4.1   ENGINES—OVERHAUL STANDARDS

4.1.1   Engines must be overhauled in accordance with the following:

a. the manufacturer's recommendations as to tolerances and procedures are to be followed;

b. parts from the engine being overhauled which are serviceable or are within reconditioning tolerances must be replaced (reinstalled) in that particular engine; the practice of pooling parts is not permitted;

c. All replacement parts must be either:

(i) new, with documentary proof of compliance with the standards for the new type, or

(ii) used parts from other engines, if subjected to a close inspection for condition, are within the manufacturer's recommended tolerances and have a known history as to the engine model and the total number of hours in service at the time of installation in the overhauled engine, the details of which must be recorded in the engine log or overhaul data sheet;

d. all replacements for lifed parts must be new;

e. matched parts (wherever components of an assembly are not available individually from the manufacturer) must be replaced as an assembly;

f. magnetos and carburetors of fuel injectors must be overhauled and tested with the overhauled engine;

g. the overhauled engine must be run in and tested before being certified;

h. the log entry shall include details of the overhaul, a list of all parts replaced, modifications incorporated, and a record of the run-in test."

F

Used Airplane Prices

THE PRICES OF USED AIRPLANES FLUCTUATE, depending on market demand. Newly manufactured airplanes will see yearly drops until a plateau is reached. Drastic price reductions will be noted on planes experiencing expensive maintenance problems or costly ADs. As seen earlier in this book, many aircraft have actually shown an increase in value due to the increased cost of new airplanes and the general lack of availability of good quality used airplanes over the past several years. Remember, in many cases they aren't making them anymore!

VALUE RATING SCALE

The value rating scale of 1 to 10 is used in the airplane market. The following is an explanation of the rating scale:

10: Showroom new with no signs of use.

9: A few small blemishes and marks from careful use. A 9 airplane looks like a 10 from 50 feet away. The avionics are state of the art.

8: Sound and solid appearing. No obvious paint problems except around the cowling and on the leading edges. Very minor window crazing or haziness. Interior used, but undamaged. No oil or fuel leaks. Looks like a 9 from 50 feet away. The avionics are modern and working. An 8 represents an average to good value.

7: Sound and solid appearing with many blemishes and signs of use. The paint might be oxidized, fiberglass parts might show crazing, and there will probably be small dents and chips along leading edges. Windows show crazing and haziness with some scratches. Interior shows some wear and damage. No oil or fuel leaks. Looks like an 8 from 50 feet away. The avionics are usable, but old. You can brag about the minor work you will be doing to improve the airplane.

6: Appearance is poor and the plane needs a paint job. Side windows are usable and the windshield needs polishing to remove scratches and blemishes. Interior shows abuse, tears, holes, and damage. Some oil or fuel leaks. Looks in poor condition. Most of the avionics operate well, but they are worn. This plane's owner didn't show off the airplane.

5: Paint is peeling and corrosion is evident. Windows require replacing. There are puddles of oil or fuel under engine, landing gear, or wings. The airplane is in poor overall condition. The avionics need replacing. This plane will require considerable money to be spent on it to bring it up to safe and usable standards.

4: The airplane can be rebuilt, but is unusable at present.

3: Possibility of using many parts for the repair of another similar-model airplane.

2: Indicates the possibility of using some parts in the repair of another similar-model airplane.

1: Unusable scrap metal.

Exterior Values

Fixed-gear airplanes with two or four seats:
- Add $4000 for a 9 or 10
- Deduct $2000 for a 7
- Deduct $4000 for a 5 or 6
- If it needs fabric re-cover, deduct $11,000

Complex single-engine airplanes:
- Add $5000 for a 9 or 10
- Deduct $3000 for a 7
- Deduct $6000 for a 5 or 6

Twin-engine airplanes:
- Add $7000 for a 9 or 10
- Deduct $4000 for a 7
- Deduct $8000 for a 5 or 6

Interior Values

Two-seat airplanes:
- Add $1500 for a 9 or 10
- Deduct $500 for a 7
- Deduct $2000 for a 5 or 6

Four-set (fixed-gear) airplanes:

- Add $2500 for a 9 or 10
- Deduct $1000 for a 7
- Deduct $2500 for a 5 or 6

Complex single-engine and heavy-hauler airplanes:

- Add $6000 for a 9 or 10
- Deduct $2000 for a 7
- Deduct $4500 for a 5 or 6

Twin-engine airplanes:

- Add $8000 for a 9 or 10
- Deduct $3000 for a 7
- Deduct $7000 for a 5 or 6

AIRFRAME TIME

Total airframe time influences the airplane's value, but this influence is also controlled by the particular make, model, and age of the plane. For example, it is certainly common to find Cessna 150s with over 4000 hours on them. There are, however, averages that can be applied.

Most two- and four-place airplanes that are privately owned will accumulate an average of 100 hours annual use time. Those in rental or instructional service will see 250 hours annual use time or better. Determine the average time by the hours indicated and the age of the airplane. Be aware that many airplanes have a history of training or rental service. If the airplane is now owned privately, you will have to adjust for the total times and years in service.

If the total time is over 50 percent above the average, reduce the airplane's value by 20 percent. If the total time is 50 percent under the average, increase the value by 10 percent. Remember that low hours do not always indicate a good value.

ENGINES AND ENGINE OVERHAULS

Values of engines are based on the TBO (time between overhauls), TTSN (total time since new) or TTSMOH (total time since major overhaul), and the projected cost of overhaul. The value for an average airplane with an engine in the first one-third of its TBO life should be increased by 20 to 30 percent, depending on the quality of the overhaul. An engine in the last third of its TBO reduces the value of the airplane by 30 to 40 percent.

The following is a listing of popular engines, the recommended TBO, and the average cost of overhaul with installation (some TBOs might differ from earlier specifications depending on specific application):

Continental/Teledyne Continental

Engine	TBO	Cost
A65	1800	$10,700
C85	1800	$10,700
C90	1800	$10,700
C125	1800	$12,400
C145	1800	$12,400
E185	1500	$18,600
O-200	1800	$11,900
E225	1500	$18,600
O-300	1800	$15,900
GO-300	1200	$17,000
IO-346	1500	$14,700
IO-360	1500	$18,600
TSIO-360	1400	$20,900 (FB, L, and M series are 1800-hour TBO)
O-470	1500	$17,000 (U series and later are 2000-hour TBO)
IO-470	1500	$18,000
TSIO-470	1400	$19,200
IO-520	1700	$18,600
TSIO-520	1400	$22,000 (some late models have 1600-hour TBO)

Franklin

Engine	TBO	Cost
6A4 series	1200	$20,000

Lycoming/Textron-Lycoming

Engine	TBO	Cost
O-235	2000	$13,000 (some have 2400-hour TBO)
O-290	1500	$10,700 (some have 2000-hour TBO)
O-320	1200	$13,000 (some have 2000-hour TBO)
O-360	1200	$14,100 (some have 2000-hour TBO)
IO-360	1200	$17,500 (some have 2000-hour TBO)
O-540	1200	$18,000 (some have 2000-hour TBO)
IO-540	1200	$22,600 (some have 2000-hour TBO)
TIO-540	1500	$29,400 (some have 2000-hour TBO)
IO-720-A1A	1800	$31,600

Porsche

Engine	TBO	Cost
PFM3200	2000	$30,000

AVIONICS

Avionics have value and do hold considerable influence over an airplane's capabilities. Although new or additional avionics can be purchased and added at any time, it is generally more cost-effective to have them already in place (and working) when you purchase the airplane.

Avionics value represents two costs: purchase and installation. Values of installed avionics can be estimated from a percentage base of the purchase and installation of like-new avionics:

- Deduct 40 percent from new cost for one-year-old avionics.
- Deduct 50 percent from new cost for two-year-old avionics.
- Deduct 60 percent from new cost for three- to four-year-old avionics.
- Deduct 70 percent from new cost for over five-year-old avionics.

An alternative means of valuing avionics is to ascertain the current price for a serviceable piece of equipment of the same make and model. Add to that the cost of installation and add the total to the value of the airplane.

Avionics values also vary depending on the specific type of airplane, e.g., a two-place, easy flyer, or complex. The more costly the airplane, the more value can generally be added for avionics, due to both the quantity and the quality of what is installed. An IFR-certified airplane ready to be used would add up to the following:

Two- or four-place easy flyer: $9000

Complex or heavy hauler: $16,000

Affordable twin: $20,000

In general, the following avionics will increase the airplane's overall value by the amount indicated (if the airplane is currently IFR-certified and the installed equipment is less than six years old):

DME: $1200

GPS: $2400

LORAN: $400

RNAV: $900

Storm scope: $2000

An autopilot does not have to be as recent as the associated avionics to hold value. An autopilot's value on the average used complex airplane, heavy hauler, or affordable twin is $2400. Note, however, that autopilots can become a maintenance problem as they age. Autopilots add more value when new.

DAMAGE HISTORY

The value of an airplane is reduced from 5 to 25 percent for historical damage, which is structural damage resulting from an accident such as a hard landing, ground loop, flipover, struck object, being struck, etc. It also includes storm damage.

Generally, old damage (over ten years) will reduce the airplane's value by only 5 percent, as long as the damage was minor and properly repaired. Deduct up to 15 percent for major damage within the past ten years on simple airplanes such as the Cessna 100 or Piper PA-28 series of fixed-gear airplanes. Deduct up to 25 percent from the overall value for recent damage on complex airplanes.

CONVERSIONS, MODIFICATIONS, AND RESTORATIONS

Conversions, modifications, and restorations made to an airplane will change its value. However, rarely will the value of the airplane be increased to an amount equal to the price paid for the conversion, modification, or restoration. In general, this type of work clouds the value issue. Investments for commonly found modifications are:

Vortex generators: $600

Large fuel tanks: $2500

Engine horsepower increase: $10,000 (30-hp gain)

Engine horsepower increase: $20,000 (50-hp gain)

Taildragger conversion: $2500

STOL conversion: $2000 to $4000 (leading-edge cuffs and gap seals)

Wing tips: $500

Floats: From $14,000 to over $20,000

ANNUAL INSPECTION

There is no real means of valuing a required annual inspection. Either the airplane is in or out of annual. In other words, it can be legally flown or cannot be legally flown. Purchasing an airplane out of annual is very risky, as the annual inspection might harbor some very expensive surprises.

Inspections are certainly a point for negotiation when setting the airplane's value. A recommended procedure would be for the prepurchase inspection mechanic to estimate the cost of a complete annual inspection (including repairs) and deduct it from the airplane's overall value.

THE AVERAGE AIRPLANE

For the purposes of this book's valuation charts, an average airplane is considered to have a value rating of 8, have midtime engine(s) (middle third of TBO), no serious damage history, working avionics appropriate to the particular

```
┌─────────────────────────────────────────────────────────────┐
│ Prepurchase Inspection Checklist for:                        │
│                                                              │
│ Make_____Model_____   │
│                                                              │
│ Year_____N-number_____   │
│                                                              │
│ Ser. number_____Color _____    │
│                                                              │
│ Owner_____Phone _____   │
│                                                              │
│ Location_____   │
│                                                              │
│ Airframe_____          Avionics/instruments_____    │
│         appearance_____        appearance_____   │
│         paint_____         NAVCOM_____    │
│         corrosion_____         ADF_____    │
│         doors_____          Loran_____    │
│         side windows_____        GPS_____    │
│         windshield_____        G/S_____    │
│         control surfaces____      intercom_____    │
│         landing gear_____   Engine(s)_____    │
│         tires_____         appearance_____   │
│         brakes_____         baffles_____    │
│         mechanical_____        wiring_____    │
│ Cabin_____                 hoses_____    │
│         appearance_____        battery_____    │
│         seats_____      Propeller(s)_____    │
│         seat tracks_____       appearance_____   │
│         carpets_____        condition_____    │
│         headliner_____                                   │
│         side panels_____   *Using value ratings from 1 to 10, average │
│         seat belts_____  each category then average all the categories │
│                               to reach a final value rating.  │
│ Remarks                                                      │
│                                                              │
│                                                              │
│                                                              │
└─────────────────────────────────────────────────────────────┘
```

Fig. F-1. *Worksheet.*

airplane, and be fully airworthy and operational, with an annual due in about six months. A worksheet aids in the tally-up of the value ratings for an airplane (Fig. F-1).

In the event that a value rating is above or below the average of 8, apply the following:

- Add 25 percent for a 10.
- Add 15 percent for a 9.
- Deduct 15 percent for a 7.
- Deduct 25 percent for a 6.

In practice, the selling price can be defined as the sum mutually agreed on by the seller and the buyer, with regard to the need to sell, the desire to buy, and the actual value of the airplane. The latter is often the least significant figure.

VALUE LIST

The following value list indicates dollar values for most general aviation airplanes commonly seen on today's market. The values are based on actual selling prices as reported during interviews with airplane dealers and brokers and *The Aircraft Bluebook Price Digest*, compared to asking prices as seen in various aviation publications, including *Trade-A-Plane*, *Flyer*, and various Internet web sites.

You can use these dollar values with the previous guidance information to determine a fair market value of a used airplane, considering airframe condition, engine value, avionics, age, etc.

Note that classic airplanes, those built from the middle 1940s through the early 1950s, are very difficult to place fair value on. Asking prices can be as much as 50 percent above the average value. The key word is "asking."

Aero Commander/Rockwell/Twin Commander

AERO COMMANDER 500/500A

1958–1963	$77,000

ROCKWELL 100 DARTER

1965	$18,000	135-hp (Volaire)
1966–1969	$12,500	150-hp

ROCKWELL 100 LARK

All	$23,000	180-hp

Aeronca/Champion/American Champion/ Bellanca

7AC

All	$17,000

7ACA

All $15,500

7EC/ECA SERIES (90-, 100-, 115-HP)

1955–1965	$21,500
1966–1971	$24,500
1972–1976	$31,000
1977–1984	$38,000

7FC

All $18,500

7GC SERIES (140-/150-HP)

1959–1967	$27,000
1967–1970	$28,500
1970–1973	$32,000
1974–1976	$35,000
1978–1984	$43,000

7GC SERIES (160-HP)

1994–1996	$61,500
1997–1998	$73,000

7KCAB CITABRIA

1967–1973	$36,000
1974–1977	$42,500

8GCBC SCOUT

1974–1978	$42,500
1978–1984	$53,500
1993–1996	$76,000
1996–1998	$83,000

8KCAB DECATHLON (150-HP)

1971–1974	$41,000
1975–1977	$45,000
1978–1980	$59,500
1992–1994	$69,000

8KCAB DECATHLON (180-HP)

1977–1979	$64,000

1980–1984	$69,000
1992–1996	$74,000
1997–1998	$84,500

11AC

| All | $16,000 |

15AC

| All | $24,500 |

Aviat

HUSKY

1988–1990	$69,500
1991–1992	$77,000
1993–1995	$88,500
1996–2000	$113,000

Beechcraft/Raytheon

TWIN BEECH 18

| All | $88,000 |

SPORT 19

| 1966–1973 | $22,500 |
| 1973–1978 | $29,000 |

MUSKETEER 23/CUSTOM 23

| 1963–1968 | $23,500 |
| 1969–1973 | $33,000 |

SUNDOWNER 23

| 1974–1977 | $39,000 |
| 1978–1983 | $49,000 |

MUSKETEER 24

| 1966–1969 | $31,000 |

SIERRA 24 SERIES

1970–1975	$50,000
1976–1979	$58,000
1980–1981	$62,000

1982-1983	$69,000

DEBONAIRE 33 SERIES

1960-1964	$69,000
1964-1967	$89,000

BONANZA 33 SERIES

1968-1970	$93,000
1971-1973	$116,000
1974-1977	$138,000
1977-1980	$153,000
1981-1983	$173,000
1983-1985	$183,000
1986-1990	$205,000
1990-1994	$225,000

BONANZA 35 SERIES

1947-1951	$35,000
1952-1960	$50,000
1961-1963	$63,000
1964-1968	$85,000
1970-1974	$107,000
1975-1978	$135,000
1979-1982	$160,000

BONANZA A36 SERIES

1968-1971	$127,500
1972-1975	$140,000
1976-1978	$159,000
1979-1981	$175,000
1982-1984	$196,500
1985-1987	$233,000
1987-1991	$263,000
1992-1995	$297,000
1996-1999	$360,000

BONANZA A36TC

1979-1981	$200,000

BONANZA B36TC

1982-1986	$244,000
1986-1990	$277,000
1991-1995	$327,000

TWIN BONANZA 50

1952–1957	$58,000
1958–1960	$75,000
1961–1962	$82,000

BARON A/B55

1961–1968	$90,000
1968–1973	$117,000
1974–1976	$143,000
1976–1980	$175,000
1981–1982	$207,000

BARON E55

1966–1968	$109,000
1969–1973	$133,000
1974–1977	$165,000
1978–1980	$190,000
1981–1982	$234,000

BARON 56TC

1967–1970	$104,000

BARON 58

1970–1973	$160,000
1974–1977	$185,000
1978–1983	$262,000
1984–1987	$365,000

BARON 58TC

1976–1979	$235,000
1980–1984	$310,000

DUCHESS 76

1978–1980	$118,000
1981–1982	$130,000

SKIPPER 77

All	$23,000

TRAVELAIR 95

1958–1965	$63,000
1966–1968	$68,500

Bellanca

14-13

All	$25,000

14-19 AND 14-19-2

All	$27,000

14-19-3

All	$34,000

VIKING 17-30/31 SERIES

1967–1972	$45,000
1973–1975	$61,000
1976–1980	$76,000
1984–1991	$134,000
1992–1997	$182,000

Cessna

120

All	$20,000

140

All	$22,000

150

1959–1969	$18,000	
1970–1974	$20,000	Add $2000 for Aerobat
1975–1977	$22,000	Add $2000 for Aerobat

152

1978–1980	$24,000	Add $3000 for Aerobat
1981–1983	$28,000	Add $3000 for Aerobat
1984–1985	$36,000	Add $3000 for Aerobat

170

All	$32,000

172

1956–1964	$24,000
1965–1968	$32,500

1969–1972	$36,500
1973–1976	$47,500
1977–1980	$53,000
1981–1984	$68,000
1885–1986	$78,000
1997–1999	$110,000
2000–2002	$128,000
2002–2004	$145,000

172 SKYHAWK XP

| 1977–1981 | $74,000 |

172 CUTLASS Q

| 1983–1984 | $74,000 |

172RG

| 1980–1982 | $69,000 |
| 1983–1985 | $78,000 |

175

| All | $23,500 |

177

1968–1971	$41,000
1972–1975	$56,000
1976–1978	$55,000

177RG

| 1971–1974 | $56,500 |
| 1971–1978 | $63,500 |

180

1953–1968	$72,500
1969–1973	$84,000
1974–1976	$91,000
1977–1979	$94,000
1980–1981	$121,000

182

1956–1961	$48,000
1962–1966	$54,000
1967–1970	$63,000

1971–1973	$72,000	
1974–1977	$85,000	
1978–1980	$103,000	
1981–1983	$128,000	Add $6000 for turbo
1984–1986	$139,000	Add $9000 for turbo
1997–2000	$180,000	
2001–2004	$265,000	

182RG

1978–1980	$116,000	Add $4000 for turbo
1981–1983	$135,000	Add $6000 for turbo
1984–1986	$161,000	Add $8000 for turbo

185

1961–1967	$82,000
1968–1971	$93,000
1972–1977	$117,000
1978–1980	$137,000
1981–1983	$155,000
1984–1985	$175,000

190/195

| All | $64,000 | Add for restoration |

205

| 1963–1964 | $51,500 |

P206

| 1965–1970 | $89,000 | Add $4000 for turbo |

U206 & H

1964–1967	$87,000	
1968–1973	$99,000	Add $6000 for turbo, add $6000 for 300-hp
1974–1978	$129,000	Add $7000 for turbo, add $6000 for 300-hp
1979–1982	$150,000	Add $9,000 for turbo
1983–1986	$170,000	Add $10,000 for turbo
1998–2000	$240,000	Add $30,000 for turbo
2001–2002	$290,000	Add $30,000 for turbo

207

| 1969–1972 | $98,500 | Add $9,000 for turbo |
| 1973–1976 | $121,000 | Add $15,000 for turbo |

| 1977–1980 | $143,000 | Add $15,000 for turbo |
| 1981–1984 | $176,000 | Add $20,000 for turbo |

208 CARAVAN

| 1985–1996 | $825,000 |

210

1960–1961	$42,000	
1962–1965	$48,000	
1966–1969	$72,000	Add $4000 for turbo
1970–1972	$85,000	Add $5000 for turbo
1973–1976	$101,000	Add $9000 for turbo
1977–1978	$123,000	Add $10,000 for turbo
1979–1981	$151,000	Add $11,000 for turbo
1982–1983	$197,000	Add $13,000 for turbo
1984–1986	$235,000	Add $19,000 for turbo

210P

1978–1980	$175,000
1981–1983	$210,000
1985–1986	$270,000

336

| 1964 | $37,000 |

337

1965–1969	$61,000	Add $3000 for turbo
1970–1973	$72,000	Add $3000 for turbo
1974–1977	$88,500	Add $3000 for turbo
1978–1980	$98,000	Add $5000 for turbo

337P

1973–1976	$106,000
1977–1978	$121,000
1979–1980	$138,000

310

1955–1958	$48,000	240-hp
1959–1963	$58,000	260-hp
1964–1965	$68,000	260-hp
1966–1968	$82,500	260-hp
1969–1970	$98,500	260-hp; add $5000 for turbo
1971–1974	$124,000	260-hp; add $5000 for turbo

1971–1974	$130,000	285-hp; add $6000 for turbo
1975–1977	$149,500	285-hp; add $8000 for turbo
1978–1981	$190,000	285-hp; add $8000 for turbo

320

| 1962–1965 | $65,000 | 260-hp |
| 1966–1968 | $85,000 | 285-hp |

Commander

112

| 1972–1975 | $65,000 |
| 1976–1977 | $78,000 |

112TC

| 1976–1979 | $88,500 |

114

| 1976–1978 | $140,000 |

114A

| 1979 | $150,000 |

114B

| 1992–1995 | $180,000 | Add $30,000 for turbo |

114TC

| 1996–1998 | $290,000 |

Commonwealth

SKYRANGER

| All | $19,000 |

Diamond

KATANA DA20-A1

1995–1997	$44,000
1998–2000	$88,000
2001–2003	$179,000

KATANA DA20-C1

| 1998–2000 | $120,000 |

2001–2003	$180,000

Ercoupe

415

All	$18,000

ALON A2

All	$24,500

MOONEY M10 CADET

All	$20,000

Gulfstream/American General

AA-1 SERIES

1969–1976	$17,500
1977–1978	$21,000

AA-5 150-HP TRAVELER/CHEETAH

1972–1977	$33,000
1978–1979	$40,000

AA-5 180-HP TIGER

1975–1979	$55,000
1990–1993	$94,000

GA7

All	$103,000

Lake

LA-4

1957	$17,000	150-hp
1958–1964	$35,000	
1965–1968	$41,000	
1969–1971	$45,000	

LA-4-200

1970–1974	$50,500	VFR
1975–1979	$63,000	VFR
1980–1982	$75,000	VFR

1983–1986	$96,000	VFR
1987–1989	$170,000	
1990–1993	$230,000	Add $25,000 for turbo

LA-250 RENEGADE

1984–1986	$160,000	IFR
1987–1989	$192,000	
1990–1992	$264,000	

Luscombe

8A/B/C/D

| All | $18,500 | Add for restoration |

8E

| All | $19,000 | Add for restoration |

8F

| All | $20,000 | Add for restoration |

SEDAN

| All | $28,000 | Add for restoration |

Maule

M-4

1962–1966	$24,000	145-hp
1965–1973	$31,000	210-hp
1967–1973	$33,000	220-hp
1970–1971	$34,000	180-hp

M-5

1974–1977	$42,000	210-hp
1974–1975	$42,500	220-hp
1977–1981	$50,000	235-hp
1979–1988	$45,000	180-hp
1985–1986	$57,000	235-hp
1987–1988	$62,000	235-hp

M-6

| 1981–1985 | $60,000 | 235-hp |
| 1986–1988 | $62,000 | 235-hp |

1989–1991	$70,000	235-hp

M-7

1984–1987	$65,000	235-hp
1988–1994	$78,000	235-hp

MX-7 SERIES

1985–1989	$61,000	180-hp
1985–1989	$67,000	235-hp
1990–1992	$72,000	180-hp; add $3000 for trigear
1990–1992	$72,000	235-hp; add $3000 for trigear
1992–1995	$81,000	235-hp; add $3000 for trigear
1992–1996	$67,000	160-hp; add $3000 for trigear
1995–1999	$84,000	160-hp; add $3000 for trigear
2000–2003	$97,000	160-hp
2000–2003	$106,000	180-hp
2000–2003	$130,000	235-hp

Meyers

200

All	$95,500	260-hp
All	$115,000	285-hp

Mooney

M20 SERIES

1955–1957	$21,000	
1958–1960	$24,000	
1961–1965	$43,000	Deduct $10,000 for master M-20D
1966–1967	$50,000	
1968–1969	$50,500	
1970–1976	$60,000	
1977–1978	$69,000	

201 M-20J

1977–1979	$92,000	
1980–1982	$100,000	
1983–1985	$109,000	Deduct $15,000 for L/M
1986–1987	$115,000	Deduct $15,000 for L/M
1988–1990	$126,000	Deduct $15,000 for L/M or AT
1991–1992	$150,000	Deduct $20,000 for AT or Limited, add $20,000 for MSE

| 1993–1995 | $180,000 |
| 1996–1998 | $205,000 |

231 M-20K

| 1979–1981 | $101,000 | |
| 1982–1985 | $120,000 | Deduct $14,00 for L/M model |

252 M-20K

1986–1988	$175,000	
1989–1990	$200,000	
1997–1998	$264,000	Encore

PFM M-20L

| 1988–1989 | $115,000 | Porsche 217-hp engine |

TLS 20M

1989–1990	$180,000
1992–1994	$210,000
1995–1998	$290,000
1999–2002	$370,000

M20R OVATION

| 1994–1996 | $235,000 |
| 1997–1999 | $275,000 |

MARK 22

| All | $72,000 |

Navion

NAVION

| 1946–1951 | $36,000 | Various engines; add for restoration |
| 1958–1964 | $44,000 | Various engines |

RANGE MASTER

| 1967–1970 | $56,000 |
| 1975–1976 | $66,000 |

Piper

J3 SERIES

| All | $25,000 | Add for restoration |

J4 SERIES

All $21,000   Add for restoration

J5 SERIES

All $23,000   Add for restoration

PA-11 SERIES

All $28,000   Add for restoration

PA-12 SERIES

All $26,000   Add for restoration

PA-14/16

All $22,000   Add for restoration

PA-15/17

All $19,000   Add for restoration

PA-18

1950–1954	$32,500	Add for restoration
1955–1961	$46,500	Deduct $12,000 for 95-hp, add for restoration
1962–1974	$58,000	Add for restoration
1975–1981	$64,000	
1982–1983	$69,000	
1988–1990	$76,000	
1991–1994	$92,000	

PA-20 PACER

All $22,000   Add for restoration

PA-22 COLT

All $15,000   Add for restoration

PA-22 TRI-PACER

All $23,000   More for restoration

PA-23 APACHE

1954–1957	$35,000	150-hp
1957–1961	$37,000	160-hp
1962–1965	$55,000	235-hp

PA-23 AZTEC

1960–1963	$48,000	
1964–1966	$57,000	Add $4000 for turbo
1967–1970	$77,000	Add $5000 for turbo
1971–1974	$90,000	Add $7000 for turbo
1975–1976	$111,000	Add $9000 for turbo
1977–1979	$132,000	Add $11,000 for turbo
1980–1981	$139,000	Add $11,000 for turbo

PA-24 COMANCHE

1958–1964	$41,000	180-hp
1958–1964	$59,500	250-hp
1964–1965	$120,000	400-hp
1965–1969	$84,000	260-hp
1970–1972	$99,000	Add $3000 for turbo

PA-280-140

1964–1972	$26,500
1973–1977	$29,500

PA-28-150/160

1962–1965	$24,500	Add $1250 for 160-hp
1966–1967	$29,000	Add $1500 for 160-hp

PA-28-151 WARRIOR

1974–1977	$40,000

PA-28-161 WARRIOR

1977–1979	$43,000	
1980–1982	$46,500	
1983–1985	$52,000	
1986–1988	$60,000	
1989–1990	$67,000	
1989–1990	$41,000	Cadet VFR
1991–1992	$83,000	
1991–1992	$53,000	Cadet VFR
1993–1994	$88,000	
1995–1999	$113,000	
2000–2002	$130,000	

PA-28-180 CHEROKEE

1963–1967	$37,500

| 1968–1973 | $47,000 |
| 1974–1975 | $55,000 |

PA-28-181 ARCHER

1976–1979	$63,500
1980–1982	$71,500
1983–1986	$79,000
1987–1988	$86,000
1989–1990	$92,000
1991–1994	$107,000
1995–1996	$120,000
1997–1999	$150,000
2001–2003	$180,000

PA-28-235 CHEROKEE

1964–1969	$53,500
1970–1974	$60,000
1975–1977	$69,000

PA-28-236 DAKOTA

1979	$82,000	200-hp turbo
1979–1981	$101,000	235-hp
1982–1985	$108,000	
1986–1989	$131,000	
1990	$139,000	
1993–1994	$150,000	

PA-28R ARROW 180

| 1967–1969 | $48,500 |
| 1970–1971 | $51,000 |

PA-28R ARROW 200

| 1969–1971 | $53,000 |
| 1972–1976 | $62,000 |

PA-28R ARROW 201

1977–1978	$75,000	Add $5000 for turbo
1979–1982	$90,000	Add $8000 for turbo
1983–1985	$96,000	Add $8000 for turbo
1986–1988	$114,000	Add $8000 for turbo
1989–1990	$130,000	Add $12,000 for turbo
1991–1994	$144,000	

1995–1996	$160,000
1997–2000	$190,000
2001–2003	$217,000

PA-30 AND PA-39 TWIN COMANCHE

1963–1967	$88,000	Add $5000 for turbo
1968–1970	$109,500	Add $5000 for turbo
1971–1972	$114,000	Add $5000 for turbo

PA-32 CHEROKEE SIX

1965–1970	$73,000	260-hp
1965–1970	$81,000	300-hp
1971–1973	$81,000	260-hp
1971–1973	$87,000	300-hp
1974–1976	$90,000	260-hp
1974–1976	$95,000	300-hp
1977–1979	$103,000	260-hp
1977–1979	$112,000	300-hp

PA-32 SARATOGA

1980–1982	$144,000	Add $6000 for turbo
1983–1984	$157,000	Add $7000 for turbo
1985–1988	$170,000	Add $5000 for turbo
1989–1990	$190,000	Add $5000 for turbo
1995–1996	$260,00	

PA-32R/300 LANCE

| 1976–1979 | $100,000 | Add $3000 for turbo |

PA-32R/301 SARATOGA

1980–1981	$163,000	Add $8000 for turbo
1982–1984	$173,000	Add $8000 for turbo
1985–1987	$194,000	Add $8000 for turbo
1988–1990	$212,000	Add $9000 for turbo
1992–1995	$250,000	Add $20,000 for turbo
1995–2000	$320,000	Add $35,000 for turbo

PA-34 SENECA

1972–1974	$79,000
1975–1977	$115,000
1978–1979	$130,000
1980–1981	$150,000

1981–1983	$177,000	Start Seneca III
1984–1985	$200,000	
1986–1988	$218,000	
1989–1993	$264,000	
1994–1995	$300,000	
1996–1997	$330,000	
1998–2000	$370,000	

PA-38 TOMAHAWK

| 1978–1982 | $16,000 |

PA-44 SEMINOLE

1979–1981	$98,000	
1981–1982	$115,000	Turbo
1989–1990	$150,000	
1995–2000	$260,000	
2001–2002	$312,000	

PA-46 MALIBU/MALIBU MIRAGE

1984–1988	$313,000
1989–1992	$390,000
1993–1995	$430,000
1995–1996	$470,000
1996–1999	$500,000

Pitts

S2-B AND S MODELS

1983–1987	$78,000
1988–1990	$87,000
1991–1995	$96,000
1996–1999	$130,000

Republic

SEABEE

| All | $70,000 | Add for refurbished/modified |

Socata

TB-9 TAMPICO CLUB

| 1990–1992 | $62,000 |
| 1993–1995 | $89,000 |

| 1996–1999 | $125,000 |
| 2000–2002 | $155,000 |

TB-10 TOBAGO

1986–1988	$63,000	
1989–1991	$73,000	
1992–1993	$89,000	Add $8000 if XL
1994–1995	$97,000	Add $10,000 if XL
1996–1998	$145,000	Add $10,000 if XL
1999–2001	$170,000	Add $20,000 if XL

TB-20 TRINIDAD

1984–1985	$115,000	
1986–1988	$120,000	Add $20,000 for turbo
1989–1990	$133,000	Add $20,000 for turbo
1991–1992	$150,000	Add $20,000 for turbo
1993–1995	$172,000	

TB-21 TC TRINIDAD

1986–1987	$140,000
1988–1990	$164,000
1991–1993	$185,000
1994–1995	$225,000
1996–1998	$270,000

Stinson

| All | $26,000 | Add for restoration |

Swift

| All | $33,000 | Add for restoration |

Taylorcraft

BC12-D

| All | $18,000 | Add for restoration |

F19

| All | $21,000 | Add for restoration |

F21

| 1980–1984 | $21,000 |
| 1985–1990 | $34,000 |

Varga/Shinn

2150

All $32,000   150-hp

2180

All $36,000   180-hp

G

Makes and Models in the FAA Registry

THE FOLLOWING LIST INDICATES the number of active aircraft, according to the FAA registry, as of December 2004, shown by make and model:

Aero Commander/Rockwell

500/500A	113
100 Darter	188
100 Lark	133

Aeronca/Champion/American Champion/Bellanca

7 series Aeronca	3,377
11 series Aeronca	1,229
15AC Sedan	240
7 series American Champion	278
8 series American Champion	375
7 series Champion	1,937 (including Citabrias)
7 series Bellanca	1,496 (including Citabrias)
8 series Bellanca	703

Aviat

Husky	510

Beechcraft/Raytheon

Fixed landing gear series	3,317
Bonanza 33 series	2,485
Mentor T34 (A45)	321
Bonanza 35 series	8,478
Bonanza 36 series	2,812
Twin Bonanza 50	199
Baron 55 & Travelair 95	2,505
Duchess 76	268
Skipper 77	206

Bellanca

1413 series	356
1419 series	265
1730/31 series	687

Cessna

120	1,046
140	2,781
150	13,604
152	4,075
170	2,835
172	24,515
172RG	739
175	1,430
177	1,790
177RG	984
180	3,125
182	12,917
182RG	1,493
185	1,837
190/195	684
206 series	2,318
207	348
210 series	5,312
336	86
337 series	1,138
310	3,835
320	322

Commander

112 series	429
114 series	381

Commonwealth

Skyranger	102

Diamond

Katana	504

Ercoupe

415 series	1,821
Alon A2	223
Mooney M10 Cadet	57

Gulfstream/American General

AA-1 series	1,194
AA-5 150-hp series	1,125
AA-5 180-hp series	980
AG-5B	164
GA-7	56

Lake

All models	224

Luscombe

8A/B/C/D	1,729
8E	490
8F	180
Sedan	40

Maule

M-4	291
M-5	485
M-6	23
M7 series	338

Meyers

200 series	36

Mooney

M20 series all	6,616
201 M-20J	1,587
231/252 M-20K	898
PFM M-20L	32
M-20M	261
M-20S	61
M-20R	304
Mark 22	20

Navion

All models	1,375

Piper

J3 series	5,485
J4 series	370
J5 series	487
PA-11 series	543
PA-12 series	1,708
PA-14	130
PA-16	451
PA-15/17	349
PA-18	4,289
PA-20 Pacer	561
PA-22 Colt	1,071
PA-22 Tri-Pacer series	4,526
PA-23 Apache	1,083
PA-23 Aztec	2,431
PA-24 Comanche series	3,280
PA-28 series	20,952
PA-28 140/150/160	6,885
PA-28-161 Warrior	2,078
PA-28-180	4,263
PA-28-181	2,578
PA-28-235	1,065
PA-28-236	523
PA-28R Arrow series	3,850
PA-30 Twin Comanche	1,226

PA-32 series	2,514
PA-32R series	2,326
PA-34 Seneca	1,959
PA-38 Tomahawk	1,021
PA-39 Twin Comanche	91
PA-44 Seminole	448

Pitts

S2-B and S2-S	219

Republic

SeaBee	302

Socata

TB-9 Tampico Club	71
TB-10 Tobago	69
TB-20 Trinidad	209
TB-21 Trinidad	68

Stinson

108 series	2,638

Swift

GC-1B	578

Taylorcraft

BC12-D	2,255
F19 series	122
F21 series	38

Varga/Shinn/Morrisey

2150	138
2180	14

Index